Also in the Variorum Collected Studies Series:

MENSO FOLKERTS
The Development of Mathematics in Medieval Europe
The Arabs, Euclid, Regiomontanus

GAD FREUDENTHAL
Science in the Medieval Hebrew and Arabic Traditions

PAUL KUNITZSCH
Stars and Numbers
Astronomy and Mathematics in the Medieval Arab and Western Worlds

RAYMOND MERCIER
Studies on the Transmission of Medieval Mathematical Astronomy

VERN L. BULLOUGH
Universities, Medicine and Science in the Medieval West

MENSO FOLKERTS
Essays on Early Medieval Mathematics
The Latin Tradition

BRUCE S. EASTWOOD
The Revival of Planetary Astronomy in Carolingian and Post-Carolingian Europe

Y. TZVI LANGERMANN
The Jews and the Sciences in the Middle Ages

EMMANUEL POULLE
Astronomie planétaire au Moyen Âge latin

THOMAS F. GLICK
Irrigation and Hydraulic Technology
Medieval Spain and its Legacy

WESLEY M. STEVENS
Cycles of Time and Scientific Learning in Medieval Europe

GUY BEAUJOUAN
Science médiévale d'Espagne et d'Alentour

DAVID A. KING
Astronomy in the Service of Islam

VARIORUM COLLECTED STUDIES SERIES

Studies in Medieval Astronomy and Optics

To the memory of my father

José Luis Mancha

Studies in Medieval Astronomy and Optics

Routledge
Taylor & Francis Group

LONDON AND NEW YORK

First published 2006 by Ashgate Publishing

2 Park Square, Milton Park, Abingdon, Oxfordshire OX14 4RN
711 Third Avenue, New York, NY 10017

Routledge is an imprint of the Taylor & Francis Group, an informa business

First issued in paperback 2018

ISBN 978-0-86078-996-3 (hbk)
ISBN 978-1-138-38255-8 (pbk)

British Library Cataloguing in Publication Data
Mancha, J.L.
    Studies in medieval astronomy and optics. – (Variorum collected studies series ; no. 852)
    1. Levi ben Gershom, 1288–1344 2. Astronomy, Medieval
    3. Astronomy, Arab
    I.Title
    520.9'2

Library of Congress Cataloging-in-Publication Data
Mancha, J.L.
    Studies in medieval astronomy and optics / J.L. Mancha.
        p. cm. – (Variorum collected studies series ; 852)
    Includes index.
    ISBN–13: 978–0–86078–996–3 (alk. paper)
    ISBN–10: 0–86078–996–9 (alk. paper)
    1. Astronomy, Medieval. 2. Optics – History – To 1500. 3. Levi ben Gershom, 1288–1344. 4. Astronomy, Arab.
    I. Title. II. Series: Collected studies ; CS852.

QB23.M335 2006
520.9'02–dc22   2006016140

VARIORUM COLLECTED STUDIES SERIES CS852

CONTENTS

Preface   vii–x

Acknowledgements   xi

ASTRONOMY AND OPTICS

I   Egidius of Baisiu's theory of pinhole images   1–35
Archive for History of Exact Sciences 40. Heidelberg, 1989

II   Astronomical use of pinhole images in William of
Saint-Cloud's *Almanach planetarum* (1292)   275–298
Archive for History of Exact Sciences 43. Heidelberg, 1992

THE ASTRONOMY OF LEVI BEN GERSON

III   The Latin translation of Levi ben Gerson's *Astronomy*   1–25
*Studies on Gersonides: A Fourteenth-Century Jewish
Philosopher-Scientist, ed. G. Freudenthal. Leiden/New York/
Köln: E.J. Brill, 1992, pp. 21–46*

IV   Levi ben Gerson's astronomical work: chronology
and Christian context   1–23
Science in Context 10. Cambridge, 1997, pp. 471–493

V   Heuristic reasoning: approximation procedures in
Levi ben Gerson's *Astronomy*   13–50
Archive for History of Exact Sciences 52. Heidelberg, 1998

VI   The Provençal version of Levi ben Gerson's tables for
eclipses   269–353
*Archives Internationales d'Histoire des Sciences 48. Rome,
1998*

VII   Right ascensions and Hippopedes: homocentric models
       in Levi ben Gerson's *Astronomy*, I: First anomaly   264–283
       Astronomy and Astrology from the Babylonians to Kepler:
       Essays Presented to Bernard R. Goldstein on the Occasion
       of his 65th Birthday, ed. P. Barker, A.C. Bowen, J. Chabás,
       G. Freudenthal, Y.T. Langermann (= Centaurus 45).
       Copenhagen: Blackwell Munksgaard, 2003

ARABIC ASTRONOMY IN WESTERN TEXTS

VIII   Ibn al-Haytham's homocentric epicycles in Latin
       astronomical texts of the XIVth and XVth centuries   70–89
       Centaurus 33. Copenhagen, 1990

IX   On Ibn al-Kammād's table for trepidation   1–11
       Archive for History of Exact Sciencs 52. Heidelberg, 1998

X   A note on Copernicus' 'correction' of Ptolemy's mean
       synodic month   221–229
       Suhayl 3. Barcelona, 2002–2003

XI   Al-Biṭrūjī's theory of the motions of the fixed stars   143–182
       Archive for History of Exact Sciences, 58. Heidelberg, 2004

Index of Names   1–3

Index of Manuscripts   1–2

This volume contains xii + 338 pages

PREFACE

Most of the research interests reflected in these articles grew out of my doctoral dissertation, an edition with translation and commentary of two related fourteenth-century texts, Henry of Hesse's *De reprobatione eccentricorum et epiciclorum* (1364) and Julman's *De reprobationibus epiciclorum* (1377). As a criticism of Ptolemy's astronomy, Hesse's work is a collection of heterogeneous arguments of unequal force, not original but very likely taken from still unidentified sources. Historically, however, it is of great interest: it contains, for instance, a sound (although irrelevant with respect to Ptolemy) criticism of the anonymous *Theorica planetarum* called *communis*, used later by Regiomontanus – author of one of the preserved copies of Hesse's work – in his *Disputationes* (1475); and also the first occurrence in Western astronomy of arguments against epicycles based on the discordance between observation and Ptolemy's theory of planetary distances and sizes. Hesse's objection to the model for the Moon – the lunar apparent diameter at quadratures ought to appear twice that at syzygies – was repeated by Regiomontanus (*Epitome*, v.22) and through him reached Copernicus (*De revolutionibus*, iv.22). Article VIII is the only one in this volume related to Hesse. It shows that, although the corresponding Arabic texts were apparently never translated, Ibn al-Haytham's Eudoxan device to account in a physically acceptable way for the oscillatory motions of the epicycle diameters in Ptolemy's latitude theory was known by Latin astronomers, as attested by the aforementioned works by Hesse and Julman, and by Albert of Brudzewo's commentary to Peurbach's *Theoricae novae*, written around 1485.

Different reasons directed my quest for Hesse's sources towards the then undated and unexplored Latin translation of Levi ben Gerson's *Astronomy*, which is the object of the articles of section II. Its Hebrew version predated Hesse's work by about 25 years and was the first known Western text addressing against Ptolemy arguments based on the unobserved variations of planetary apparent diameters; moreover, the model concentric with equant which Hesse proposed for the Sun was also described by Gersonides in chapter 20 of his *Astronomy*. Lastly, the relationship between Gersonides' work and Hesse was made plausible through Philippe de Vitry, who made a request to Levi for his *De numeris harmonicis*, and Nicole Oresme, who could mediate between Vitry and Hesse.

Article III describes the characteristics of the Latin version of Gersonides' *Astronomy* – in fact an *editio* of the text with numerous variants, redacted by Levi himself close in time to the Hebrew version, probably with the collaboration of Peter of Alexandria, the Hermit friar who translated with Levi his two works dedicated to pope Clement VI, the *Prognosticatio* and the *Tractatus instrumenti astronomie* (chapters 4–11 of the *Astronomy*). Article IV describes the main stages of Levi's astronomical research as well as the chronology of the redaction of the Hebrew and Latin versions. Article V concerns also the history of mathematics and explains the method (iterative application of the rules of false position and two false positions) used by Gersonides in his derivation of planetary parameters when dealing with problems whose solution cannot be obtained by rigorous Euclidean proofs – i.e., problems of the kind that Ptolemy also solved in the *Almagest* by iterative procedures (to find the chord of an angle α from the chord of an angle 3α or the eccentricity and apsidal line of a superior planet from three oppositions). Article VI presents an edition, translation and commentary of the Provençal version of Levi's tables for finding syzygies, earlier than the Hebrew version and composed ca. 1336 'at the request of many great and noble Christians'.

My interest in the medieval Latin contributions to the theory and astronomical use of the pinhole camera was due to Gersonides' work on the Jacob staff. The theoretical problem of the pinhole images was to provide a geometrical answer to the question 'why does light after passing through apertures of whatever shape always reproduces the shape of the luminous source?' The practical problem, which still troubled Tycho Brahe at the end of the sixteenth century, demanded a correct procedure to derive the apparent diameters of the luminaries from the measures of the images cast onto the screen. These are, respectively, the subjects of items I and II. Egidius of Baisiu's *Improbatio* contains a detailed refutation of Pecham's analysis of the problem of pinhole images and an explanation, geometrically equivalent to Kepler's, of the formation of images cast by light through triangular apertures. Saint-Cloud explains that finite and point apertures require different geometrical treatment for measuring apparent solar diameters (in the first case, the diameter of the aperture must be subtracted from the diameter of the solar image on the screen) and suggests that the pinhole camera could be used to derive a value for the solar eccentricity from apparent solar diameters at apogee and perigee. An appendix to item II contains the only known medieval derivation of the solar eccentricity using this procedure, made by Levi ben Gerson in 1334, although there is no evidence that Gersonides was aware of Saint-Cloud's suggestion.

The two main problems about the scientific status of astronomy in the Middle Ages provide a clue to understand Levi ben Gerson's concern for homocentric theory. The first was the inadequacy between some accepted

Aristotelian physical principles – that the heavenly motions are circular, uniform, and around the Earth, for instance – and some features of Ptolemy's theory – the eccentric, the epicycle, and the equant. Its solution required a new and coherent assembling of physics and astronomical theories, namely to modify the physical principles or the models or both. The second, that astronomical theories provided only *a posteriori* explanations, which proceeded from the effects to the causes and were not able to achieve true scientific demonstrations – this agreeing with Aristotle's distinction in his *Posterior Analytics* between *propter quid* and *quia* demonstrations.

Levi ben Gerson's purpose was to unravel these problems, and the rationalistic attitude with which he faced them was ingenious and inevitably ingenuous. According to him, the first could be solved by an empirical proof demonstrating, for instance, that the eccentricity of the heavenly spheres is a fact, not a hypothesis: this is the meaning of the derivation of the solar eccentricity with a pinhole camera from the apparent diameters of the Sun at apogee and perigee, already mentioned. The second, by enumerating first *all the possible models* and rejecting later those unable to account for the features of the observed motions in longitude and anomaly. If this search enables one to find the right model (the cause), it will also allow one to deduce the phenomena from that model, achieving thus a true *demonstrative* astronomy.

Article VII contains an edition, with translation and commentary, of one of the two chapters devoted by Gersonides in his *Astronomy* to investigate the ability of homocentric models to reproduce the observed planetary motions. It follows from it, first, that the working of Eudoxan couples (pairs of concentric spheres rotating in opposite directions around inclined axes) was not forgotten in the Middle Ages, as it is usually assumed; second, that Ptolemy's procedure in the *Almagest* for the computation of right ascensions was used to deal with the geometry of the curve which results from these couples (in our words, to calculate the width of Simplicius' hippopede) and that, since Gersonides did not claim originality for himself, this method was a medieval standard procedure to deal with homocentric theory.

Article XI provides a confirmation of this point showing that the procedure was used by al-Biṭrūjī in the chapter of his *Kitāb fi'l-haya* devoted to the fixed stars, providing thus a proof that the model underlying al-Biṭrūjī's text is indeed a Eudoxan couple and that the eight-shaped curve accounts for the so-called trepidation of the stars, which results according to al-Biṭrūjī from a true motion in declination and another, merely apparent, in accession and recession.

Article IX is a complement to the study, published in 1994 by Chabás and Goldstein, on the Latin translation made by John of Dumpno in 1260 of Ibn al-Kammād's *zīj al-Muqtabis*. It presents a different reconstruction of the trepidation table of the set, based on a thirteenth-century Castilian text which probably translates a chapter of Ibn al-Kammād's *zīj al-Amad^c alā al-abad*. Article X deals with an accidental discovery made when preparing item VI: that the correction to Ptolemy's value for the mean synodic month which supposedly was first made by Copernicus, was indeed a commonplace among medieval astronomers, inheritors on this point to al-Ḥajjāj's Arabic version of the *Almagest* through Cremona's Latin translation. It shows that the lack of critical editions of sources available to medieval and early modern scholars cannot be filled by critical editions of ancient texts.

JOSÉ LUIS MANCHA

Sevilla
December 2005

ACKNOWLEDGEMENTS

For kind permission to reproduce material in this volume I thank Springer Science and Business Media (I, II, V, IX, and XI); Cambridge University Press (IV); Koninklijke Brill N.V. (III); Blackwell-Munksgaard Publishing (VII and VIII); Robert Halleux and Julio Samsó, editors of *Archives Internationales d'Histoire des Sciences* (VI) and *Suhayl* (X), respectively. Colleagues and friends with whom I am in debt are mentioned further on; here, I wish to express my gratitude especially to Bernard R. Goldstein and John D. North, who read most of these papers before their first publication, to Gad Freudenthal, for suggesting to assemble them in a Variorum volume, and to John Smedley for kindly accepting it.

PUBLISHER'S NOTE

The articles in this volume, as in all others in the Variorum Collected Studies Series, have not been given a new, continuous pagination. In order to avoid confusion, and to facilitate their use where these same studies have been referred to elsewhere, the original pagination has been maintained wherever possible.

Each article has been given a Roman number in order of appearance, as listed in the Contents. This number is repeated on each page and is quoted in the index entries.

I

Egidius of Baisiu's Theory of Pinhole Images

Nec invidebo melius pertractanti
sed ut didascalum venerabor.
PECHAM, *Perspectiva*, rev. ed., I, 7.

Introduction

In a series of articles published in this *Archive*[1], D. C. LINDBERG demonstrates that the problem of pinhole images constituted an important element of study into the nature and propagation of light in the Middle Ages. The problem was approached repeatedly and fruitlessly throughout the thirteenth century by a number of leading researchers and compilers of optical theory (BACON, WITELO, PECHAM) in the same terms used in the *Problemata* attributed to ARISTOTLE: Why does the sun produce circular images after passing through angular apertures? Although IBN AL-HAYTHAM (Latinized as ALHAZEN, in his *Sūrat al-Kusūf (On the shape of the eclipse)*[2], 965 ca. 1040) had provided an analysis for the case of the images formed through circular apertures during a partial solar eclipse these medieval authors were unacquainted with it since the treatise was apparently never translated into Latin. Thus, being unable to discover a geometrical explanation for the problem, they finally accepted the case, implicitly or explicitly, as an exception to the principle of rectilinear propagation of light.

Dealing with the problem in several works, in each case PECHAM arrives at the conclusion that, since it is capable of reproducing only the shape of the aperture, rectilinear propagation alone cannot account for the phenomena of pinhole images. PECHAM'S conclusion was widely known because of his *Perspectiva*, used as a textbook in universities from the fourteenth until the sixteenth century and called *communis* like that other readily available text, the anonymous *Theorica planetarum*. As his commentators (LANGENSTEIN and PARMA[3], for example, in the fourteenth century and, after them, a long line of authors, known or anonymous, published or unpublished) were even less sucessful in applying EUCLID'S *Elements*, they generally accepted PECHAM'S position without dispute. In doing so, they illustrate a common occurrence in the history of medieval science: the

---

[1] LINDBERG [1], [2], [3].
[2] WÜRSCHMIDT [1]; WIEDEMANN [2]; SABRA [1], pp. 195–6.
[3] LINDBERG [3], pp. 299–300 and 308–25.

persistence of error due to technical incompetence. Other approaches to the problem, more "physical" than geometrical, were tried, but a generalized solution (that is, one valid for any shape of aperture and light source) was not reached until KEPLER (1604) and MAUROLICO (1611)[4], both apparently working independently of IBN AL-HAYTHAM.

In one of his articles[5], LINDBERG calls attention to a text on pinhole images entitled *Improbatio cuiusdam cause que solet assignari quare radius solis transiens per foramen quadrangulare facit figuram rotundam in pariete* ("Disproof of a certain cause by which is usually explained why a solar ray passing through a quadrangular aperture produces a round figure on a wall"). Composed by Brother EGIDIUS OF BAISIU probably at the end of the thirteenth century or at the beginning of the fourteenth, the work is preserved in only one known copy, apparently incomplete, from the first half of the fourteenth century, in MS. 569 of the Jagiellonian Library in Cracow[6]. In his summary of the text, LINDBERG emphasizes the clear expression of some of the concepts found in the *Improbatio*[7]. Such clarity is noteworthy in a moment in history when not only a correct analysis of the problem was unattainable, but also the basic phenomena are sometimes ignored[8]. LIND-

---

[4] KEPLER [1], ch. II, pp. 37–50; MAUROLICO [1], Theorem XXII, pp. 16–9. On the genesis of KEPLER's theory of pinhole images, see STRAKER [1].

[5] LINDBERG [3], pp. 301–3.

[6] Codex in fol. (27×20), parchment, French origin, old signature D.D. IV. 19. Descriptions of its content can be found in WISŁOCKI [1], p. 178; CURTZE [1], pp. X–XII; BJÖRNO & VOGL [1], pp. 132–4 and 172, and HAJDUKIEWICZ [1], pp. 287–8.

[7] After making reference to the passage where EGIDIUS affirms that luminous rays passing through an aperture always acquire the shape of the luminous source, independently of the shape of the aperture, LINDBERG ([3], p. 303) writes: "This is surely the clearest expression of the phenomena of pinhole images (or those of the phenomena relating to shape) that we find in the entire Middle Ages".

[8] PECHAM, for instance, in the first version of the *Perspectiva communis* does not know that, when the luminous source is the partially eclipsed sun, the incidence on the screen has the same shape as the source, but inverted (LINDBERG [4], pp. 70–1). This is indicative both of his limited attention to the observational data of the problem and of his ignorance of the geometry implied by the idea of luminous cones. In the fourteenth century, HENRY OF LANGENSTEIN, in his *Questiones super perspectivam* does not even seem to know that the image formed behind the aperture during an eclipse will reproduce the shape of the eclipsed sun since he always attributes circularity to the image independently of the shape of the source. If this assertion is in a rather obscure passage in the *Questiones* (LINDBERG [3], p. 314), LANGENSTEIN states the same opinion without any ambiguity in his *Tractatus de reprobatione eccentricorum et epiciclorum* (1364), which was probably revised after the writing of the *Questiones*: "Istud etiam est falsum, quoniam si sol esset quadratus vel cuiuscumque alterius figure eque bene per foramen rotundum vel angulare faceret incidentiam rotundam. Igitur, abscisa quacumque parte solis, residua eque bene faciet incidentias rotundas etiam cuiuscumque figure fuerit; quia quod ille incidentie sint rotunde non est ex parte figure luminosi, ut experientia patet de candela translucente per foramen triangulare. Patet etiam hoc in commento quinte propositionis *Perspective communis*, ubi huiusmodi incidentiarum rotunditas nature lucis attribuitur et non figure luminosi" (I am preparing an edition of this text, based on the nine extant copies, and I quote from it; but see, for example, MS. Paris, Bibl. Nat., 16401, f. 65v).

BERG'S summary is not sufficient, however, to give us an accurate idea of the relevance of EGIDIUS' work.

The object of this paper is to provide the Latin text of EGIDIUS OF BAISIU'S *Improbatio*, with a translation and commentary. Although the extant copy is apparently incomplete and in spite of the additional difficulties arising from a copyist who was obviously unfamiliar with the subject matter[9], the manuscript supplies enough information to establish EGIDIUS' contribution to the problem of pinhole images. The text contains a detailed refutation of PECHAM'S analysis of the problem, an analysis which is merely a false geometrical argument based on an incorrect figure (*i.e.*, a ψευδογράφημα, to use ARISTOTLE'S term)[10]. In the course of this refutation, EGIDIUS offers a solution, *geometrically equivalent* to KEPLER'S, for the circular image cast by light passing through a triangular aperture.

The manuscript lacks figures; those which accompany the translation are a reconstruction based on the text[11]. Editorial insertions are enclosed in square brackets. Sentence numbers have been added to the translation also in square brackets for ease of reference. I am indebted to Dr. G. ROSIŃSKA and the Jagiellonian Library for supplying me with photographs of MS. 569 and to Dr. F. SOCAS of the University of Seville for his suggestions in the interpretation of the Latin text.

Text

[120r] *Improbatio cuiusdam cause que solet assignari quare radius solis transiens per foramen quadrangulare facit figuram rotundam in pariete, quam improbationem facit Egidius de Baisiu.*

Queritur causa quare lux transiens per foramen triangulare vel alterius figure habet in pariete distante figuram rotundam. Et dicunt quidam huius esse causam perfectionem figure circularis, affirmantes, cum lux maxime sit activa, talem sibi figuram acquirere, cum alia mundi corpora minus activa hanc sibi acquirant, ut patet de aqua roris super herbam. Nituntur etiam hoc astruere destruendo causam aliam, que assignatur intersectionem, scilicet radiorum, ex directo processu sic:

"Radii solis qui applicant se lateribus foraminis angularis maioris sunt dila-

---

[9] The codex is due to a single copyist. CURTZE'S judgement of the copy of AL-NARIZI'S *Commentary* to EUCLID'S *Elements* (ff. 1r–37r) is: "Qui scripsit librum, eum de mathematica pauca vel nihil scivisse verisimillimum est" (CURTZE [1], p. XII).

[10] ARISTOTLE [1], pp. 274–7 (101 a 15–17); [2], pp. 60–1 (171 b 12–18).

[11] The work begins in f. 120r, immediately after a copy of MESSAHALA'S *Liber motus orbis et nature eius* (*De orbe*) and ends with f. 120v. Folio 121r begins with the last words of Proposition 49 of Book VII of IBN AL-HAYTHAM'S *Optics* (*Kitāb al-Manāzir*): "comprehendet illam rem visam in figure armille. [50ª] Si vero BGDZ fuerit in corpore ...", which ends on f. 122v: "... causa est reflexio. Nunc autem terminemus hunc tractatum qui est finis libri. Explicit VII liber Alhacen de aspectibus". The title of EGIDIUS' work is on the right margin on f. 120*r*, in rubric. In spite of this, a later hand, probably in the fifteenth century, wrote in the margin, as well as in other places of the codex, the title of the copied work; in this case, confused by the *explicit* on f. 122v, this other hand writes "Alhacen de aspectibus" on the right margin and "Incipit Alhacen" on the left one.

tationis quam aliquis aliorum per foramen transeuntium; sed incidentia lucis figuram habet a radiis extremis et maxime dilatatis; ergo si illi procedunt linealiter, ut dicis, si in foramine sunt triangularis figure, si in infinitum protrahantur semper erunt triangularis.

"Quod autem isti sint maxime dilatati probo. Radii procedentes a sole, a quolibet puncto solis procedunt in omnem partem medii et toto sole procedunt in quemlibet[1] punctum medii in quo intersecant se et intersecti ulterius procedunt; igitur radiorum procedentium ab eodem puncto solis et per foramen transeuntium planum est maxime dilatationis eos esse[2] qui lateribus foraminis se applicant. Idem probo de aliis, quoniam omnium radiorum concurrentium tanto ad maiorem angulum concurrunt quanto breviores sunt. Igitur cum alique piramides radiose concurrant inter solem et foramen, alique ultra foramen, ille maiorem angulum faciunt que inter solem et foramen concurrunt, quia ille breviores sunt per 21 primi Euclidis; sed [quanto] obtusiorem angulum constituunt tanto latius intersecando se extendunt, quia anguli contra se positi sunt equales per 15[3] primi Euclidis; quanto ergo piramides breviores sunt tanto amplius dilatantur.

"Sed inter omnes per foramen transeuntes ac inter solem et foramen concurrentes magis dilatatur illa piramis cuius latera foramini applicantur, quia omnes alie cadunt intra illos. Igitur necesse est lineas radiosas conformari foraminibus per que transeunt quantum est de vi irradiatonis." Ergo irradiatio non est causa rotunditatis.

Habent etiam figuram per se, quod in mane et in vespere, quando solis irradiatio est debilior propter diversas fractiones, non acquirit lux per triangulare transiens ita complete et ita cito figuram rotundam, quod dicunt esse[4] propter debilitatem sue virtutis. Sed quia hoc primo contradicit auctoritati, contradicit sensui, contradicit demonstrationi, ideo penitus tanquam erronea reputanda est.

Contradicit inquam auctoritati quia — cum auctor *Perspective*[5] dicat et experimentis manifestis demonstret, rectum incessum radiorum nullatenus[6] posse mutari nisi variato medio, in quo etiam omnes alii principium recte sentientes concordant — si aer sit eiusdem dyphaneitatis a foramine usque ad parietem, qui dicunt ipsum, ut figuram spericam acquirunt, incessum [120v] variare, contradicunt auctoribus et destruunt principia perspective, que tota super hoc inmutabile principium est fundata.

Contradicit etiam rationi, quia omnis natura fortius agit, completius assequitur propriam operationem cum est sue origini propinquior; ergo si naturaliter radii hanc figuram acquirunt tanquam aliquod completum intentum, cum iuxta foramen sint propinquiores sue origini quam multum ultra foramen, statim post foramen acquirerent hanc figuram. Si forte diceres quod redditur distantia proportionaliter, quero qua causa; preterea [si] radii transeuntes per magnum foramen sint plures et plus habeant de virtute, citius acquirerent hanc figuram, quod patet esse falsum.

---

[1] *ms.*: quem est.
[2] *ms.*: erunt.
[3] *ms.*: 51.
[4] *ms. add.* quod esse.
[5] *ms.*: Perspectiva.
[6] *ms.*: nulla.

Preterea si hanc figuram acquirerent tanquam sui aliquod completum, in qua scilicet partes magis unirentur et virtus curvaretur et fortificaretur, tunc completiorem[7] haberent actionem cum illam figuram acquisissent[8], quod patet esse falsum.

Etiam est contra sensum; si enim corpus a quo procedunt radii fuerit alterius figure, ut semicircularis seu novacularis, post foramen, in tanta distantia in quanta a corpore [circulari] luminoso veniens acquireret[9] rotundam, acquiret talem qualis est superficies corporis a quo venit, ut patet in partiali eclipsi solis. Ad quod experimentum quidam falso nituntur respondere. Dicunt enim quod in tali incidentia tempore eclipsis sunt due partes distinguibiles: una interior, angularis, ad modum foraminis disposita, que habet lucem primariam seu radiosam; alia[10], que complet figuram novacularem, minus claram, quasi mediam inter primariam lucem et secundariam. Et per hoc credunt evadere, sed ad pauca inspicientes et non interius intendentes aliquid enuntiant unde possint sue ignorantie solatium querere. Unum siquidem patet experienti: quod non solis tunc sed in omni lucis incidentia meritum post foramen lux est fortioris in medio quam circa extrema, cuius causa potest dicere. Sed, quid de hoc ad propositum? Si enim dicas quod illa pars lucidior sola est ex directa processione radiorum, tunc deberent acquirere figuram circularem, cuius contrarium concessisti. Iterum, illa lux secundaria unde venit? Si a radiis primariis, tunc similiter ab eis procedet. Cum luminosum est orbiculare quero ergo: illa incidentia rotunda que est in pariete, [post] foramen triangulare, aut est ex illa luce secundaria simul et ex primaria, aut ex altera solum; si ex utraque et nec — scilicet eclipsis debet fieri orbicularis — ex utraque; si ex altera ergo et nec ex altera; quod falsum est. Preterea si lux accidentalis sive ab illa primaria oritur, debet oriri undique equalem, cum talis [sit] natura multiplicationis lucis, ut probat auctor[11] Perspective. Unde ergo figura novacularis? Si autem sit lux primaria, tunc habeo propositum, scilicet quod directa irradiatione acquiritur talis figura qualis est corpus luminosum. Si autem neutra, tu ponis novam speciem lucis et facis auctorem[12] Perspective insufficientem, qui non ponit nisi radios rectos, reflexos ad corpora solida et [refractos] ad diversa dyaphana, et lucem accidentalem sive secundariam. Et preterea quero: unde venit et quare talem figuram habet et cornua situm habent determinatum opositum, scilicet contra corpus luminosum?

Et ut non inveniant mendacia de illo experimento quia eclipsis raro contingit, docebo hoc semper exper[iendo] et de nocte quando luna lucet et de die ad solem. Accipias tabulam parvam subtilem et facias in eam foramen rotundum vel cuiusque figure volueris. Et ponas illud in loco radii intrantis per fenestram ad situm ad quem veniant radii solum ab una portione solis, ita tamen quod totum foramen intra radium perfecte includatur. Et videbis in pariete opposito[13] foramini lumen in tali portione qualis est portio corporis luminosi, a qua veniunt radii in loco in

---

[7] ms.: completionem.
[8] ms.: acquisiscent.
[9] ms.: acquirerent.
[10] ms.: aliam.
[11] ms.: aut.
[12] ms.: auctore.
[13] ms.: apposito.

quo est foramen per in[14] tabula, et in situ[15] opposito; cuius causa post patebit. Et similiter est de luna sive foramen sit rotundum sive non. Si etiam ponas parvum foramen rotundum in loco in quo veniat lumen solum ab una parte solis, triangularis figure, fieret post in pariete figura triangularis; unde manifestum est quod in omnibus modis consequitur figuram sue originis.

Erubescant novi presumptores[16] qui causam veram attingere nec valentes nec verentur mendacia fingere, quibus aures simplicium delundutur; quod si nec verentur maiorum dictis contradicere, perspective fundamenta negare, vereantur[17] saltem omnium visui contradicere manifeste. Solutio autem huius[18] questionis difficilis est, et in eo patet veritas irradiationis.

Notandum igitur quod a quolibet corpore luminoso concurrit una piramis in quemlibet punctum medii; inter autem piramides que concurrunt inter foramen et solem, omnium scilicet[19] de quibus aliud transit per foramen, illa est maxima que post intersecationem applicat suas extremitates extremitatibus foraminis, cuius scilicet basis in foramine est circularis circumscribens figuram foraminis; sed de illa piramide non transit per foramen nisi radii venientes a simili figura inscripta a figura corporis luminosi. Verbi gratia. Si sol dum est circularis figure irradiat per foramen triangulare per illam piramidem cuius basis est circulus inscribens triangulum foraminis, irradiat in foramen maxima irradiatione, id est, maxime dilatata, per piramidem cuius sectio in foramine[20] seu basis est circulus circumscribens triangulare foramen; sed de illa non transeunt nisi radii qui veniunt per piramidem triangularem circumscriptam a predicta piramide, cuius basis ex parte solis est triangulus circumscriptus a circumferentia superficiei solis, maiori scilicet circulo quoad sensum. Piramis autem cuius basis est circulus circumscriptus a figura foraminis transit tota et, cum veniat a latiori origine, secundum partes propinquas medietati laterum foraminis intersecat radios prime piramidis triangularis post foramen. Et sic alie intermedie piramides similiter; unde amittunt illi radii terminum figuralem.

Ad cuius quantitatis ostensionem notandum quod, secundum quod dicit Ptolomeus[21] in *Almagesti*[22] dictione 5, capitulo 15, distantia solis a terra est 1210 partes secundum quas semidiameter terre est pars una. Et diameter solis est quincuplum ad diametrum terre et medietas eius fere, ut ostendit 16 eiusdem. Unde patet quod proportio axis piramidis cuius conus est in terra et basis superficies solis se habet ad suam basim in 111ª proportione fere. Et eadem proportio erit inter axem et diametrum basis, ubicumque piramis hec, vel equalis ei, que fit post intersectionem, equidistanti basi secetur, ut patet per secundam sexti Euclidis et etiam per Esculeum et *De pyramidibus*[23].

---

[14] *sic in ms.*

[15] *ms.*: situm.

[16] *ms.*: persumptores.

[17] *ms.*: venitur.

[18] solutio autem huius: *dubitanter statuo* (*ms.*: solo ā h' h').

[19] *ms.*: est.

[20] *ms. add.* est.

[21] *ms.*: Apolonius.

[22] *ms.*: Alimagesti.

[23] *ms.*: de py^{b3}.

Intelligamus ergo foramen hic triangulare laterum equalium, 4 scilicet digitorum, quod vocemus palmam. Igitur diameter [quadratus] circuli ipsum circumscribentis erit productum [lateris] sesquitertium[24] per 8 decimotertii et 47 primi[25] Euclidis. Si[26] igitur quadrato lateris, scilicet 16, addideris tertiam partem extraserisque radicem, habebis diametrum circuli 4 digitos et 37 minuta et 7 secunda, quod vocemus[27] 4 pollices. Erunt ergo conus piramidis huius ante foramen 444 pollices, id est, 40 pedes et 4 pollices; tantumdem igitur ultra foramen erit diameter circuli circumscribentis basim piramidis triangularis que transit ex hac circulari piramide, 8 pollicum. Sed si a centro circuli circumscribentis ducatur perpendicularis ad latus triangulare equilateri, equalis medietati[28] semidiametri ut probatur super 8 decimotertii Euclidis per quartam primi et primam partem tertie tertii et per penultimam primi, ducta hac linea ad circumferentiam et ab utroque termino eius ad unde de[29] angulis lateris secti ductis lineis rectis, igitur post foramen 40 pedibus et 4 pollicibus, a centro axis piramidis triangularis usque ad angulos 4 pollices, usque ad medietatem laterum duo.

Accipiamus igitur piramidem aliam cuius basis circumscripta a triangulo foraminis et erit eius diameter 2 pollices. Per eadem igitur conus[30] eius ante foramen 222 pollices erit, id est, 20 [pedes] et 2 pollices. Igitur quando veniet ad locum signatum piramidis in pariete, scilicet quadraginta pedes et [4] pollices ultra foramen, erit diameter eius fere 6 pollices. Ergo a centro axis communi huic et triangulare usque ad huius basis circumferentiam erunt 3 pollices. Sed usque media latera prime nec fuerunt nisi duo; ergo excedit latera prioris uno pollice et excedetur ab angulis uno. Si autem prima, scilicet triangularis, adhuc per 40 pedes et 4 pollices processerit, ita quod distet a foramine per 80 pedes et 8 pollices, diameter circuli eius erit duodecim pollices, et a centro usque ad media laterum trianguli erunt 3 pollices. Si autem secunda, scilicet rotunda, usque ibi procedat, diameter eius erit 10 pollices fere, id est, excessa 11ᵃ[31] parte unius pollicis fere. Cum igitur a centro communi usque ad circumferentiam huius sint quinque pollices, excedit medium lateris triangularis 2 pollices et exceditur ab angulis uno. Et sic procedendo, si igitur accipiuntur piramides[32] medie inter has duas, replebunt circularibus portionibus illum excessum, quo excedunt anguli hanc minorem piramidem. Et erit lumen circulare, sed non erit penitus equalium diametrorum.

Post hec accipiamus tres piramides quarum coni sint in tribus angulis foraminis, quarum axes egrediuntur a centro solis. Cum igitur veniant ad locum priorem, id est, 80 pedibus et 8 pollicibus post foramen, uniuscuiusque[33] diameter [erit] 8 pollicum, et axes earum[34] distinguntur et distant secundum triangulum cuius

---

[24] *ms.*: sexquitertium.
[25] *ms.*: 4 secundi.
[26] *ms.*: scilicet.
[27] *ms.*: volemus.
[28] *ms. add. in mg.*
[29] *sic in ms.*
[30] *ms.*: corpus.
[31] *ms.*: 55.
[32] *ms.*: piramidis.
[33] *ms.*: utriusque.
[34] *ms. rep.* distant.

circulus circumscribens habet diametrum plusquam quatuor pollicibus[35] in proportione que octoginta pedes [et octo pollices] ad distantiam foraminis a sole. Et reli[n]quatur: diameter utriusque est 4 pollices. Igitur circa duodecim pollices habebit diameter piramidis composite ex hiis tribus. Sunt igitur attingenda[36] latera et transeunda prioris piramidis triangularis, ita[37] quod etiam angulos attingunt, quippe cum idem sit radius angularis prioris piramidis triangularis et extremus in circumferentia unius illarum trium.

Translation

[1] *Disproof of a certain cause by which it is usually explained why a solar ray passing through a quadrangular aperture produces a round figure on a wall; disproof effected by Brother Egidius of Baisiu.*

[2] We ask why light passing through a triangular aperture, or any shape whatsoever, produces a round figure on a distant wall. [3] And some say the cause of this is the perfection of the circular figure, asserting that, since light is most active, it acquires such a figure for itself, as other less active substances of the world also acquire said figure for themselves, as is evident in dewdrops on the grass. [4] Furthermore, they endeavour to prove this by rejecting, in the following way, the explanation which attributes the phenomenon to the intersection of the rays produced as a consequence of their rectilinear propagation:

[5] "The rays of the Sun falling on the sides of the angular aperture suffer a greater dilatation than any other passing through the aperture; but the incidence of light acquires its shape as a function of the outer and most dilated rays; therefore, if these rays are propagated in straight lines, as it is asserted, [and] if they are triangular shaped in the aperture, they will remain triangular if they are prolonged to infinity.

[6] "That these rays are the most dilated, I prove thus: rays issuing from the sun propagate from any point on the sun toward any part of the medium and they propagate from the whole sun to any point in the medium where they intersect and, once the intersection is produced, they continue to propagate still further; therefore, among the rays propagating from the same point of the sun and passing through the aperture, it is evident that the most dilated will be those which come into contact with the sides of the aperture. [7] The same I prove with respect to the rest, given that, among all the converging rays, the shortest have the greatest angles. Hence, as some radiant pyramids have their vertices between the sun and the aperture and others beyond the aperture, those formed between the sun and the aperture produce a greater angle since they are shorter, according to Proposition 21 of Euclid's Ist Book; [8] but the more obtuse the angle they form, the more they expand after intersection, given that vertically opposite angles are equal, according to Proposition 15 of Euclid's Ist Book; therefore, the shorter the pyramids, the more they dilate.

---

[35] *ms.*: pedibus.
[36] *ms.*: attingentia
[37] *ms.*: y^a.

[9] "But, among all the pyramids passing through the aperture and converging between the sun and the aperture, the pyramid whose sides touch the aperture dilates the most, as all others are included in these rays. So, it is necessary, with respect to the strength of rectilinear radiation, that rays of light adapt their shape to the aperture through which they pass." [10] Therefore [they conclude] rectilinear radiation is not the cause of roundness.

[11] Thus the rays acquire this figure for themselves, since in the morning and in the afternoon, when the sun's radiation is weaker due to the different refractions, the light which passes through a triangular aperture does not acquire a round figure as completely or as quickly; this occurs, they assert, due to the weakness of the light. [12] But, as this opinion, in the first place, contradicts the authorities on the subject, contradicts experience, and contradicts demonstration, it must consequently be rejected as entirely erroneous.

[13] I say it contradicts all authority on the subject for, as the author of the *Perspective* states and proves with unquestionable experiments and all those who correctly understand the principle are also in agreement, the rectilinear trajectory of the rays cannot be modified in any way except by a variation in the medium; thus, if the air has the same diaphaneity from the aperture to the wall, anyone who claims that the trajectory of the rays is modified to assume a round figure, is contradicting the major authors and destroying the principles of optics, which are entirely based upon this immutable principle.

[14] It also contradicts reason, since every substance acts more intensely and accomplishes its corresponding activity more completely, the nearer it is to its origin; [15] therefore, if rays acquire this round figure naturally, as the plenitude of their intentionality, [then, it follows that] they should acquire the aforesaid figure immediately after the aperture, since they would be nearer their origin close to the aperture than at a great distance from it. [16] If one were to assert that such a round figure is recovered in proportion to the distance, I ask by what reason; [17] moreover, in this case, rays passing through a large aperture would acquire this round figure more quickly since their number is greater and they have a greater capacity for doing so, but this is evidently untrue. [18] Furthermore, if these rays, as a function of their own plenitude, were to acquire a round figure, a figure in which the parts would unite and their virtuality would curve and increase in strength, then they would accomplish their most complete action only when they had acquired this round figure, which is evidently untrue.

[19] It is also contradicts experience for if the body from which the rays originate had another figure, semicircular or crescent-shaped, for example, the image cast would acquire the same shape as the surface of the body from which it originates at the same distance behind the aperture that the light from a circular luminous body acquires a round figure, as is evident in a partial solar eclipse. [20] This is an experiment against which some try in vain to argue. For they say that during the time of the eclipse, two parts are distinguishable in said incidence: the interior angular part, with the same shape as the aperture, possesses a primary or radiant light; the other, which produces the crescent-shaped figure, is less clear, intermediate, so to speak, between the primary and secondary light. [21] And, with this expedient, they believe they avoid the problem, although, taking into account only a few things and without examining them deeply they are actually

only expounding something which can serve as a consolation to their ignorance. [22] For the following is evident to anyone who does experiments: that, beyond the aperture, the light is more intense in the centre [of the image] than near the edges, not only when it proceeds from the sun, but in any incidence of light. The reason for this can be explained. But, what has this to do with what interests us? [23] Since if it is asserted that only this more luminous part is produced as a consequence of the rectilinear propagation of the rays, then [the rays] should take on a circular figure, the contrary of which is admitted. [24] Moreover, where does this secondary light originate? If it proceeds from primary rays, then it will propagate from them in an identical form. [25] When the luminous body is circular, I then ask: is either the round incidence formed on the wall behind the triangular aperture produced from this secondary light as well as the primary one, or is it a consequence of only one of them? [26] We cannot [reasonably] assume it was produced from both kinds of light, since in the case of an eclipse the incidence would be round. [27] If it is assumed to be produced by only one of them, neither can it be produced by only one of them, which is untrue. [28] Furthermore, if this accidental light is generated by or emanates from the primary light, it must be generated equally everywhere, since this is the nature of the multiplication of light as the author of the *Perspective* proves. [29] Then, where does the crescent-shaped figure originate? For if it issues from primary light, then I have proved my point, which is that the incidence assumes the shape of the luminous body as a consequence of rectilinear radiation. [30] But if it is not due either to one or to the other, then a new kind of light is assumed and thus the author of the *Perspective*, who accepts only straight rays, reflected by solid bodies and refracted by different transparent media, and accidental or secondary light, must be considered incomplete on this point. [31] And still I ask: where does [the figure] originate and why does it have that shape and why, if the figure is crescent-shaped, are its horns set in opposite places, that is, inverse to those of the luminous body?

[32] And, so that false ideas cannot be imagined, given that eclipses occur so infrequently, I shall illustrate the case with experiments which can be undertaken at any moment, at night by moonlight or during the day by sunlight. [33] You can take a small thin panel, make a round aperture in it, or whatever shape desired. And you can place the aperture in such a way that a ray passing through a window falls on it, in a position where the rays falling on the panel proceed from only a part of the sun, being careful, however, that the whole aperture is covered by the ray. [34] And on the wall opposite the aperture, the incidence of light will be seen to possess the same figure of the luminous body from which the rays are propagated to the site where the aperture in the panel is found, but inversely placed; the cause of which will be evident later. [35] And the same happens with the moon, whether the aperture is round or not. [36] Furthermore, if a small round aperture is placed in a position where it is struck by light proceeding only from a triangular-shaped part of the sun, a triangular figure will be produced on the wall behind; [37] hence, it is obvious that, in every case, light will reproduce the figure of its luminous source.

[38] Let them be ashamed, all those pretentious moderns, who are unable to find the true cause and are not afraid of inventing falsehoods to deceive the ears of fools; for if they are not afraid of contradicting the words of the ancients

or denying the foundations of optics, at least let them fear contradicting openly the testimony of common vision. But the solution of this problem is difficult and therein clearly lies the truth of [the assertion that] rectilinear radiation [is the true cause].

[39] For it must be admitted that from any luminous body a pyramid converges towards any point in the medium; [40] but of all the pyramids whose vertices are between the aperture and the sun, that is, all those from which some light passes through the aperture, the largest is the one which, after intersection, touches the edges of the aperture with its outermost rays, [and] whose base in the aperture is a circle circumscribing the figure of the aperture; [41] although the only rays from that pyramid to pass through the aperture are those issuing from a figure similar [to that of the aperture] inscribed in the figure of the luminous body. [42] If, for example, the sun, possessing a circular figure, radiates through a triangular aperture by means of the pyramid whose base is the circle circumscribing the triangle of the aperture, it radiates towards the aperture with the greatest radiation, that is, with the greatest dilatation, by means of the pyramid whose section or base in the aperture is the circle circumscribing the triangular aperture; [43] but the only rays from that pyramid to pass through the aperture are those originating from the triangular pyramid circumscribed by the aforementioned [circular] pyramid, whose base in the sun is a triangle circumscribed by the circumference of the sun's surface, that is, by a perceptibly greater circle. [44] But, the pyramid whose base is the circle circumscribed by the figure of the aperture, passes through in its entirety and, as it is generated by a larger base [than that of the triangular

Fig. 1

pyramid], once behind the aperture, it intersects the rays of the first triangular pyramid near the midpoints of the sides of the aperture. [45] And the same occurs with the other intermediate pyramids; as a result, the rays [of the triangular pyramid] lose the form of the figure of the aperture.

[46] To demonstrate this in a quantitative way, it is necessary to note that, according to Ptolemy's assertion in his *Almagest* (Book V, Chapter 15), the distance between the sun and the earth is 1210 parts, the radius of the earth being equivalent to one of those parts. [47] And the diameter of the sun is approximately five and a half times greater than the diameter of the earth, as [Ptolemy] shows in Chapter 16 of the same work. [48] Hence, it is evident that the ratio of the axis of the pyramid, whose vertex is on the earth ad whose base is the surface of the sun, to its base is approximately 111/1. [49] And the same proportion will exist between the axis and the diameter of the base, wherever the pyramid, produced after intersection, or another equal to it, is formed by a base parallel [to that of the first], as is evident according to Proposition 2 of Euclid's Book VI, and also according to Esculeus and [the book] *On pyramids*.

[50] Therefore let us consider, in this case, a triangular aperture whose sides are equal, that is, 4 fingers each, a measurement we shall call a palm. [51] Consequently, the square of the diameter of the circle which circumscribes the triangular aperture will be 4/3 of the square of one side [of the aperture], according to Proposition 8 of Euclid's Book XIII and number 47 of Book I. [52] Therefore, if the square of the side, which is 16, is increased by a third part of that quantity and [from the result] the square root is extracted, it will be found that the diameter of the circle is 4 fingers, 37 minutes and 7 seconds, which we will call 4 thumbs. [53] The vertex of this [circular] pyramid will be situated 444 thumbs in front of the aperture, that is, 40 feet and 4 thumbs; [54] consequently, at an equivalent distance behind the aperture, the diameter of the circle circumscribing the base of the triangular pyramid passing through the aperture and generated by the circular pyramid, will be 8 thumbs. [55] But if a perpendicular line is drawn from the center of the circumscribing circle to one side of the equilateral triangle (the line being equal to half the radius of the circle, as proved by Proposition 8 of Euclid's Book XIII, by means of Proposition 4 of Book I, the first part of Proposition 3 of Book III and the next to the last proposition of Book I), [56] once this line is extended to the circumference and straight lines are drawn from each end of the radius to the angles of the side intersected by the perpendicular line, then [the distance from the aperture to the base] will be 40 feet and 4 thumbs, [the distance] from the centre of the axis of the triangular pyramid to the angles 4 thumbs and to the midpoints of the sides 2 thumbs.

[57] But let us take the other pyramid, the one whose base is the circle circumscribed by the triangle of the aperture, and whose diameter will be 2 thumbs. [58] Therefore for the same reason, its vertex will be located 222 thumbs, that is, 20 feet and 2 thumbs, in front of the aperture. [59] Thus, when it reaches the fixed position of the pyramid on the wall, that is, 40 feet and 4 thumbs beyond the aperture, its diameter will be approximately 6 thumbs. [60] Therefore, [the distance] from the center of the axis, common to both this pyramid and the triangular one, to the circumference of its base, will be 3 thumbs. [61] But [the distance] to the midpoints of the sides of the first [*i.e.*, the triangular one] was only 2 thumbs;

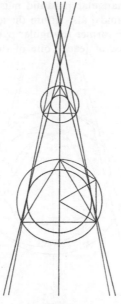

Fig. 2

[62] therefore the circular pyramid exceeds the sides of the first by 1 thumb and is surpassed by the angles of the triangular pyramid by 1 thumb. [63] But if the first, that is, the triangular one, is extended 40 feet and 4 thumbs further, in such a way that the base is located 80 feet and 8 thumbs from the aperture, the diameter of the circle [which cirumscribes it] will be 12 thumbs and the distance between the centre and the midpoints of the sides of the triangle will be 3 thumbs. [64] But if the second that is, the circlar pyramid, is extended to the same distance, its diameter will be approximately 10 thumbs, that is, 10 thumbs and approximately 1/11 of a thumb. [65] Thus, given that there would be 5 thumbs from the common center to its circumference, it surpasses the distance from the midpoint of the side of the triangle by 2 thumbs and is surpassed by the angles by 1 thumb. [66] And proceeding in this way, if the intermediate pyramids between these two are considered, their circular portions will fill in the excess by which [in the aperture] the angles surpass the smallest pyramid. [67] And the light [of the incidence] will be circular although the diameters will not be entirely equal.

[68] Moreover, let us consider the three pyramids whose vertices are located on the three angles of the aperture [and] whose axes issue from the centre of the sun. [69] Thus, when they reach the aforementioned place, that is, 80 feet and 8 thumbs behind the aperture, the diameter of each will be 8 thumbs. [70] And their axes will separate, forming a triangle whose circumscribing circle has a diameter which is greater than 4 thumbs in the same proportion maintained by the 80 feet and 8 thumbs with regard to the distance between the aperture and the sun. [71] But let both diameters be 4 thumbs. [72] Then the diameter of the pyramid composed of those three [pyramids] will be approximately 12 thumbs. [73] Therefore,

the sides of the former triangular pyramid must be attained and exceeded in such a way that [these pyramids] also attain the angles, given that [the endpoint of] the angular radius of the former triangular pyramid and the endpoint [of the radius] on the circumference of [each] one of those three coincide.

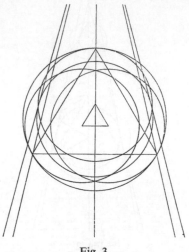

Fig. 3

Commentary

Ad [3]. In three works dealing with the problem of pinhole images (*Tractatus de sphera* and both versions of *Perspectiva communis*) PECHAM concludes that the circular incidence produced by light passing through an angular aperture cannot be explained by the rectilinear propagation of light which, according to his reasoning, would always give rise to an image with the same shape as the aperture. He, then, goes on to attribute the roundness of the incidence to a so-called "natural tendency" of light toward circularity. The first part of the *Improbatio* (Sections 2–37) demonstrates the inconsistency of such an analysis. Although neither PECHAM nor his works are mentioned by name, the extensive quotation of his arguments leaves no doubt as to where EGIDIUS' criticisms are directed.

PECHAM's notion that light naturally tends to assume a circular shape seems to be derived by analogy from the orbicular diffusion of light from any luminous source through the medium (*Perspectiva communis*, I, 6). In this section, EGIDIUS makes a direct reference to the *Tractatus de perspectiva*, where PECHAM writes: "But, given that light is active in the highest degree, it acquires a circular figure for itself"[1]. The same assertion is restated in the first version of the *Perspectiva communis*: "Then, it must be known that the spherical figure, which in its interior

---

[1] "Quia igitur lux est maxime activa, acquirit sibi figuram circularem" (LINDBERG [1], p. 168, n. 40). I give my own translation for PECHAM's Latin texts edited by LINDBERG. When I quote from the *Perspectiva communis*, the notation "I, 5, 106–10", for instance, means: first part, Proposition 5, lines 106–10.

unites all its parts in the most perfect manner, is the shape in accordance with light and is in harmony with all earthly bodies since it safeguards its nature to the maximum; hence even a raindrop achieves roundness. Thus, light naturally seeks this [figure] and acquires it by degrees as the distance increases"[2]. And again in the *Tractatus de sphera*: "Similarly the rays of the incidence passing through a plane angular aperture, the length of whose sides is not too great, acquire a round shape on the facing wall; the cause of which is that light, being strongly active, easily acquires the form to which it is naturally inclined"[3].

In the revised version of the *Perspectiva communis* PECHAM appears to have altered his position. After reminding the reader that some people attribute the circularity of the incidence to the natural tendency of light to assume a spherical figure, the most suitable for simple bodies, he adds that if this were true, the incidence during an eclipse would be circular rather than crescent-shaped. As he continues, however, it becomes clear that he has, in fact, only changed the presentation of his arguments. Thus, after apparently refuting the claim that primary light propagating in straight lines can explain the circularity of the image on the wall, he concludes: "Owing to which it seems to some [*i.e.*, to PECHAM] that the roundness of the incidence originates from both; which is to say, [the incidence] possesses the magnitude of the dilatation as a consequence of the intersection [of straight rays] and the perfection of roundness as a consequence of the nature of light."[4]

Ad [4]. PECHAM maintains that the natural tendency of light towards circularity explains the formation of pinhole images after having refuted an attempt at a geometrical solution which he attributes to others (see my commentary on Sections 39–45). Furthermore, he considers that his refutation is equivalent to *a proof of the impossibility of any geometric explanation of the phenomenon*. Thus, in the *Tractatus de sphera*, he writes: "That this roundness cannot be produced by any intersection of the radiation, but rather only by diffusion, is proven in the following manner ..."[5]. (The corresponding passages in both versions of the *Perspectiva communis* are: I, 5, 118–9, and I, {7}, 201–5.)

Ad [5]–[9]. These sections are taken almost word for word from the *Tractatus de sphera* (*cf.* MACLAREN [1], pp. 98–100). The passage on pinhole images from PECHAM'S *Tractatus de sphera* can be found also in DUHEM ([1], pp. 524–9), who used only MS. 419 of the Bibliothèque Municipale of Bordeaux. The text quoted

---

[2] "Sciendum igitur figuram spericam esse luci cognatam et omnibus mundi corporibus esse consonam utpote nature maxime salvativam, que omnes partes suo intimo perfectissime coniungit, unde et stilla in rotunditatem incidit. Ad hanc igitur naturaliter lux movetur et eam protelata distantia paulatim acquirit" (LINDBERG [4], I, 5, 106–10).

[3] "Similiter radii incidentie cadentis per foramen planum angulare, cujus tamen latera non multum se extendant, in pariete opposita rotundantur; cujus ratio est quod lux ut potissime activa hanc formam facilius consequitur ad quam naturaliter inclinatur" (MACLAREN [1], p. 97; DUHEM [1], p. 526).

[4] "Propter quod videtur non nullis quod rotunditas incidente causatur ex utroque ut videlicet per intersectionem habeat dilatationis magnitudinem sed per naturam lucis rotunditatis perfectionem" (LINDBERG [4], I, {7}, 218–20).

[5] "Quod autem nulla radiatione vel radiorum intersectione possit causari illa rotunditas, sed sola diffusione, probatur sic ..." (MACLAREN [1], p. 98; DUHEM [1], p. 527).

by EGIDIUS is on pp. 527–8. The corresponding passages in the *Perspectiva communis* are: I, 5, 70–104, and I, {7}, 171–200.

Ad [7]. EUCLID, *Elements*, I, 21; HEATH [1], p. 289.

Ad [8]. EUCLID, *Elements*, I, 15; HEATH [1], pp. 277–8. This second reference to EUCLID is not in the *Tractatus de sphera* (*cf.* MACLAREN [1], p. 99; DUHEM [1], p. 528).

Ad [11]. The view that the circularity of the image is due not only to the natural tendency of light towards roundness, but is also dependent on the strength of the rays, can be found in PECHAM's *Tractatus de perspectiva*: "... since at sunrise, when the rays are extremely weak because at that time they incline greatly as they are passing through vapours, they do not acquire roundness for themselves as they do at noon"[6]. BACON offers a similar explanation in *De multiplicatione specierum* (II, 8): "And in the same way, we observe that rays proceeding from the sun or from any other [luminous body], passing through the same aperture produce a more circular image at noon than in the morning. Hence, there is no need [to assume] that although the aperture is always the same, for this reason the characteristics of the image will be the same in every case, near as well as far off, at one hour or another, according to one trajectory of the rays or another; because it is not so. Since rays fall at oblique angles through vapours in the morning, this causes them to be weaker, producing a weaker image; such conditions are not present at noon and so, given that [the rays] are stronger, they assume a round figure at noon"[7].

Ad [13]. "*Auctor Perspective*" is the usual way of refering to IBN AL-HAYTHAM (ALHAZEN) in the thirteenth century (*cf.* LINDBERG [5], p. xiii). Here EGIDIUS may be making reference to Propositions 14, 17 or 18 of Book I, or Proposition 2 of Book VII of ALHAZEN's *De aspectibus* (*K. al-Manāzir*). Proposition 17 of Book I, for example, states that "... light extends through a diaphanous body in straight lines as long as the diaphanous body maintains a similar diaphaneity, and if it were to encounter another body with a different diaphaneity than the preceding body through which it had extended, it would not traverse it along the straight lines by which it had previously extended, except in the case that these lines were perpendicular to the surface of the second diaphanous body"[8].

---

[6] "... quia in ortu solis cum radii valde debiles sunt, tum quia/multum declinantes tum quia per medium vaporum transeuntes, non acquirunt sibi ita rotunditatem sicut in meridie" (LINDBERG [3], p. 302, n. 15).

[7] "Et similiter videmus, quod radii solis vel alterius cadentis in meridie transeuntes per idem foramen faciunt speciem magis rotundam quam in mane. Unde non oportet quod licet sit idem foramen, quod propter hoc sit omnibus modis eadem conditio speciei, et prope et longe, et in una hora et in alia, et ad unum casum radiorum et alium; non enim ita est. Nam in mane cadunt radii ad angulos obliquos per medium vaporum, propter quas causas debiliores sunt, facientes speciem debiliorem, quae causae non accidunt in meridie, et ideo acquirunt sibi figuram rotundam in meridie, quoniam fortiores sunt" (BRIDGES [1], p. 493).

[8] "... lux extenditur per corpus diaphanum secundum lineas rectas, dum corpus diaphanum fuerit consimilis diaphanitatis: & cum occurrerit corpus aliud diuersae diaphanitatis a diaphanitate corporis praecedentis, in quo extendebatur, non pertransibit secundum rectitudinem linearum, super quas extendebatur ante, nisi quando illae lineae fuerint perpendiculares super superficiem secundi corporis diaphani ..." (ALHAZEN [1] I, 5, p. 9).

For EGIDIUS, PECHAM'S conclusion clearly implies a violation of the principle of rectilinear propagation of light (also see Section 38 of the text). It is not easy, however, to assume, as does LINDBERG ([7], p. 50), that PECHAM was forced into such a position by "an unwelcome, but inescapable, teaching of the observational data". I would suspect that PECHAM probably did not conduct any experiments on the passage of light through small apertures (see commentary on Sections 39–45). In any case, it is clear that PECHAM'S inability to provide a geometrical explanation for the problem of pinhole images was not due to observational data.

Ad [15]. Refering to arguments based on the distance between the aperture and the incidence, EGIDIUS uses PECHAM'S own logic against him. Refuting those who claim that the circular incidence is caused by the circular shape of the sun (since during an eclipse the incidence is crescent-shaped), PECHAM argues rather ineffectively that such an explanation is insufficient for if this were the case, the incidence should be circular *near* the aperture as well as *far* from it: "Some simply attribute it to the sun's roundness, so that, as the ray [is generated] by the sun, so roundness [is generated] by roundness, using for this an observational argument, since in the case of the solar eclipse the rays become crescent-shaped, according to the portion of the sun cut off by the moon; for this reason roundness seems to proceed from roundness. But, if this cause were sufficient, the incidence would acquire roundness both near and far from the aperture, which is the contrary of what we observe"[9]. The same argument is repeated in the *Tractatus de sphera* (MACLAREN [1], p. 97; DUHEM [1], p. 527). If this argument is valid, then, it must also be valid, as EGIDIUS points out, when applied to PECHAM'S own conclusions: the natural tendency of light towards circularity should produce a circular image not only at a distance from the aperture, but also immediately behind it.

Ad [16]. To claim this natural tendency functions in proportion to distance (*cf.* Note 2), EGIDIUS appears to conclude, is *post hoc, ergo propter hoc* reasoning and as such requires further explanation.

Ad [20]. Although reference is made to a distinction between *lux primaria* and *lux secundaria* in the Latin version of IBN AL-HAYTHAM'S *K. al-Manāzir* (ALHAZEN [1], IV, 5, p. 104; VII, 1, p. 231) and in the *Perspectiva* of WITELO ([1], def. 4, p. 61; V, 4, p. 192), PECHAM is apparently the first to use this distinction to explain the formation of pinhole images. In the first version of the *Perspectiva communis*, he states: "However, during an eclipse the aforementioned crescent shape is produced because the accidental and secondary diffusion of the light is obstructed in that part where the solar rays are cut off; thus, lacking principal light, secondary light must be also lacking"[10]. In other words, when the sun is full, secondary light fills out the triangular image cast by rectilinear radiation (*i.e.*, primary light),

---

[9] "Quidam simpliciter solari tribuunt rotunditati ut sicut radius a sole sic rotunditas a rotunditate, ad hoc argumentum sensibile assumentes quia tempore eclipsis solaris huius incidentie fiunt novaculares secundum portionem quam abscindit luna de sole, propter quod rotunditas videtur esse a rotunditate. Sed si hec causa sufficeret incidentia ista rotunditatem acquireret sic prope foramen sicut longe a foramine, cuius contrarium videmus" (LINDBERG [4], I, 5, 63–9).

[10] "In tempore autem eclipsis fit novaculatio predicta quia impeditur accidentalis et secundaria lucis diffusio in parte illa qua radii solares abscinduntur, deficiente enim principali lumine necesse est secundarium deficere" (LINDBERG [4], I, 5, 110–3).

producing a circular incidence. But, during a solar eclipse, the lack of secondary light does not permit the formation of a circular image so the resulting incidence is crescent-shaped. PECHAM is clearly unaware of the inconsistency inherent in his argumentation. If secondary light contributes nothing to the crescent-shaped image, then primary light conforms to the shape of the aperture in one case and to the shape of the luminous source in the other. The contradiction involved is all too obvious. On the other hand, if a combination of primary and secondary light produces a circular incidence when the sun is full, but a crescent-shaped one when primary and secondary light are diminished during an eclipse, then (taking in consideration PECHAM's claim that primary light always assumes the angular shape of the aperture), the effects of secondary light are different in each case. The reason for this, however, is left unexplained.

Although PECHAM's views are more developed in the *Tractatus de sphera* and in the revised version of the *Perspectiva communis* than in the first version, it is surprising to discover that the inconsistency, merely inherent in his former explanation, is now explicitly stated. He goes so far, in fact, as actually to defend the view that primary light produces the crescent-shaped incidence during an eclipse. Thus, in the *Tractatus de sphera*, he asserts: "Let them listen, those who so speak, that in such an incidence, two parts are distinguishable to the eye, of which the interior is angular like the aperture and the exterior, which is said to be less clear and of intermediate brightness, so to speak, between primary and secondary light, is circular. I call the radiating light "primary light" and the light which is outside the incidence of the rays, as in houses open only to the north when the sun is in the south, "secondary light". Therefore, this crescent-shaped configuration is not in the exterior incidence, which does not assume the shape of the aperture, but rather in that incidence [*i.e.*, the interior] which is generated by the lengthening of the radiating light. The incidence, thus, reproduces the shape of the darkening of the sun; that is, the incidence produced by intersection [*i.e.*, the interior], not the one which externally tends towards roundness as a consequence of natural diffusion, except by accident. The roundness of the incidence is not caused by the roundness of the sun, given that when the sun is said to be crescent-shaped because of the interposition of the moon, the roundness of the accidental light continues to exist as is evident to the eye. Since radiating light is nevertheless the principal one and the cause of the accidentally diffused light, thereby, if radiating light is hindered, secondary light is consequently also hindered so that, during the eclipse, its roundness is not easily discerned"[11].

In the revised version of the *Perspectiva communis*, PECHAM continues to stress

---

[11] "Attendant igitur qui sic locuuntur quia in tali incidentia lucis due sunt partes sensui distinguibiles, quarum una interior angularis in modum foraminis, altera exterior circularis, que minus clara dicitur et quasi medii splendoris inter lucem primariam et secundariam. Dico lucem primariam, lucem radiosam; secundariam, illa que est extra incidentiam radiorum, ut in domibus apertis ad aquilonem tantum, sole existente in austro. Ergo in exteriori incidentia, que modum foraminis non sequitur, non est illa novaculatio, sed in illa que est ex radiosa luminis protensione generata. Sequitur ergo incidentia modum obumbrationis solaris, scilicet illa que intersectione gignitur, non illa que exterius naturali diffusione in rotunditatem deducitur nisi per accidens. Non est igitur rotunditas incidentie ex solari rotunditate, quia sole ex interpositione lune quasi nova-

this point: "Solar light is diffused in two manners, that is, in a direct and radiating form, which is called primary light, and in another manner, indirectly in all directions and outside of the rays such as, with the sun located over the horizon, a house is full of light although no solar ray has penetrated it, and this is called secondary or accidental light (...). Moreover, since the circle of this incidence of the light is often of a lesser brightness than that part of the light proportioned to the aperture through which it passes, this is also why the answer to the objection raised concerning the incidence during an eclipse is evident, for the most intense light assumes the shape of a crescent and nevertheless this same crescent shape is surrounded by a certain minor light. Furthermore, it should be known that the most important part of the radiating light consists of vertical rays issued from the center of the sun, which were spoken of above. And since a considerable part of these rays is interrupted during an eclipse, otherwise there would be no eclipse, quite rightly the incidence appears crescent-shaped and the light of the other rays is reduced. But through any aperture, no matter how small, passes as much of the first class of rays [*i.e.*, those from the pyramids with base on the sun and vertex on the plane of the aperture] as of the third class [*i.e.*, those from the pyramids with base on the sun and vertex behind the aperture] or the second [*i.e.*, those from the pyramids with base on the sun and vertex between the sun and the aperture], and thus, when there is no eclipse, the light is stronger and more capable of acquiring the shape of that [*i.e.*, the aperture] than during an eclipse"[12].

EGIDIUS, KEPLER and MAUROLICO[13] base their approach to the problem

culato manet rotunditas luminis accidentialis, sicut sensui patet. Quia tamen lumen radiosum est principale et causa luminis accidentaliter diffusi, ideo, impedito lumine radioso, per consequens impeditur lumen secundarium, ut non faciliter discernatur, in tempore eclipsis, eius rotunditas" (MACLAREN [1], pp. 97 8; DUHEM [1], pp. 527–8). I have modified MACLAREN'S and DUHEM'S wording and punctuation in some places; consequently, the English translation is mine.

[12] "... lux solaris dupliciter se diffundit, scilicet, directe et radiose, et hec dicitur lux primaria, item alio modo indirecte in omnem directionem et extra radios, sicut sole existente super orizontem domus est plena luce quamvis nullus eam intret radius solaris, et hec dicitur lux secundaria vel accidentalis (...). Ad hoc autem quoniam frequenter circulus huius lucis incidentie est debilioris luminis quam sit pars luminis proportionata angulari foramini quod transit, et per hoc patet responsio ad obiectum de incidentia tempore eclipsis, quoniam lux intensior novaculatur, et tamen ipsa novaculatio luce quadam minori superducitur. Preterea sciendum est quod gloria luminis radiosi consistit in radiis verticalibus a centro solis procedentibus, de quibus dictum est supra. Et quia tempore eclipsis abscinditur pars magna istorum radiorum, alioquin non appareret eclipsis, merito apparet incidentia novaculata, et lumen aliorum radiorum minoratur. Sed per omne foramen quantumcumque parvum transit tantum de primo genere radiorum quantum de tertio vel secundo, et ideo fortius est lumen extra tempus eclipsis et potentius ad figuram huius acquirendam quam tempore eclipsis" (LINDBERG [4], I, {7}, 151–5 and 237–49).

[13] See Section 37 of EGIDIUS' text. KEPLER [1], p. 39: "Patuit itaque concurrere ad problema demonstrandum, rotunditatem non radii visorii, sed ipsius solis, non quia haec perfectissima sit figura, sed quia haec lucentis corporis figura sit in genere". MAUROLICO [1], p. 18: "Concludimus ergo, quod quo magis à quocumque lucido per qualemcumque foramen radiante, processerint radij; eo magis in planum, quod lucido parallelum est, profectum lumen ad ipsius lucidi similitudinem accedit ...".

of pinhole images on explaining why light passing through an angular aperture reproduces the shape of the luminous source *at all times* provided the distance between aperture and image is sufficient. PECHAM, however, never formulates the problem in these terms. He seems to be more concerned with explaining *the lack of circularity* during an eclipse than the crescent-shaped incidence actually observed. This, in part, may account for the incoherence of his arguments in the *Perspectiva communis*.

Ad [23]. The passage from the *Tractatus de sphera* quoted above poses several problems of interpretation, due not only to a certain obscurity in the text, but also to the inconsistency of PECHAM's solution. First, exactly what kind of light produces this less clear part of the incidence whose brightness is somewhere between primary and secondary light? And second, what is the relationship between these interior and exterior areas of light and the formation of pinhole images?

At first, LINDBERG ([1], p. 174) assumes that the exterior part of the incidence (*i.e.*, that of intermediate brightness) is produced by secondary light. Yet, later on, taking for granted that PECHAM has used the treatise *De speculis comburentibus* attributed to BACON, LINDBERG ([2], p. 222) maintains that the exterior part of the incidence is primary light. Refering to the passage quoted from the *Tractatus de sphera*, he states that PECHAM "attributes circularity of the solar image formed by a triangular aperture alternately to exterior primary radiation (following BACON) and to secondary radiation; in his revised version of the *Perspectiva communis* he restricts himself to secondary radiation". But, if it is difficult to figure out which type of light, primary or secondary, is the one whose brightness, according to PECHAM, is intermediate, it is no easier to understand how the circularity can be caused *alternately* by one type of light ("exterior primary radiation") and then another.

To assume that the exterior part of the incidence is produced by primary light simply increases the lack of coherence already present in PECHAM's views. To begin with, PECHAM does, in fact, seem to attribute the circularity of the image exclusively to secondary radiation not only in the first version of the *Perspectiva communis*, but also in the revised version (see Note 10). If PECHAM argues that the scarcity or absence of secondary light explains why the image is crescent-shaped during an eclipse rather than circular, it is surely because he believes that secondary light is responsible for the round image produced when the sun is full.

Second, there is little evidence to support LINDBERG's claim that, in the *Tractatus de sphera*, PECHAM identifies the exterior part of the incidence with "BACON's circular pyramid converging to an apex at the central point [E] of the aperture [DF]" (LINDBERG [2], p. 222; see Figure 4). In the *Tractatus de sphera* (MACLAREN

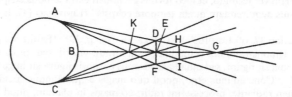

Fig. 4. LINDBERG [2], p. 219.

[1], pp. 99–100; Duhem [1], p. 528) and both versions of the *Perspectiva communis* (Lindberg [4], I, 5, 80–119; I, {7}, 185–200), Pecham explicitly states that the pyramid ACE is absorbed by the pyramid whose vertex is at point K. Pecham also maintains in the *Tractatus de sphera* that rectilinear radiation can only produce triangular incidences. Thus, if the circularity of the image is due to "exterior primary radiation", this radiation ought to be outside the pyramid represented by lines AKF and CKD and this is clearly impossible.

Finally, there is nothing in the first version of the *Perspectiva communis* or in the passage quoted from the *Tractatus de sphera* which would lead us to believe that Pecham attributes primary light with any means of propagation other than rectilinear. Thus, if this "exterior radiation" is primary and primary light can only reproduce the shape of the aperture, then how can it cast a circular image? In the revised version of the *Perspectiva communis* (see Note 12), Pecham repeats his conclusion: when the sun is full, the most intense light is not responsible for the circular incidence as it is restricted to reproducing the shape of the aperture; during an eclipse, however, this same intense light does produce the crescent-shaped incidence. It is more than a little difficult to reconcile Pecham's views with the notion that this intermediate light is primary. Consequently, it seems that the only way to resolve the contradiction between the *Tractatus de sphera* and the first version of the *Perspectiva communis* on this point is by assuming, as Egidius does, that this intermediate light is secondary light (or a third type of light). The fact that Egidius (Section 20) attributes to the *Tractatus de sphera* a statement not contained in the text (*i.e.*, during an eclipse, intermediate light gives the angular incidence produced by primary light its crescent shape) can, in my opinion, be explained as an attempt to make some kind of sense of Pecham's exasperating contradictions.

Ad [24]. According to Ibn al-Haytham (*cf.* Sabra [1], p. 191; Rashed [1], pp. 274–6), secondary light radiates in straight lines in all directions from any point in the medium accidentally struck by primary light. Thus, if secondary light proceeds from primary rays, to use Egidius' words, then, assuming its effects were visible on the screen, it should cast a nearly triangular shape (as in Figure 5) when the beam from which it emanates has a triangular base. According to Sabra ([1], p. 197), the known Latin manuscripts of Ibn al-Haytham's *Optics* do not contain the first three chapters of the Arabic text. It is in these missing chapters that Ibn

Fig. 5

AL-HAYTHAM explains in detail the difference between primary and secondary light and the rectilinear propagation of both. This may explain why PECHAM assumes that the propagation of secondary light is not rectilinear or, at least, that the effects of secondary light are not compatible with rectilinear propagation.

In the revised version of the *Perspectiva communis*, PECHAM appears to have given his assumptions some additional thought. Thus, he notes that in the case of a circular incidence "wherever the eye rests on the incidence, the body of the sun can be seen since vision occurs only by means of straight lines"[14]. He goes on to conclude that whatever the shape of the image, it is determined by straight lines. This does not seem to pose any special problems for PECHAM as he continues to maintain that the circularity is not due to primary light propagated in straight lines. He simply adds that "the light lacking little to be round takes on its connatural shape due to the proximity of another light"[15], by which he probably means secondary light. In other words, the circular image on the screen is formed by primary light which produces a triangular image and by the rectilinear rays of secondary light which rounds out the triangular incidence. The deflection of the secondary light rays, possibly produced in the aperture, is justified by PECHAM using a weak analogy. As he affirms in Proposition 40 of the *Perspectiva communis* (Proposition 43 in the revised version), the rays which produce vision must modify their trajectory when penetrating the vitreous humour in order to converge in the optic nerve. This deflection is produced "more in agreement with the law of the spirits than in agreement with the law of diaphaneity"[16]. Without realizing that he is contradicting not only ALHAZEN ([1], I, 5, p. 9 ff.), but also his own proposition 29 (32 in the revised version) in the *Perspectiva communis*, PECHAM claims it is an analogous process which explains the trajectory of secondary light in the case of pinhole images. EGIDIUS' affirmation of the rectilinear propagation of secondary light implies if not a knowledge of the first three chapters of the Arabic text, at least a more coherent reconstruction of IBN AL-HAYTHAM's theory of primary and secondary light than PECHAM was able to provide.

Ad [26]–[27]. EGIDIUS takes PECHAM's arguments to their logical and absurd conclusion. To claim that the circular incidence is produced by a combination of primary and secondary light, is incompatible with the crescent-shaped image cast during an eclipse which, if such were the case, should also be round. If, on the other hand, the circularity is due to *only* one type of light, it cannot be primary light which is supposedly limited to reproducing the shape of the aperture, nor can it be secondary light which exists simply as a consequence of primary light. That the round image cannot be caused by only one of these types of light is obviously false because it is, in fact, produced by primary light.

Ad [28]. EGIDIUS stresses that if secondary light comes from primary light, it should, assuming its effects on the screen are visible, take on the same shape in

---

[14] "... ubicumque ponatur oculus in incidentia potest videre corpus solis, visio autem non est nisi per lineam rectam" (LINDBERG [4], I, {7}, 144–6).

[15] "... lux deficiens modice a rotunditate per vicinitatem alterius lucis formam iudicat [vindicat?] connaturalem" (LINDBERG [4], I, {7}, 220–2).

[16] "... secundum legem spirituum magis procedit quam secundum legem diaphonei-tatis" (LINDBERG [4], I, 40 {43}, 783–4).

all cases and not circular in some and crescent-shaped in others. I have been unable to locate EGIDIUS' reference to ALHAZEN in RISNER's edition of the Latin text.

Ad [30]. EGIDIUS has demonstrated that PECHAM's explanation based on the two types of light allowed is inconsistent. If the shape of the image cannot be derived without contradiction either from primary light or secondary or from a combination of the two, then one must suppose there exists a third. But the possibility of a third type of light contradicts the principles of optics and the authority of ALHAZEN ([1], IV, 1, p. 102; IV, 5, p. 104; VII, 1, p. 231).

Ad [31]. The observation that the image on the screen is inverted is not a trivial one. As we have seen, PECHAM is unaware of it in the first version of the *Perspectiva communis*: "Furthermore, if the roundness were produced by intersection, then if the sun were eclipsed in the eastern part, the incidence would be diminished in the western part and not in the same part ..."[17].

Ad [39]–[45]. Earlier I qualified as $\psi\varepsilon\nu\delta o\gamma\varrho\acute{\alpha}\phi\eta\mu\alpha$ the geometric argument PECHAM uses to prove that rectilinear radiation would necessarily produce a triangular image (quoted by EGIDIUS in Sections 5–9). This is the term ARISTOTLE uses to describe the attempts of HIPPOCRATES, ANTIPHON and BRYSON to square the circle. It is used not only to label false reasonings "which do not start from the true principles of the science, but also those which do start from those principles but go wrong in some particular point" (HEATH [2], p. 77; see also ALLMAN [1], pp. 62–75). Without refuting *in extenso* PECHAM's paralogism, EGIDIUS simply substitutes it for an analysis without erroneous premises. In doing so, he is following ARISTOTLE's rule :"... one solves the correctly reasoned arguments by demolishing them, the apparent reasonings by making distinctions" (ARISTOTLE [2], pp. 100–1).

The false proof from which PECHAM derives his conclusion is repeated without any important changes in both versions of the *Perspectiva communis* and in the *Tractatus de sphera*. That PECHAM manages from beginning to end to cling so resolutely to his belief in the validity of his argument is, indeed, puzzling. But, even more puzzling is the increasing frequency of his geometric errors. He, himself, realizes that his figures are not the most appropriate for visualizing the phenomena he is trying to analize ("... as is evident in the figure", he writes, "insofar as a plane can properly represent three-dimensional figures")[18]. But the term $\psi\varepsilon\nu\delta o\gamma\varrho\alpha\phi\acute{\omega}\nu$ is used by ARISTOTLE ([1], pp. 274–5) not only to mean one who draws or employs an incorrect figure, but also one who elaborates an incorrect proof. In the latter case, the figure is inappropriate because it is generated by false reasoning. KEPLER was aware that in the *Perspectiva communis* the error lay in the concepts as well as in their geometric representation when he wrote: "Years ago a certain light shone for me in the darkness of PISANUS [*i.e.*, PECHAM]. Since

---

[17] "Amplius si rotunditas esset ex intersectione tunc si sol eclipsaretur in parte orientali deficeret incidentia in parte occidentali et non in eadem parte ..." (LINDBERG [4], I, 5, 113–5).

[18] "... sicut patet in figura quatenus planities potest figuras solidas declarare ..." (LINDBERG [4], I, 5, 89–90).

I couldn't understand the obscure meaning of the words from his plane figure, I resorted to personal observation in three dimensions"[19].

In Sections 39–45 EGIDIUS demonstrates the erroneousness of PECHAM'S conclusion that rectilinear radiation can produce only triangular images. There are two crucial, logically related points not taken into account by the author of the *Perspectiva communis*. First, the triangular pyramid which, according to PECHAM, behind the aperture absorbs all others, has a *triangular* base on the sun. Second, there are, however, other pyramids which pass through the aperture whole, pyramids having their vertices between the sun and the aperture and whose bases on the sun are *circular*. The intersection of these pyramids with the triangular-based one causes the latter to lose its shape. In order to appreciate the extent of PECHAM'S failure to consider these points, a step by step analysis of the arguments developed in the *Perspectiva communis* (those presented in the *Tractatus de sphera* being basically the same) is particularly useful.

Both versions begin with the same mistaken claim. If the screen were at the same distance from the aperture as the aperture is from the sun, the incidence would be the same size as the sun (*Perspectiva communis*, I, 5, 56–61; I, {7}, 125–30). From here on, the two versions differ.

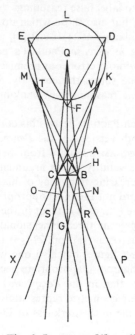

Fig. 6. LINDBERG [4], p. 69.

[19] "Mihi ante annos complures ex Pisani tenebris aliqua lux affulsit. Cum enim sensum verborum adeo obscurum ex schemate in plano comprehendere nequirem: confugi ad αὐτοψίαν in solido" (KEPLER [1], p. 39).

1. In the *first version*, PECHAM writes that "some people" try to explain the circularity of the image in the following manner. Triangle DEF (Figure 6) on the solar disk is the base of the luminous pyramid whose vertex behind the aperture is point G. KLM is a circumference which is also inside the solar disk. The only circular-based pyramid able to pass through the aperture whole is the one whose vertex is in the middle of the aperture at point H. Since the outermost rays of this pyramid form a larger angle than the one formed by DEF converging at G, the former will absorb the latter intersecting at points O and N. It will also absorb the triangular-based pyramid whose origin and vertex are at point Q on the surface of the sun intersecting, in this case, at points S and R. PECHAM goes on to refute this proof developed by "others" using the following arguments:

1.1. If this explanation were correct, the luminous pyramid, triangular directly behind the aperture, would suddenly become circular. The reason for this is, as PECHAM writes, that "everything beyond NO or, at least, beyond RS would be completely round [and] what came before triangular; the contrary of what we clearly observe since we can see that this light acquires roundness little by little"[20].

1.2. Furthermore, "rays VX and TP adapt themselves to the sides of the aperture and adopt its form, and it is certain that these rays include all the rest which could produce the roundness by means of rectilinear radiation"[21].

2. In the *revised version*, PECHAM rejects the explanation that the circularity of the image is due simply to the natural tendency of light toward roundness since it is only acquired "to the extent that it is caused by radiating lines"[22]. This is the reason, he continues, that "some people" claim that the cause is the intersection of the rays and the phenomenon is produced in the following way. The shortest of all the pyramids passing through the aperture whose vertices are in the aperture, is KLM (Figure 7). This pyramid, whose base on the sun is KL and whose vertex is in the middle of the aperture at point M, is the shortest because its axis, which is perpendicular to the plane of the aperture, is shorter than the axes of the rest. If it is the shortest, the angle formed by its outermost rays will be greater. Thus, once past the aperture, pyramid KLM will end up absorbing all the others whose vertices are in the aperture. PECHAM adds to this explanation the following commentaries:

2.1. In addition to the luminous pyramids having their bases on the sun and their vertices in the aperture, one must consider those whose vertices are between the sun and the aperture, and those whose vertices are behind the aperture.

2.2. As a result, the outermost rays of the pyramid with its vertex at point O (Figure 8) form an angle larger than the one produced by pyramid KLM. The pyramid with vertex O is, therefore, shorter and the lines of its outermost rays

---

[20] "... quicquid esset ultra NO vel ad minus RS esset rotundum complete, quicquid citra triangulare, cuius contrarium manifeste videmus, quia videmus lumen ipsum paulatim rotunditatem acquirere" (LINDBERG [4], I, 5, 101–4).

[21] "... radii VX et TP applicant se lateribus foraminis et sequuntur figuram eius et certum est quod isti omnes alios includunt qui rotunditatem possint radiositate recta generare" (LINDBERG [4], I, 5, 115–8).

[22] "... nisi quatenus causatur a lineis radiosis" (LINDBERG [4], I, {7}, 148).

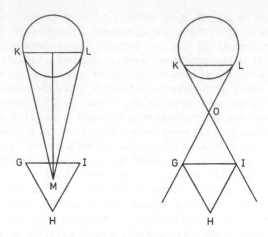

Figs. 7 and 8. LINDBERG [4], p. 77.

"will absorb, prolonged in straight lines, the lines and the incidences of all the rest of the longer pyramids"[23].

2.3. To re-enforce his refutation of the explanation developed by "some people", PECHAM adds a surprising argument. Let's suppose, he says, that the passage of the light were partially blocked by an opaque disk inscribed in the triangle of the aperture. The rays passing through the three resulting apertures would form three circles on the screen. If the disk were taken away, the rays, passing whole through the aperture, would form a larger circular image (larger than each of the three previous circles, we can assume since the text is not explicit). Thus, "it is impossible that the three circular images previously mentioned are parts of the last. Therefore, it is impossible, by [rectilinear] radiation, that the small beams passing through the three corners of the aperture outside the circumference of the circle inscribed in the aperture should converge on the same circular surface as the light passing through the concavity of the circle inscribed in the aperture"[24].

It seems fairly clear from the content of some of PECHAM'S objections that he does not understand correctly the statements he is trying to refute. This would lead one to think that the explanation PECHAM attributes to "others" is only a self-made rhetorical device to introduce his own conclusions. If he really understood Figure 6, would he state that the argument of his opponents implies that before NO (or, at least, RS) the incidence on the screen would be triangular and com-

---

[23] "... absorbent, in continuum et directum producte, omnium aliarum pyramidum longiorum lineas et incidentias" (LINDBERG [4], I, {7}, 198–200).

[24] "... cuius partes impossibile est esse predictorum trium circulorum portiones. Ergo impossibile est quod per viam radiationis radioli transitantes per tres angulos ysopleuri extra circumferentiam circuli inscripti concurrant in eadem superficie circulari cum lumine traseunte per concavitatem circuli inscripti ysopleuro" (LINDBERG [4], I, {7}, 212–7).

pletely round directly afterwards? PECHAM'S confusion is complete. After attributing to his opponents an error which is really his own (the instantaneity of the transition from triangular to circular figure), he refutes the conclusion of "others" saying that experience proves that the transition is gradual. This fact, however, is not excluded either by the explanation of "others" or by Figure 6.

PECHAM'S answer (2.2) in the revised version of the *Perspectiva communis* is another example of his poor understanding of EUCLIDES. If such were not the case, would he consider correct the absurd statement that the cone KLM (Figure 7), whose axis is perpendicular to the plane of the aperture, ends up absorbing all other cones also having their vertices in the aperture *because* it is the "shortest"? It is evident that all these cones have the same height despite having axes of different lengths. Their bases behind the aperture are, therefore, equal and it is impossible that KLM would absorb the rest.

PECHAM'S reasoning is equally defective in his attempt (1.2) to prove that rectilinear radiation can only produce images with the shape of the aperture. Looking at Figure 6, one wonders exactly where points T and V are located as they are easily confused with points M and K. This confusion implies (and the text does not allow any other interpretation) that PECHAM considers lines VX and TP to be tangent to circle KLM. Thus, since the intersection of VX and TP is nearer the sun than is point H, points T and V are nearer H than points M and K. But PECHAM does not seem to realize that the pyramid whose outermost rays are VX and TP and whose base is triangular behind the aperture ABC, also has a triangular base on the sun. Thus, while K and M must be the endpoints of the diameter of a circle, T and V can only be the end-points of one side (*i.e.*, the side corresponding to BC in the aperture) of an equilateral triangle inscribed in that circle. It is clear, therefore, that the angle formed by the outermost rays VX and TP is not larger than all others which can be formed having a vertex between the sun and the aperture.

In the revised version of the *Perspectiva communis*, the same errors are repeated. PECHAM seems unaware that KL (Figure 8) can only be one side of a triangle inscribed in the sun similar to GHI and *not* a solar diameter. Nor does he take into account that pyramids from the sun, whose angles are greater than KOL, can pass whole through the aperture. PECHAM also explicitly excludes the possibility that the circular-based pyramids extend beyond the sides of the triangular pyramid without reaching the angles however: "It can be seen, therefore, that by rectilinear radiation the image [on the screen] fails to be round proportionally the same amount as the aperture fails to be round"[25]. The same conclusion can be found in the *Tractatus de sphera*: "If [the rays] have a triangular shape in the aperture, will they always be triangular if prolonged to infinity"[26].

Although PECHAM denies it, Figure 6 obviously implies that the image formed by the triangular pyramid with vertex G and the circular one with vertex H *imme-*

---

[25] "Ergo videtur quod quantum deest foramini de rotunditate tantum deest incidentie proportionaliter de rotunditate per viam directe radiationis" (LINDBERG [4], I, {7}, 203–5).

[26] "... si in foramine sunt triangularis figure, si in infinitum procedant, semper erunt triangulares" (MACLAREN [1], p. 99; DUHEM [1], p. 528).

diately before NO (or *immediately before* RS according to the pyramids with vertices at Q and H) will be a mixture of the shape of the luminous source and the shape of the aperture as EGIDIUS explains in Sections 44–45.

In the revised version of the *Perspectiva communis*, argument 2.3 simply produces perplexity and illustrates the scarce use PECHAM makes of experience, as the premises of the argument are contrary to observation. Clearly, the light passing through the partially blocked aperture produces at the proper distance *only one* circular image on the screen and this image is exactly the same size as the one formed by the unblocked triangular aperture.

Ad [46]–[47]. In the *Almagest* (V, 15 and 16) the solar distance is 1210 terrestrial radii and the sun's radius is exactly 5;30 t.r. (*cf.* PTOLEMY [1], pp. 255–57).

Ad [48]. Although $1210/11 = 110$, throughout most of the text EGIDIUS uses the value 111 which he probably found more convenient for purposes of calculation. In his *Kayfiyyat al-aẓlāl*, of which a Latin translation is not known to exist, IBN AL-HAYTHAM uses the ratio 1210/11 to determine the length of a shadow (*cf.* WIEDEMANN [2], pp. 239–40).

Ad [49]. *Elements*, VI, 2; HEATH [1], II, pp. 194–5. ESCULEUS (also known as ESCULEIUS, ASCULEUS, ASSICOLAUS or ACEFALUS) is, in fact, HYPSICLES (second century B.C.), author of the so-called "Book XIV" of the *Elements* and the Ἀναφορικός, translated by GERARD OF CREMONA with the title *De ascensionibus signorum* (*cf.* HEATH [1], I, pp. 5–6; III, pp. 438–9 and 512–9; BULMER-THOMAS [1]). GERARD OF CREMONA's translation of the *Elements* mentions the author of "Book XIV" (*cf.* CLAGETT [1], p. 28; MURDOCH [1], pp. 283 and 301) and this is the translation which EGIDIUS uses. The reading *pyramidibus* is doubtful and I have been unable to find any medieval text with that title. Another possible reading would be *proportionibus*, but that seems even less likely.

Ad [51]. *Elements*, XIII, 12, and I, 47; HEATH [1], III, pp. 466–7, and I, pp. 349–50. EGIDIUS refers to Proposition 12 of Book XIII as "Proposition 8" because the order of the propositions in that book had been changed in most of the medieval Latin translations in accordance with the Arabic versions they were based on (*cf.* HEATH [1], I, p. 80).

Ad [52]. $d = \sqrt{21, 33} = 4; 37, 7, 41, 24$.

Ad [55]. *Elements*, XIII, 12; I, 4; III, 3; I, 47; HEATH [1], III, pp. 466–7; I, pp. 247–8; II, pp. 10–11; I, pp. 349–50.

Ad [59]. The expression "approximately 6 thumbs" can be taken as a sign that here EGIDIUS has used the exact value of 110 for the ratio 1210/11 instead of the approximate one of 111. The diameter of the pyramid's base would then be $666 : 110 = 6, 054$.

Ad [64]. *Excessa 11ᵃ parte* is a correction of *excessa 55ᵃ parte* found in the manuscript ($1110 : 110 = 10, 09$). Note that here EGIDIUS again uses the exact value for the ratio 1210/11.

Ad [67]. According to Figure 3, these diameters would be 1) the diameter of the circular incidence whose base in the aperture is the circle inscribed in the triangle of the aperture, and 2) the diameter of the circle circumscribing the triangular incidence produced by the pyramid whose base in the aperture is the circle which circumscribes the triangle of the aperture.

Ad [68]. EGIDIUS starts off his solution by considering the effects on the screen of those cones whose axes issue from the center of the sun and whose vertices are in the three corners of the aperture. LEVI BEN GERSON (1288–1344) also uses these same cones to explain the rounding off of the angles of the image formed by light passing through a quadrangular aperture. LEVI writes (*Liber Bellorum Dei*, V, 1, ch. 5): "It is evident from this demonstration that if the ray [of light] passes through a window formed by straight lines, said ray will not be received on the opposite wall as a straight-lined shape in the parts of the angles because it will dilate in any part of any angle only in the quantity corresponding to the radius of the circle of the luminous body and not more; and the radius will reach the angle with the form of a fourth of the circumference of the circle whose center is the vertex of the angle. And this is what we observe with the rays which come from the sun and the moon through windows formed by straight lines"[27].

LEVI's main concern is not the problem of pinhole images, but rather the use of angular or circular apertures to determine the apparent diameters of bodies such as the sun and moon. The passage quoted above derives from a proposition in which LEVI establishes that the incidence is larger than the aperture by twice the angle formed on the screen by the radius of the luminous body[28]. In other words, to calculate the apparent diameter of the sun or moon, an amount equal to the size of the aperture must be subtracted from the diameter of the image formed on the screen. The direction of LEVI's interest is confirmed by Book V, part 1, Chapter 9 of *Liber Bellorum Dei* where he provides another geometric proof of the same Proposition[29]. Nevertheless, a solution to the problem of pinhole images can be deduced

---

[27] "Ex ista demonstratione est manifestum quod si radius transeat per fenestram factam a lineis rectis, ipse radius non recipietur in pariete obiecto in forma linearum rectarum in partibus angulorum, quia dilatabitur in qualibet parte cuiuslibet anguli solum in quantitate circuli semidyametri corporis radiosi et non plus; et veniet radius ad angulum in forma quarte partis circumferentie circuli, cuius centrum est punctum anguli. Et hoc videmus ad sensum in radiis que veniunt a sole et luna per fenestras lineas rectas habentes" (Vat. Lat. 3098, f. 7ra; see also an English translation of the Hebrew text in GOLDSTEIN [1], pp. 48–9).

[28] "Radius solis vel lune vel cuiuscumque corporis radiosi qui intrat per quamcumque fenestram vel per quodcumque foramen, terminatus ad aliquid obiectum distans a predicto foramine, est latior quantitate foraminis ex qualibet parte in quantitate in qua terminat angulum semidiametri corporis radiosi in loco fenestre" (Vat. Lat. 3098, f. 6vb; GOLDSTEIN [1], pp. 48 and 141–2).

[29] "Ad cuius probationem ponamus diametrum radiosi lineam AB et diametrum foraminis parallelum lineam CD et diametrum radii in tabula secunda recepti parallelum eisdem lineam EF. Et protrahamus lineas ADF et BCE rectas que intersecabunt se in puncto G. Manifestum est quod angulus diametri radiosi est angulus AGB; dico quod

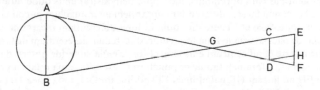

Fig. 9

from LEVI's text. The starting points of LEVI's and EGIDIUS' arguments are the same. The only difference is that what is *implicit* in LEVI's text (*i.e.*, at the proper distance, the bases of the luminous cones are superimposed forming a circle) is totally *explicit* in EGIDIUS'. Is it possible that EGIDIUS knew LEVI's work? As the extant copy of the *Improbatio* does not allow us to date EGIDIUS' work with exactitude, the possibility cannot be rejected out of hand. There are chronological considerations, however, which would lead us to suspect that EGIDIUS' solution was original and independent of LEVI's.

What do we know about the dates of the Latin translation of LEVI's *Astronomy*? Chapters 4–11 of the first part of Book V of *Milḥamot Adonai*, dedicated to trigonometry and the instrument known as the Jacob Staff, were translated by PETRUS OF ALEXANDRIA in 1342 under the title *De sinibus, chordis et arcubus, item instrumento revelatore secretorum* (*cf.* CURTZE [2], pp. 226–32, for an edition of Chapter 3 of *De sinibus* according to MS. 7293, Bibliothèque Nationale, Paris). Another Latin translation of the Hebrew text, preserved among others in MS. Vat. Lat. 3098 cited above, is far more complete as it includes, with some gaps, 110 of the 136 chapters of the first part of Book V. But, it is unlikely that this more complete version can be dated before 1342. A collation of three of the existing copies of *De sinibus* (Paris, Bibl. Nat., 7293; Vienna, Nat. Bibl., 5277, and Klagenfurt, Bischöfliche Bibl., XXX.b.7) with MS. Vat. Lat. 3098 allows us to conclude that we are dealing with two slightly different versions by the same translator, PETRUS OF ALEXANDRIA, and not two independent translations. Also, in his dedicatory *Epistola* of *De sinibus*, LEVI states that the text had not been translated into

---

ille angulus erit scitus. Ad cuius probationem protrahemus de puncto D in linea CD ad punctum H in linea EF lineam DH parallelam linee CE. Sequitur quod linea DH est equalis linee CE, que est equalis linee DF; sequitur quod linea DH est equalis linee DF et quod linea HF est differentia diametrorum, quia linea EH est parallela et equalis linee CD. Set linea HF est scita per experientiam, et longitudo columne que ueniret a puncto D in lineam HF est scita, quia ipsa est longitudo baculi. Sequitur quod quadratus dicte columne coniunctus cum quadrato medietatis differentie est scitus, quorum quadratorum simul radix est scita, que est equalis linee DH; quod angulus HDF est scitus, quia qualem proportionem habet HF ad HD talem habet corda illius anguli ad 60 gr. Dico quod angulus HDF, qui est scitus, est equalis angulo AGB, quia manifestum est quod angulus EGF est equalis angulo AGB; set quia angulus EGF est equalis angulo HDF, quia linea HD est parallela linee EC, sequitur quod angulus AGB est equalis angulo HDF, qui est scitus; sequitur quod angulus diametri radiosi in puncto G est scitus, quia punctus G est in superficie terre. Experientia punctaliter requiretur quod distantia semidiametri terre coniungeretur cum distantia solis radiosi ad punctum G et scieretur quem angulum faceret in centro terre; set quia nondum habemus distantiam radiosi ad punctum G, set in sequentibus eam docebimus, ideo tunc demonstrabimus quod additio semidiametri terre ad eam non faceret differentiam notabilem inter angulum puncti G et angulum centri terre in Ioue, Sole et Venere. Set quia in luna differentiam faceret notabilem, ideo in loco in quo de hac materia in luna loquemur, docebimus differentiam distantiarum lune ad superficiem terre et ad centrum eiusdem, et ibi scientia superior poterit dari complete. Distantia tamen inter foramen tabule et punctum G est scita, quia qualem proportionem habet linea FH ad lineam HE habet linea FD ad lineam DG, quia linea HD est parallela linee EG trianguli FEG. Hoc igitur est aiutorium quod possumus a nostro instrumento habere" (Vat. Lat. 3098, f. 9va–b; see also GOLDSTEIN [1], pp. 70–1).

Latin before 1342 ("Et licet praedictum sacramentum in iamdiu fuit revelatum, ut apparebit inferius, et annotatum hebraicis litteris, et quod alijs verbis sine debito ordine ex ore meo prolatus [sic] forte fuerit aure tenus, scilicet, partialiter repraesentatum, *nunquam tamen ordinate translatum fuerat in latinum*"; Vienna, Nat. Bibl., 5277, f. 40v). Since the copy of the *Improbatio* is assumed to be from the first or second quarter of the fourteenth century (LINDBERG [3], p. 301, n. 11), to conclude that EGIDIUS was familiar with LEVI'S text would mean that the *Improbatio's* date of composition would have to be moved forward to some time after 1342. Finally, the extent to which EGIDIUS' refutation follows the line of PECHAM'S arguments should logically place it chronologically nearer the Englishman's work rather than later.

Indeed, the starting point of EGIDIUS' solution can even be found in the revised version of the *Perspectiva communis,* where PECHAM examines the possibility only to reject it later, using, as is his custom, inadequate arguments: "Perhaps it might seem to someone that this roundness is caused by the pyramids whose vertices are in contact with the sides of the aperture since these pyramids are round, given that they come from the roundness of the sun's surface. But, this can be refuted in the same manner as above since the shorter pyramids absorb the longer. Furthermore, three round pyramids whose axes are very separated cannot by means of rectilinear radiation reunite in a round pyramid, as every pyramid has, each one, only one axis"[30]. Once again, PECHAM has forgotten that his figures must represent the facts. His conclusion is invalid since it does not take into account the variables of the problem: the size of the luminous source and the aperture, the distance between the two and the distance between the aperture and the screen.

Ad [70]. The numerical example EGIDIUS gives in Sections 46–47 is not superfluous. It is used to point out that due to the distance between the sun and the aperture, the axes of the cones whose vertices are in the three corners of the aperture are *sensibly* parallel behind it. The copy of the *Improbatio* clearly deteriorates in the last lines of the text (see the critical apparatus of the edition) and the passage refered to here presents some minor problems. Obviously, what EGIDIUS means is that the length of the diameter of this circle exceeds 4 thumbs by an amount proportional to

$$\frac{888 \text{ thumbs}}{1210 \text{ t.r.}},$$

that is, if we let AC (Figure 10) be the diameter of the circle circumscribing the triangular aperture and DH be the diameter of the circle circumscribing the triangle formed on the screen by joining the end points of the axes of the cones, then,

---

[30] "Forte posset aliqui videri quod ista rotunditas causaretur ex pyramidibus applicantibus conos suos lateribus ysopleuri, quia huius pyramides sunt rotunde utpote a rotunditate solaris superficiei procedentes. Sed illud improbatur ut supra, quia breviores pyramides absorbent longiores. Preterea tres pyramides rotunde non possunt per viam irradiationis convenire in unam pyramidem rotundam quarum maxime axes differunt et continue disgregantur, omnis enim pyramidis unius est axis una" (LINDBERG [4], I, {7}, 257–64).

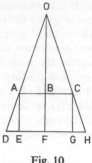

Fig. 10

given that $OB = 1210$ t.r. and $BF = 888$ thumbs,

$$\frac{2DE}{AC} = \frac{AE}{OB};$$

hence

$$2DE = (4 \times 888 \text{ thumbs}) : 1210 \text{ t.r.} \approx 4 \times 2{,}05309 \times 10^{-9} \text{ fingers},$$

using the values for a foot and a thumb given by EGIDIUS and assuming, as is usual in the Middle Ages, that 1 t.r. $= 3250$ *miliaria*, 1 *miliare* $= 4000$ *cubitos* and 1 *cubitus* $= 2{,}5$ feet.

The sentence *habet diametrum plusquam quatuor pollicibus in proportione que octoginta pedes et octo pollices ad distantiam foraminis a sole* is a rather odd way of expressing the proportion

$$\frac{2DE}{4 \text{ thumbs}} = \frac{888 \text{ thumbs}}{1210 \text{ t.r.}}.$$

I interpret *que* as a *locus corruptus*, not worth making conjectures about since EGIDIUS' concept is correct.

Ad [73]. Since, for all practical purposes, the axes of the three pyramids are parallel, the radii of their bases increase much more quickly behind the aperture than does the diameter of the circle circumscribing the triangle formed by their axes.

Conclusion

As EGIDIUS' solution for the passage of light through triangular apertures is correct, his contribution to the problem of pinhole images can, without a doubt, be considered superior to that of any other known medieval scholar in Europe before KEPLER, with the exception of LEVI. An overall evaluation of this contribution with respect to KEPLER clearly depends on the extent to which EGIDIUS' explanation can be generalized. The apparently unfinished state of the copy preserved in the Jagiellonian Library makes any definitive conclusion difficult.

Assuming the copy is icnomplete, I think it likely that EGIDIUS would have generalized his solution for any type of angular apertures and (according to Sections 36 and 37) for partial solar eclipses and other possible shapes for the luminous source, proving (according to Section 38) that all cases can be explained by the principle of rectilinear propagation of light. That such conjectures are well-founded does unfortunately not constitute proof of what was actually contained in the *Improbatio*.

One can only hope that the discovery of another copy of the work would provide a clear answer to the question of its content and the date of its composition. Such a discovery would certainly help to mitigate the sharp contrast between the dozens of known copies of PECHAM'S *Perspectiva communis* (not to mention the commentaries on the text) and this single testimony of EGIDIUS' work. It is a situation which brings to mind BACON'S words in *The Great Instauration* (BACON [1], p. 15): "Time is like a river, which has brought down to us things light and puffed up, while those which are weighty and solid have sunk".

Note added in proof: Ad [49]. If the reading *pyramidibus* is correct, it is probably a reference to *De curvis superficiebus*, a tract in the tradition of AR-CHIMEDES' *Sphere and Cylinder*, known also in the thirteenth centure as *De (rotundis) pyramidibus* (cf. J. L. HEIBERG, *Archimedis opera omnia cum commentariis Eutocii*, vol. III, Leipzig, 1915 (repr. 1972), pp. XCVI–XCVIII, and M. CLA-GETT, *Archimedes in the Middle Ages, V: Quasi-Archimedean Geometry in the Thirteenth Century*, Philadelphia, 1984, pp. 7–8. The text is edited in M. CLAGETT, *Archimedes in the Middle Ages, I: The Arabo-Latin Tradition*, Madison, 1964, pp. 439–557.

References

ALHAZEN [1]. *Opticae Thesaurus Alhazeni Arabis libri septem ... Item Vitellonis Thuringo-poloni libri X, omnes instaurati ... a F. Risnero*, Basileae, 1572.
ALLMAN [1]. G. J. Allman, *Greek Geometry from Thales to Euclid*, Dublin, 1889; repr. New York, 1976.
ARISTOTLE [1]. Aristotle, *Topica*, trans. by E. S. FORSTER, London, 1960.
ARISTOTLE [2]. Aristotle, *On Sophistical Refutations*, trans. by E. S. FORSTER, London, 1965.
BACON [1]. *The Works of F. Bacon*, coll. and ed. by J. SPEEDING, R. L. ELLIS & D. D. HEATH, vol. IV, London, 1860.
BJÖRNO & VOGL [1]. A. A. Björno & S. Vogl. "Alkindi, Tideus und Pseudo-Euklid: Drei Optische Werke", *Abhandl. zur Geschichte der math. Wissenschaften*, XXVI, 3 (1912), 1–176.
BRIDGES [1]. J. H. Bridges, *The 'Opus Maius' of Roger Bacon*, vol. II, London, 1900; repr. Frankfurt M., 1964.
BULMER-THOMAS [1]. I. Bulmer-Thomas, "Hypsicles", *Dictionary of Scientific Biography*, 6 (1981), 616–7.
CLAGETT [1]. M. Clagett, "The Medieval Latin Translations from the Arabic of the *Elements* of Euclid, with Special Emphasis on the Versions of Adelard de Bath", *Isis* 44 (1953), 16–42.
CLAGETT [2]. M. Clagett, *Archimedes in the Middle Ages, I: The Arabo-Latin Tradition*, Madison, 1964.

CLAGETT [3]. M. Clagett, *Archimedes in the Middle Ages, V: Quasi-Archimedean Geometry in the Thirteenth Century*, Philadelphia, 1984.

CURTZE [1]. M. Curtze, *Anaritii in decem libros priores Elementorum Euclidis Commentarii ex interpretatione Gherardi Cremomensis*, Leipzig, 1889.

CURTZE [2]. M. Curtze, "Die Dunkelkammer", *Himmel und Erde* 13 (1901), 225–36.

DUHEM [1]. P. Duhem, *Le Système du Monde*, vol. III, Paris, 1958.

GOLDSTEIN, [1]. B. R. Goldstein, *The Astronomy of Levi ben Gerson (1288–1344)*, New York-Berlin, 1985.

HAJDUKIEWICZ [1]. L. Hajdukiewicz, *Biblioteka Macieja z Miechowa*, Wrocław, 1960.

HEATH [1]. T. L. Heath, *The Thirteen Books of Euclid's Elements*, 2^n ed., 3 vols, Cambridge, 1926; repr. New York, 1956.

HEATH [2]. T. L. Heath, *Mathematics in Aristotle*, Oxford, 1949; repr. New York, 1980.

HEIBERG [1]. J. L. Heiberg, *Archimedis opera omnia cum commentariis Eutocii*, vol. III, Leipzig, 1915 (repr. 1972).

KEPLER [1]. J. Kepler, *Ad Vitellionem Parallipomena quibus Astronomia pars optica traditur* ... , Francofurti, 1604.

LINDBERG [1]. D. C. Lindberg, "The Theory of Pinhole Images from Antiquity to the Thirteenth Century", *Archive for History of Exact Sciences* 5 (1968), 154–76; repr. in LINDBERG [6].

LINDBERG [2]. D. C. Lindberg, "A Reconsideration of Roger Bacon's Theory of Pinhole Images", *Archive for History of Exact Sciences* 6 (1970), 214–23; repr. in LINDBERG [6].

LINDBERG [3]. D. C. Lindberg, "The Theory of Pinhole Images in the Fourteenth Century", *Archive for History of Exact Sciences* 6 (1970), 299–325; repr. in LINDBERG [6].

LINDBERG [4]. D. C. Lindberg, *John Pecham and the Science of Optics. Perspectiva communis*, Madison, 1970.

LINDBERG [5]. D. C. Lindberg, "Introduction" to the reprint of ALHAZEN [1], v–xxxiv, New York, 1972.

LINDBERG [6]. D. C. Lindberg, *Studies in the History of Medieval Optics*, London, 1983.

LINDBERG [7]. D. C. Lindberg, "Laying the Foundations of Geometrical Optics: Maurolico, Kepler, and the Medieval Tradition", in D. C. LINDBERG & G. CANTOR, *The Discourse of Light from the Middle Ages to the Enlightenment*, pp. 1–65, Los Angeles, 1985.

MACLAREN [1]. B. R. MacLaren, *A Critical Edition and Translation, with Commentary of John Pecham's 'Tractatus de sphera'* (unpublished dissertation), University of Wisconsin, 1978.

MAUROLICO [1]. F. Maurolico, *Photismi de lumine, & umbra ad perspectivam* ..., Neapoli, 1611.

MURDOCH [1]. J. Murdoch, "Euclides Graeco-Latinus. A Hitherto Unknown Medieval Latin Translation of the *Elements* Made Directly from the Greek", *Harvard Studies in Classical Philology* 71 (1966), 249–302.

PTOLEMY [1]. *Ptolemy's Almagest*. Translated and Annotated by G. J. TOOMER. London, 1984.

RASHED [1]. R. Rashed, "Optique Géometrique et Doctrine Optique chez Ibn Al-Haytham", *Archive for History of Exact Sciences* 6 (1970), 271–98.

SABRA [1]. A. I. Sabra, "Ibn al-Haytham", *Dictionary of Scientific Biography*, 6 (1981), 189–210.

STRAKER [1]. S. Straker, "Kepler, Tycho, and the *Optical Part of Astronomy*: the Genesis of Kepler's Theory of Pinhole Images", *Archive for History of Exact Sciences* 24 (1981), 267–93.

WIEDEMANN, [1]. E. Wiedemann, "Über eine Schrift von Ibn al Haitam", *Sitzungsberichte der Physikalisch-medizinischen Sozietät in Erlangen* **39** (1907), 226–48.
WIEDEMANN [2]. E. Wiedemann, "Über die *Camera obscura* bei Ibn al Haitam", *Sitzungsberichte der Physikalisch-medizinischen Sozietät in Erlangen* **46** (1914), 155–69.
WISŁOCKI, [1]. W. Wisłocki, *Catalogum codicum manuscriptorum Bibliothecae Universitatis Jagiellonicae Cracoviensis*, Vol. I, Cracovia, 1877.
WITELO [1]. See ALHAZEN [1].
WÜRSCHMIDT [1]. J. Würschmidt, "Zur Theorie der *Camera obscura* bei Ibn al Haitam", *Sitzungsberichte der Physikalisch-medizinischen Sozietät in Erlangen* **46** (1914), 151–4.

ADDENDA

p. 28, ad [49]: On this passage and my edition and translation of sentence 51 of Egidius' text, H.L.L. Busard wrote in *Mathematical Reviews*: «In his commentary ad [49] the author says: "Gerard of Cremona's translation of the *Elements* mentions the author of 'Book XIV' and this is the translation which Egidius uses." In the reviewer's opinion it is very unlikely that Egidius used Gerard's translation, for the usual edition referred to in those days is that of Campanus. That Egidius did use the Campanus edition is indicated by his mentioning of the propositions *4 secundi* (not *47 primi*), *quartam primi, primam partem tertie tertii, et penultimam primi*, all of which Campanus gives in his proof of XII.8».

p. 31: On the date of the Latin version of Levi ben Gerson's *Astronomy* see IV, esp. pp. 14–16.

II

Astronomical Use of Pinhole Images in William of Saint-Cloud's Almanach Planetarum *(1292)*

1. Introduction

There are two aspects of the so-called problem of pinhole images during the Middle Ages. The first (why does light passing through apertures of whatever shape always reproduces at a certain distance the shape of the luminous source?) concerns optical theory (or "*perspectiva*") directly and requires an answer to the question of the formation of these images. The second one refers to the astronomical use of pinhole images, that is, to the possibility of casting images of the full or eclipsed solar disk through an aperture onto a screen, and it demands a method for a correct measurement of apparent solar diameters. According to the final passage of a commentary on astronomical tables entitled *Theorica planetarum* and attributed to ROGER OF HEREFORD (ca. 1178)[1], this procedure seems to have been well known at least since the XIIth century.

In the second half of the XVIth century, the same procedure came into increasing use among professional astronomers, as is attested in the works of ERASMUS REINHOLD (1511–1553), GEMMA FRISIUS (1508–1555) and TYCHO BRAHE

---

[1] ROGER OF HEREFORD's text reads as follows: "Si autem die qua sol eclipsabitur totumque eclipsim conspicere uolueris absque oculorum lesione, hoc est, quando incipit et quanta sit et quamdiu durat solis eclipsis, obserua casum solaris radij per medium alicuius rotundi foraminis. Et circulum clarum quem perficit radius in loco super quem cadit diligentius inspice; cuius circuli rotunditatem cum in aliqua parte uideris deficere scias quod eodem tempore deficit claritas in corpore solis ex parte opposita illi parti; nam cum in circulo claro incipit rotunditas deficere [ex parte orientis], tunc incipit sol eclipsari in parte occidentis. Et semper dum decrescit rotunditas circuli clari crescit eclipsis et proportionaliter secundum quantitatem; quot enim digiti diametri solis eclipsantur, tot pereunt digiti circuli clari quem figurat radius solis in loco casus sui postquam transierit per medium foramins rotundi" (Paris, Bibliothèque Nationale, lat. 15171, f. 157v; Oxford, Bodleian Library, Digby 168, f. 84vb). See also DUHEM [1], vol. 3, p. 505, and STRAKER [1], pp. 116–122.

(1546–1601)[2]. Although a general solution of the problem of the formation of images cast by light behind small apertures obviously implies a satisfactory account of the method for calculating apparent solar diameters with a pinhole camera, there is some textual evidence to suggest that astronomers were faced with the astronomical use of the camera without dealing at the same time with the problem of the formation of images; it seems that a method, though incorrect, for calculating apparent lunar and solar diameters during eclipses may have been developed by (or transmitted to) TYCHO BRAHE without relating it to a general theory of pinhole images[3].

---

[2] "Quomodo item solarium defectuum quantitates, augmenta, decrementa, initia, atque exitus, sine ulla offensione oculorum, etiam cum non aspicias coelum, obseruari ac considerari possint, illud quoque optime lector te non caelabo, Nec dubito, quin vbi semel fuerit hac usus ratione, qua nihil potest esse simplicius, eam reliquis modis omnibus, quorum descriptiones quidem extant, commoditate, & certitudine, iucunditate denique longe sit antelaturus. Ea est huiusmodi, ne multis te detineant. Quando calculus monet futurum defectum solis, recipe te siue sub tectum altae domus, siue in cubiculum minus humile, aut quamuis contignationem, quae quo est altior, eo aptior erit ad hoc negotium. Sit denique hic locus, in quo instituis obseruationem omnis expers lucis, quantum fieri potest. Etiamsi autem omnia clauseris, & obturaueris, facile tamen reliqua tibi erit seu rima seu foramen cuiuscunque figurae, in quod solis radij incidere queant. Sin minus, ipse tenue foramen ingruentibus radijs aperias. Hoc facto, si vel in area pauimenti, vel in latere quod foramini opponitur, incidens solis lumen obserues, videbis (mirabile dictu) id prorsus effigiem solis repraesentare, tantámque portionem deesse circulo luminoso, quantam ipsa luna intercedens a nostro conspectu aufert. Quare si eiusdem luminosi circuli diametrum partiaris in 12 digitos, vt vocant artifices, reliqua omnia, quae initio dixi, ante oculos tibi posita erunt, etiamsi terram non coelum aspicias. Caeterum ingeniosus obseruator, ec hac breui admonitione multo plura intelliget, & iudicabit & caetera" (REINHOLD [1], Riiiv(131v)–Riiiir(132r)). See also GEMMA FRISIUS [1], pp. 312–313; STRAKER [1], pp. 311–361; STRAKER [2], pp. 269–272. REINHOLD's and GEMMA's passages are commented by DANIEL SANTBECH in the *propositio* XIII (*Qua ratione metiamur Eclipsium magnitudines*) of his *Problematum astronomicorum et Geometricorum sectiones septem*, printed in 1561 (*cf.* REGIOMONTANUS [1], 2nd part, p. 47).

[3] TYCHO's method between 1591 and 1598 probably consisted in correcting the measured solar and lunar diameters on the screen subtracting in both cases the diameter of the aperture, instead of by addition in the case of the lunar image, producing for that reason an anomalous diminution in the apparent diameter of the moon during eclipses (*cf.* BRAHE [1], vol. 12, p. 108; STRAKER [1], pp. 336–344; STRAKER [2], pp. 275–282). The knowledge of certain conditions for optimum use of the pinhole camera for astronomical purposes, perhaps entirely empirically based, which some texts by REINHOLD and MAESTLIN (1560–1631) reveal, is equally interesting for the history of the problem. The former points out (see note 2) that the greater the distance between the aperture and the screen (*i.e.*, the smaller the aperture), the better the experiment. As for MAESTLIN, his recommendation of using great distances between the aperture and the screen (as that obtained by placing the aperture in the roof of a cathedral) and apertures as small as compatible with a visible image is equivalent to recognizing that from very small apertures *no sensible* error follows. MAESTLIN arrived at this conclusion while trying to avoid the deformations produced by great apertures (*e.g.*, half a finger) in the image on the screen of the horns of the partially eclipsed sun, according to his letter to

The correct method for the use of the pinhole camera in astronomical observations was provided by JOHANNES KEPLER (1571–1630) in his *Ad Vitellionem paralipomena* (1604), though it was clearly explained almost three centuries before him by LEVI BEN GERSON (1288–1344) in his *Astronomy*[4]. It is generally accepted, however, that KEPLER arrived at his solution without knowledge of LEVI's analysis of pinhole images.

In their discussions, LEVI and KEPLER explain the effect of the size of the aperture on the size of the image, asserting that the solar image on the screen is a measure of the apparent size of the luminary provided the size of the aperture is subtracted from it; a statement derived from the distinction between point apertures and finite apertures, since the angular diameter of the luminary must be measured in each case from different distances.

The object of this paper is to provide an edition, with translation and commentary, of a passage on pinhole images of WILLIAM OF SAINT-CLOUD's *Prologue* to his *Almanach planetarum*, composed in 1292[5]. In this text, and in a purely

---

KEPLER of 9 October 1600: "Quae de Eclipsi Solis scribis, ea omnia dudum animaduerteram, videlicet quod lux radij ampliet Solis, et minuat diametrum. Verum si scena seu obseruationis locus sit amplior, (.cuiusmodi locum in nostro templo esse nosti.) isti impedimento egregiè prospicitur. Foramen enim factum quam fieri potest minimum (.nec opus est esse punctum mathematicum.) excludit omnem sensibilitatem ad distantiam. Ibi enim cornua extrema ☉ acutissimè cernuntur ... Sin veró foramen fuerit aliquantum maius, vtpote dimidiati digiti, vel vltra semidiametrum vel diametrum huius: fit omnino vt radius vndique amplietur, at cornua ... fiunt obtusa. Idque certissimum argumentum est, obseruationem eiusmodi esse fallacem ... Et quidem in eiusmodi obseruationibus aliquoties deprehendi diametrum ☽ minorem, quam secundum calculum esse debebat ... Praecauetur autem minutiore foramine, eoque remoto, ibi enim cornua ... tam acuta fiunt, vt de nullo errore sensibili supersit suspicio" (KEPLER [2], vol. 14, pp. 156–157).

[4] KEPLER [1], pp. 37–50; STRAKER [2], pp. 282–286. On LEVI's theory of pinhole images see CURTZE [1]; LINDBERG [1], pp. 303–308; STRAKER [1], pp. 197–222; GOLDSTEIN [3], pp. 48–50, 69–71, 140–143, and 156–157. See also MANCHA [1] for a medieval solution of the problem of the formation of images cast by light through triangular apertures.

[5] The dates of birth and death of WILLIAM OF SAINT-CLOUD are unknown. The earliest recorded date in his *Almanach* is 1 June 1285 when he observed a conjunction of Mars and the Moon. WILLIAM OF SAINT-CLOUD's works are entirely devoted to astronomy and they remain unpublished. The *Directorium* is a description of a compass sundial with a graduation in unequal hours, provided with a table for computing the duration of diurnal arcs. The *Kalendarium* (*Calendar*) is dedicated to Queen MARIE OF BRABANT, widow of PHILIP III, and it was translated into French at the request of JEANNE OF NAVARRE, wife of PHILIP IV (on MSS of these works, see POULLE [1], p. 391). The purpose of the *Almanach planetarum* is to provide the positions of the planets for a period of twenty years starting from 1292. The *Prologue* is an introductory text where WILLIAM OF SAINT-CLOUD presents an account of the observations and considerations on which his work is based, especially a criticism of the Tables of Toulouse concerning the values for mean planetary motions and a criticism of the trepidation theory attributed in the Latin West to THĀBIT IBN QURRA (836–901). The *Almanach* gives us solar and lunar positions in the ninth sphere for every day in degrees and minutes, and planetary posi-

astronomical context, SAINT-CLOUD clearly states (i) that finite and point apertures require different mathematical treatment for measuring apparent solar diameters on the screen of the camera, *i.e.*, the main point in LEVI's and KEPLER's analysis, and (ii) that the pinhole camera can be used to find the solar eccentricity from measurements of the apparent solar diameters at apogee and perigee. This procedure, which is not mentioned in the *Almagest*, was used by LEVI BEN GERSON in 1334, and no precedent of it was till now known. These two points are enough to grant to SAINT-CLOUD's text an important place in the history of the problem of pinhole images. An appendix on LEVI BEN GERSON's determination of the solar eccentricity with a pinhole camera is added at the end of the article[6].

---

tions with the same accuracy also in the ninth sphere at 10-day intervals for the superior planets and at 5-day intervals for the inferior ones. Intermediate positions, computed by interpolation, are given only in degrees. The times of mean syzygies are also given and, for these times, the mean position of the Moon in the eighth sphere, its mean argument, the solar and lunar hourly velocities, and the latitude of the Moon. Finally, the *Almanach* provides us with places, times, and magnitudes of solar and lunar eclipses for the same period. A table at the end of the *Prologue* also gives the diurnal arc, the number of equal hours of daylight, the equation of time, and the solar declination at 10-day intervals. On SAINT-CLOUD's *Almanach* see also LITTRE [1]; DUHEM [1], vol. 4, pp. 10–24; POULLE [1], pp. 389–391, and MERCIER [1], pp. 201–204.

The text of the *Prologue* of the *Almanach* is preserved in at least four MSS: Paris, Bibliothèque Nationale, n.a.l. 1242, ff. 41va–44vb; Paris, Bibliothèque Nationale, lat. 7281, ff. 141r–144v; Cues, Stiftsbibliothek, 215, ff. 24va–31va, and Utrecht, Universiteitsbibliothek, 725, ff. 201v–204r (hereafter MSS *A, B, C,* and *D,* respectively). The passage edited in this article occurs in *A,* ff. 43va–44r; *B,* ff. 143v–144r; *C,* ff. 29va–30rb, and *D,* f. 204r. The copies *A* and *C* seem to date from the end of the XIIIth century or the beginning of the XIVth. The copies *B* and *D* are, respectively, from the middle and the end of the XVth. In a marginal note to the section on eclipses of JOHN OF SICILY's commentary to the canons of the Toledan tables, the copyist of MS *B* repeats almost word for word HEREFORD's text quoted above, adding at the end: "Et ut dicit Guillelmus de Sancto Clodoaldo de quadam eclipsi solis anno christi 1285 potest idem probari per candelam ardentem". Manuscript *D,* which lacks the figure reproduced below with the text, is a very deficient copy; I have therefore omitted most of its variant readings in the critical apparatus. In the Latin texts, editorial insertions are enclosed in square brackets and sentence numbers have been added to texts and translations (also in square brackets) for ease of reference.

Microfilms of manuscripts were kindly supplied by the Directors of the Stiftsbibliothek, Cues; Universiteitsbibliothek, Utrecht; Biblioteca Nazionale, Firenze; Biblioteca Vaticana, Roma; Bibliothèque Municipale, Lyon, and Staatsbibliothek, München. I also wish to thank the Directors of the Bibliothèque Nationale and the Institut des Recherches et d'Histoire des Textes, Paris, for permission to inspect the other manuscripts used in this study and a microfilm of MS Oxford, Bodleian Library, Digby 168, respectively.

[6] I am grateful to Professor BERNARD R. GOLDSTEIN of Pittsburgh University for his collation of my transcription of the Latin copies of chapter 56 of LEVI's *Astronomy* with the Paris manuscripts of the Hebrew text and his suggestions on a draft of this paper.

2. Latin Text

[1] Quia uero intendo loca et tempora et quantitates eclipsium tam solis quam lune notare in almanach — [2] eclipsium autem solarium nulla nisi medietatem excedat potest ad oculum obseruari nisi contingat eam esse circa ortum uel occasum solis, ita quod propter oppositionem uaporum lumen solis debilitetur, quod tamen raro accidit —, [3] ideo intendo hic experimentum quodam ponere per quod poterit quelibet eclipsis solis quantumcumque parua notari, etiam si solum medietas puncti deficeret, [4] ne forte si propter defectum uisus aliqua talis eclipsis in almanach posita uideri non posset imputaretur errori, [5] ne etiam ipsas eclipses obseruantibus contingat illud quod pluribus accidit anno domini 1285, quarta die iunij, [6] scilicet quod propter fortem intuitum solis per paruam quantitatem eclipsis accidit illis qui sic solem fortiter inspexerant quedam in oculis tenebrositas, que communiter accidit intrantibus umbram postquam fuerint in claritate solis; [7] que quidem tenebrositas in quibusdam remansit per duos dies, in alijs per tres, in alijs etiam per plures, secundum quod intensius et diucius solem inspexerant, et forte etiam secundum quod plus uel minus apti erant eorum oculi ad huiusmodi tenebrositatem. [8] Unde uidetur quod intantum posset aliquis solem aspicere quod penitus excecatur, iuxta illud dictum Philosophi: excellencie sensibilium corrumpunt sensum; causa autem illius accidentis alibi declaratur.

[9] Solent autem aliqui eclipsim solis in aqua posita in pelui aspicere, sed illud non sufficit; quia ab aqua reflectitur lumen solis, licet debilius sit lumen reflexum quam lumen proprium, et etiam predictum accidens induceret licet debilius. [10] Si tamen fiat, proprie debet fieri in aqua clara et uase profundo existente in quieto loco.

[11] Ut igitur predictum accidens penitus euitetur, experimentum aliud explanctur per quod non solum eclipsis solis absque lesione oculorum poterit obseruari, sed etiam hora initij eius et finis necnon et quantitas punctaliter mensurari et etiam quedam alia que alibi locum habent.

---

1. et[1] *om.* B | et[2] + etiam C | tam *om.* B | quam AC, et B | almanac B.

2. oppositionem AC, interpositionem B | tamen *om.* B | accidit AB, contingit C.

3. eclipsis + ipsius C.

5. etiam AC, ex intuentibus B | quod *om.* B.

6. fortem + aspectum et B | per paruam quantitatem eclipsis *om.* B | illis *om.* B | intrantibus BC, intuentibus A.

7. quidem tenebrositas *om.* B | in[2] *om.* B | in[3] *om.* B | etiam *om.* B | inspexerant AC, aspiciebant B | forte etiam C, fortiter A, *om.* B | huiusmodi *om.* B.

8. unde uidetur AC, ita B | aliquis AC, quis B | illud *om.* B | excellencie AB, excellencium C.

9. posita AC, imposita B | illud *om.* B | et etiam A, et ita B, ita etiam C | predictum AC, dictum B | induceret AC, inducere posset B | licet + minus et B.

10. si + sic A | proprie *om.* B | aqua AC, aliqua B | et + in B | existente AC, posito B.

11. igitur AC, ergo B | experimentum, *om.* A | aliud *om.* B | solis *om.* B | necnon *om.*, B | punctaliter AC, punctorum B.

[12] Fiat igitur in domo clausa foramen in tecto uel fenestra uersus partem illam in qua debet eclipsis solis euenire. [13] Sit autem quantitas foraminis sicut est foramen a quo extrahitur uinum a dolijs. [14] Lumine ergo solis per huiusmodi foramen intrante, ad distantiam foraminis 20 pedum uel 30 aptetur aliqua res plana, ut pote asser unus, ita quod huiusmodi lumen solis super illius rei superficiem perpendiculariter cadat. [15] Videbitur autem lumen in suo casu super huiusmodi superficiem penitus rotundum, etiam si foramen angulare esset. [16] Erit etiam maius foramine, et quanto magis distabit huiusmodi res plana a foramine tanto lumen super ipsam cadens latius apparebit; erit tamen debilius quam prope.

[17] Et si a centro foraminis, si paruum fuerit foramen, uel a concursu extremorum radiorum solis ultra foramen, si magnum fuerit, usque ad casum luminis describatur unus circulus ita quod centrum huius circuli sit centrum foraminis uel concursus extimorum radiorum, et circumferentia eius transeat per ipsum casum luminis, [18] inuenietur lumen in loco casus proportionaliter abscindere de circulo secundum proportionem dyametri solis in celo. [19] Ita quod si dyameter solis abscindat 30 minuta in celo, dyameter etiam luminis abscindet 30 minuta de circulo; si uero plus fuerit, plus abscindet. [20] Unde per hoc uidetur posse probari ecentricitas solis ad oculum supposito quod sit rationabilius ipsum habere ecentricum quam epiciclum, cum oporteat alterum. [21] Cum enim sol in auge existens remotior sit a terra quam quando fuerit in opposito augis, minor debet apparere, similiter etiam et lumen huiusmodi cadens per foramen super planum huiusmodi minus erit.

[22] Hijs ita dispositis hora qua debet esse eclipsis obseruetur illud lumen cadens super planum. [23] Et quando eclipsis incipiet uidebitur illud lumen proportionaliter deficere secundum defectum in sole. [24] Et augmentabitur per eius augmentacionem et decrescet secundum eius decrementum. [25] In hoc solum erit differentia quod pars deficiens in lumine opposita erit parti deficienti in sole; ita quod si pars orientalis solis deficiat, in lumine occidentalis deficiet, et econuerso.

---

12. igitur *AC*, ergo *B*, om. *D* | tecto *ABD*, directo *C*, + domus *B* | uel + in *A*, + in aliqua *D* | illam + celi *B* | euenire *ACD*, apparere *B*.

13. sit autem *AC*, et sit *B* | foramen + dolij *B* | a dolijs *om. B*.

14. unus *om. B* | huiusmodi[2] *om. B* | superficiem + planam *B*.

15. huiusmodi *AC*, illam *B*.

16. etiam + lumen *B* | maius + illo *B* | quam + si *B*.

17. fuerit[1] *ACD*, sit *B* | extremorum *ACD*, extimorum *B* | huiusmodi *AC*, huius *BD* | extimorum *BC*, extremorum *AD*.

18. circulo *ABD*, oculo *A*.

19. si dyameter ... in celo *om. C* | abscindat *AC*, fuerit *BD* | 30 *BCD*, 10 *A* | celo + quantum ad uisum *B* | circulo + descripto *B* | fuerit + dyameter *B*.

20. unde *om. A* | probari + distantia solis a terra et *D* | alterum + eorum *B*.

21. huiusmodi *om. BD*.

22. ita *AC*, itaque *BD* | lumen *om. A*.

23. lumen + cadens super planum *B*.

24. augmentationem *BC*, augmentum *AD* | eius *A*, illius *B*, ipsius *CD* | decrementum + et *B*.

25. solis *om. B* | lumine + illo pars *B* | occidentalis + solis *C*.

[26] Et hoc est propter intersectionem radiorum in ipso foramine; per ipsam enim fit radius ueniens a dextra parte solis sinister et a sinistra dexter. [27] Per eandem etiam causam apparent in speculis concauis res euerse.

[28] Et hoc quidem faciliter apparet in figura. [29] Sit AB sol, C centrum foraminis, AD radius ueniens a parte orientali solis, BE radius ueniens a parte occidentali. [30] Manifestum erit quod si A deficiat in sole, D deficiet in lumine cadente per foramen, ab eo enim causatur. [31] Et si B deficiat, E deficiet simili ratione.

3. Translation

[1] But, as I intend to indicate in the almanac the places, times, and magnitudes of both solar and lunar eclipses — [2] and given that none of the solar eclipses, except when more than half the Sun is eclipsed, can be observed with the naked eye unless it occurs near the rising or setting Sun in such a way that the solar light is weakened because of the opposition of vapours, which however rarely occurs —, [3] I intend to expound here a certain experiment by which any eclipse of the Sun, no matter how small, can be observed, even if only half a point were eclipsed; [4] so that if a certain eclipse indicated in the almanac cannot be observed because of a defect of vision, it cannot be attributed to an error [of the almanac], [5] and in order to avoid befalling to those observing these eclipses what happened to many people on the day 4 of June of 1285 A.D.; [6] that is, because of their intense gazing at the Sun when the eclipsed part was still small, it came upon those who look fixedly at the Sun a certain darkness in the eyes, which usually occurs to those who enter in a place in shadow after beging in the light of the Sun; [7] darkness that lasted two days for some of them, three for others, and more for others, according to the intensity of their gazing at the Sun and also according to the predisposition of their eyes to this darkness. [8] From which it seems that someone could gaze at the Sun so much that he would be certainly blinded, according to that sentence of the Philosopher: the excellence of the sensible objects destroy the sense; but the cause of this accident is declared elsewhere.

[9] Thus, some are accustomed to observe the eclipse of the Sun in water placed in a vessel, but it is not enough; because the light of the Sun is reflected by the water, and even though the reflected light is weaker than the direct light, yet the aforementioned accident will still occur although less intensely. [10] If however it is so done, it ought to be done with clear water in a deep vessel in a still place.

[11] Thus, to avoid completely the aforementioned accident, a certain experiment will be explained by which it will be possible to observe without injury

26. est *ACD*, fit *B* | radius + solis *B* | parte *om. B* | et + ueniens *B* | dexter + et *B*.
27. etiam *om. B*.
29. radius[1] + solis *B* | occidentali + solis *B*.
30. per foramen *CD*, super planum quia *B*, *om. A*.
31. deficiat + in sole *B* | deficiet + in lumine *B*.

to the eyes not only the eclipse of the Sun, but also the hour of its beginning and end, and also to measure its magnitude and the remaining circumstances that occur.

[12] Let there be made an aperture in the roof or in a window of a closed house towards that part [of the sky] in which the eclipse of the Sun is to happen. [13] Let the size of the aperture be like that through which wine is drawn from barrels. [14] Once the light of the Sun passes through the aperture, let there be placed at a distance of 20 or 30 feet from the aperture something flat, as for example a panel, in such a way that the light of the Sun falls perpendicularly on the surface of that flat object. [15] The light that falls on this surface will be seen completely round, even if the aperture were angular. [16] The light that falls on this surface will be also greater than the aperture, and the greater the distance of the flat object from the aperture, the greater will appear the light falling on it; however the light will be weaker than if it were near.

[17] And if there be described a circle about the center of the aperture in the case where the aperture is small, or from the intersection of the outermost rays of the Sun beyond the aperture in the case where it is great, up to the surface on which the light falls, in such a way that the center of that circle is the center of the aperture or the intersection of the outermost rays, and its circumference passes through the very fall of the light, [18] the light in the place of its fall will be found to cut off proportionally from the circle according to the proportion of the diameter of the Sun in the sky. [19] In such a way that if the diameter of the Sun cuts off 30 minutes in the sky, the diameter of the light will also cut off 30 minutes of the circle; if [the diameter of the Sun] were greater, the light will cut off a greater arc of the circle. [20] From this, it seems that the eccentricity of the Sun could be proved by eye-sight, on the assumption that it is more likely that the Sun has an eccentric than an epicycle, as one of them is necessary. [21] But, since the Sun at apogee is farther from the earth than when it is in the [place] opposite apogee, it must appear smaller [at apogee]; similarly, the light falling through the aperture on the plane also will be smaller.

[22] Once these things so disposed at the time when the eclipse is to occur, observe the light falling on the plane. [23] And when the eclipse begins that light will be seen proportionally lacking according to the lack [of light] in the sun. [24] And it will increase in size according to its increasing and it will decrease according to its decreasing. [25] The only difference will be that the part lacking in the light will be opposite to the part lacking in the Sun, in such a way that if the eastern part of the Sun is lacking, in the light the western part will be lacking, and vice versa. [26] And this occurs because of the intersection

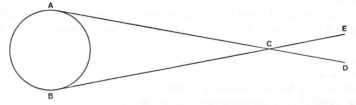

Fig. 1. *A*, 43v; *B*, 144r; *C*, 30r.

of the rays in the aperture; thus, because of this intersection, the ray proceeding from the right part of the Sun becomes left and the one proceeding from the left, right. [27] And also by the same cause objects appear inverted in concave mirrors.

[28] And this is easily apparent in the figure. [29] Let AB be the Sun, C the center of the aperture, AD a ray proceeding from the eastern part of the Sun, BE a ray proceeding from the western part. [30] It will be evident that if A is lacking in the Sun, D will be lacking in the light falling through the aperture, as it is caused by it. [31] And if B is lacking, by the same reason E will be lacking.

4. Commentary

Ad 3. In the Middle Ages, the apparent solar diameter was usually divided into 12 points or digits to measure the magnitude of a solar eclipse.

Ad 5. The solar eclipse of 4 June 1285 (OPPOLZER [1], no. 5945; $16;35.5^h$ UT) is not the only recorded observation in the *Prologue* of the *Almanach*. WILLIAM OF SAINT-CLOUD also reports the following: 1) a conjunction of Mars and the Moon *sub Aldebaran* (α Tauri) on 1 June 1285; 2) two observations of Saturn and Jupiter near conjunction on 28–29 December 1285; 3) a conjunction of Mars with the star situated *in ungula septentrionali Scorpionis* ("in the nothern nail of the Scorpion"); near Sco 21°, according to the text, on 3 March 1290; 4) a determination of the solar altitude at noon on 12 March 1290[1]; 5) a second conjunction of Mars with the aforementioned star during its retrogradation on 21 April of the same year; 6) two other determinations of the solar altitudes at summer and winter solstices, not dated in the text but probably also made in 1290; 7) an observation of Mars during its retrogradation in Capricorn on 1 July 1292 after a conjunction with the Moon. No instruments are mentioned, excepting an armillary sphere for observation no. 7 (*et inspiciendo per armillas uidebatur luna transiuisse martem quasi per unum gradum*). SAINT-CLOUD reports that the three determinations of the solar altitude (nos. 4 and 6) were realized at Paris; therefore, observations nos. 3 and 5 were probaboy made at the same place. The motivation of all these observations was purely astronomical: to test the planetary positions derived from the Toledan and Toulouse tables (observations nos. 1–3, 5, and 7), and to find the value for the obliquity of the ecliptic and to test the value attributed to the motion of the eighth sphere by the trepidation theory (observations nos. 4 and 6). WILLIAM OF SAINT-CLOUD also informs us of another conjunction of Saturn and Jupiter on 4 April 1226 as reported in the margin of a certain book read by him (*in quodam libro in margine*) and he predicts another conjunction of Mars and the Moon (with Mars at the apogee of the epicycle and the epicyclic center at the apogee of the eccentric) for the end of May or the begining of June 1293.

---

[1] From this observation derives SAINT-CLOUD's determination of the vernal equinox (16^h, Sunday 12 March 1290) repeatedly quoted by JOHN OF MURS in his *Expositio intentionis regis Alfonsii circa tabulas eius* (*cf.* POULLE [2], pp. 261–267).

Ad 8. The passage quoted by WILLIAM is ARISTOTLE's *De anima*, II, 12 (424a).

Ad 9–10. This technique, whose only purpose was probably to minimize such injuries to the eye as that described in §§ 6–7, is also mentioned by IBN YŪNUS (d. 1009), who informs us that the solar eclipse of 18 August 928 was observed in this way by ABŪ AL-QĀSIM 'ABDALLĀH B. AMĀJŪR (c. 910) (see CAUSSIN [1], pp. 120–122), and by AL-BĪRŪNĪ (d. after 1050) (AL-BĪRŪNĪ [1], p. 131). In the XVIth century the procedure is still attested by APIANUS in his *Astronomicum Caesareum*: "Postremum est, & quasi parergum, ut ecleipses quod fusissimè descripsi, oculari quoque observatione contuendas doceam. Cum multi sint, qui variè variis videndi instrumentis utantur, omnibus tamen perperàm, Alii enim in pelui aqua referta, Alii speculis, Alii simplici papiro perforata, Alii aliter obseruare ecleipses solent" (APIANUS [1], f. kIIra).

Ad 15–16. The procedure described by SAINT-CLOUD is very simple and requires only an aperture in the roof or in a window and a screen perpendicularly placed with respect to the solar rays passing through the aperture. Two features of the image on the screen are clearly stated: its round shape, even if an angular aperture is used, and its increasing size according to its removal from the aperture, although the greater the distance, the weaker the image.

Ad 17–19. These are the relevant sentences on pinhole images in SAINT-CLOUD's text. WILLIAM claims that finite apertures and point apertures require two different procedures for measuring apparent solar diameters on the screen of the camera. In the first case, *i.e.*, when the aperture is "great" (as, for example, AB in figure 2), the angle corresponding to the apparent solar radius must be measured on the screen from the intersection before the aperture of the outermost rays casting the solar image CD, that is, from point H at which lines CA and DB intersect. This is in fact equivalent to asserting that, for an aperture like AB, the apparent solar radius on the screen is half the chord of arc CKD of the circle whose radius is HC (and not half the chord CD of a circle whose radius is OC, since angle α cannot be equal to angle β).

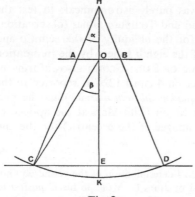

Fig. 2.

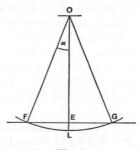

Fig. 3.

In the second case, if the aperture is like a point (*i.e.*, O in figure 3), then the apparent radius of the Sun, FE, is half the chord of arc FLG of the circle whose radius is OF. In my opinion, this is the only correct interpretation of the text despite the apparent ambiguity introduced by the word *parvum* ("small") that WILLIAM uses for referring to the aperture in this case, in so far as the outermost rays of the solar image always intersect in a point situated between point O and the Sun, except in the case of point-apertures. Otherwise, it would be necessary to attribute to SAINT-CLOUD a statement that is absurd from a geometrical point of view: that there is an aperture smaller than AB and greater than the point-aperture O (for example, MN in figure 4), for which angle α will be equal to angle δ.

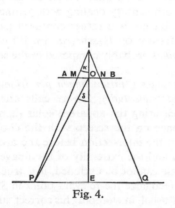

Fig. 4.

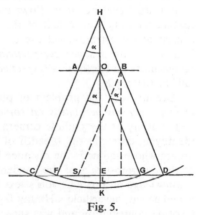

Fig. 5.

Although the text does not continue beyond this point, from the superposition of figures 2 and 3 it is obvious that measuring the half chord CE (fig. 5) of arc CKD of the circle with radius HC is equivalent to measuring the half chord FE of arc FLG of the circle whose radius is OF, since angle CHE is equal to angle FOE. It does not seem unreasonable to attribute to a highly competent astronomer like WILLIAM OF SAINT-CLOUD the elementary knowledge of EUCLID (*Elements*, VI, 2) required to arrive at this corollary.

Obviously, SAINT-CLOUD's claim is equivalent to the procedure explained forty years later by LEVI BEN GERSON in his *Astronomy*, *i.e.*, taking into account the distance between the aperture and the screen, and subtracting from the diameter of the solar image on the screen the diameter of the aperture, since angle CHD (fig. 5) is equal to angle SBD, and CS is equal to AB. This does not imply, however, that, if WILLIAM OF SAINT-CLOUD at some time used the pinhole camera for solar observations, he actually subtracted AB from CD, inasmuch as he could use the *ratio*

$$\frac{HO}{AO} = \frac{HE}{CE} = \frac{HO + OE}{CE}$$

to find HO and, once HE is known, HC is found from

$$HE^2 + CE^2 = HC^2.$$

Then angle α, the solar radius, follows directly from

$$CE = \frac{\text{chord of arc CKD}}{2}.$$

The possibility of investigating the solar eccentricity by means of a pinhole camera, mentioned by SAINT-CLOUD in § 20, suggests that he was relying on the accuracy of the procedure. Unfortunately, it is not easy to reconstruct the parameters involved in the case considered by him in §§ 13–14 in so far as the exact size of the aperture is not mentioned.

Little can be said about the possible influence of SAINT-CLOUD's text on authors of the following century. Except for LEVI BEN GERSON (hitherto there is no textual evidence that allows us to assert that he knew the *Almanach planetarum*), the only known text of the fourteenth century dealing with geometrical aspects of the astronomical use of pinhole images is a rather confused passage in the *Questiones super perspectivam* by HENRY OF HESSE (*ca.* 1325–1397), a commentary on PECHAM's *Perspectiva communis* probably composed in the second half of the 1360s.

Dealing with the problem of pinhole images (*"utrum lumen per triangulare foramen incidens per lineas ad rotunditatem reducatur"*), HENRY calls attention to the utility of the pinhole camera for measuring the apparent solar diameter, identifying correctly the breadth of the image on the screen with the chord of an arc whose radius is the distance between the intersection before the aperture of the outermost rays of the incident beam and the extremity of the image. But if knowledge of SAINT-CLOUD's text by HESSE cannot be excluded, it is true that, if he knew it, nevertheless HENRY failed to recognize the relevant point in SAINT-CLOUD's contribution and was succesful, as usual, in wrapping his correct suggestion in a cloud of obscurities and geometrical errors. HESSE writes:

"I answer this question by supposing that the incident beam [passing] through the aperture increases continuously in size as the distance to the screen [increases] because of the intersection of the rays tangents to the sides of the aperture [proceeding] from opposite [parts of the Sun], as it is commonly declared in optics. Thus, the first conclusion is that this intersection occurs between the aperture and the Sun. This is proved because [the rays of] the sides of the image, being continuously inclined towards an angle, reach the extremities of the aperture. Consequently, as they go farther away according to the rectilinear propagation [of light] from the Sun, they necessarily intersect beyond the aperture [*i.e.*, between the aperture and the Sun]. Therefore, the conclusion is true. The antecedent is evident for anyone who understands [this matter].

Secondly, if the intersection of the rays producing the increase in size of the image occurred on the other side of the aperture [i.e., between the aperture and the screen], it would follow that the image would be somewhere smaller and narrower than immediately after the aperture. It is sufficiently evident from experience that [this smaller and narrower part of the image would be] near the intersection of the rays; because in this case the rays of the sides of the image could not appear right, but refracted in the direction of [the Sun] C, as it appears in the figure, where the sides of the image after the aperture appear according to lines KE and GE, and not [according to lines] FO and DO. From this it follows that the angle to which the Sun is opposed to the eye [*i.e.*, angle NEM] is known, because it is equal to the angle formed before the aperture by the intersection of the sides of the image, as they are opposite, and this angle is known.

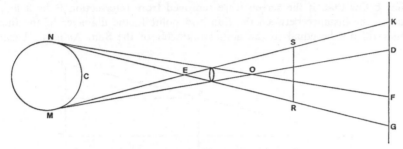

Fig. 6. HESSE [1], 50v; MS Conv. soppr. J.X.19, 60r.

From this is known the amount of the light on the screen. And the width of the image at any place is known by interposition of a staff like SR in the middle of the image. Thus, if SR and KG are known and SK is known, the angles [S]K[G] and [R]G[K] will be known. And in this way the angle KEG is known. From this, it is furthermore known how many minutes measures the arc of the sky that the Sun cuts off as seen by the eye, because KE is known, and in the same way KG [is known], which is a chord of the circle described about E by the [radius of] length EK. Therefore, KG is known in the units of this radius. Thus, the amount of the chord KG is found from the table of chords, and what we sought will be known. Moreover, it is evident that if the screen is removed from the [aforementioned] intersection of the rays by a distance equal to the distance between this intersection and the Sun, the diameter of the light [on the screen] would still not be equal to the diameter of the Sun. The reason [for this] is that [the outermost rays of] the pyramid whose sides intersect in point E are tangents to the Sun; therefore, these rays proceed from an arc smaller than a semicircle, that is, from [arc] NM, as we will see better further on. It also follows that the right-hand rays of that image proceed from the left-hand side of the Sun, and the left-hand ones from the right-hand [side]"[2].

---

[2] "Ad istam questionem respondeo supponendo quod incidentia per foramen secundum remotionem obiecti continue maioratur propter radiorum intersectionem contingentium latera foraminis ex oppositis, sicut communiter in perspectiva declaratur. Sit ergo prima conclusio quod talis intersectio est inter foramen et solem. Probatur quia latera incidentie continue declinando ad angulum pertingunt ad oppositas extremitates foraminis; igitur ex quo ulterius procedunt secundum rectam radiationem a sole necessario concurrent post foramen. Igitur conclusio uera. Antecedens patet cuilibet intuenti. Secundo, si esset huiusmodi intersectio radiorum causans maiorationem incidentie ex alia parte foraminis, sequeretur quod incidentia esset alicubi minor et strictior quam inmediate post foramen; quia circa concursum radiorum satis patet ad experientiam, quia tunc laterales radij incidentie non appare[ren]t recti, ymmo fracti uersus C,

Leaving aside HESSE's unnecessarily complicated demonstration that EK and EG are the rays of light producing the breadth of the image on the screen, it is difficult to be sure if he actually refers to calculating the value of the angle KEG from the angle KSL (figure 7), which is half the angle KEG and can be known from SK, KL, and SL. But, be that as it may, HESSE surprisingly claims below that, if the screen were removed from intersection E by a length equal to the distance between the Sun and point E, the diameter of the image would still not be equal to the actual diameter of the Sun. And this because

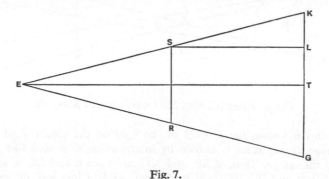

Fig. 7.

sicut apparet in figura, ubi latera incidentie post foramen apparent secundum lineas KE et GE, et non FO et DO. Ex quo sequitur quod angulus cui opponitur sol secundum uisum est notus, quia est equalis angulo ad quem concurrunt latera incidentie ultra foramen propter contrapositionem, qui est notus. Ex quo quantitas luminis in obiecto est nota. Et linea spissitudinis ubique est nota per positionem virge per medium incidentie sicut SR. Si ergo SR et KG sunt note et RK [*ms. and pr. ed.*: SK] est nota, erunt anguli G et K noti; et sic KEG angulus est notus. *Ex quo* ulterius *comprehenditur quot minutorum sit arcus celi quem sol cordat* secundum uisum, *quia KE est nota et similiter KG, que est corda circuli descripti super E secundum quantitatem EK; igitur KG est nota in gradibus suis. Inueniatur igitur quantitas arcus corde KG per tabulam cordarum* et habetur propositum. Patet item quod elongato obiecto pariete tantum ab intersectione radiorum sicut sol ex alia parte nondum diameter luminis esset equalis diametro solis. Ratio quia piramis cuius latera secant sese super E contingunt solem; ergo necessario procedunt ab arcu minori semicirculo. scilicet NM, sicut posterius uidebitur melius. Sequitur etiam quod radij destri illius incidentie exiantur a sinistro latere solis et sinistri a destro".

To establish this text I have used the printed edition of HESSE's *Questiones* (*cf.* HESSE [1], ff. 49v–50r) and one of the extant copies of the work: Firenze, Biblioteca Nazionale, MS Conv. soppr. J.X.19, f. 60r–v. The sentences in italics have been edited by LINDBERG ([1], pp. 311–312] and commented upon by STRAKER ([1], pp. 195–196). STRAKER, relying on the sentences edited by LINDBERG, points out that "HENRY's failure to tell the reader how he should determine the distance EK makes him guilty either of not understanding the geometry of image formation, or of not seriously directing his suggestion to practicing astronomers, or, what is most likely, both. His remark 'since KE is known', must be read as an implicit suggestion that E is a point in the plane of the aperture" (STRAKER [1], p. 196). As we will see, some of STRAKER's conclusions are essentially correct, in spite of the fact that HESSE actually distinguishes point E from the center of the aperture.

line NM is not a solar diameter! HESSE's text reveals not only his misunderstanding of the geometry involved in the case, but also his ignorance of the inconsistency of this conclusion, because if NM is not a diameter of the solar disk, then angle NEM and, therefore, angle KEG are not measurements of the apparent solar diameters.

It has been suggested that, facing up to the problem of pinhole images and asking quantitative questions about the relative sizes of the images cast on the screen of the camera, images seen as those of actual bodies whose apparent sizes are already known, the astronomer has a distinct advantage over the student of optics whose attention is focused directly on questions about the nature and propagation of light[3]. But, beyond the fact that astronomers could easily reject certain measured values of the apparent solar diameter on the screen as obviously incorrect in comparison with the accepted ones as asserted in astronomical tables or derived from other procedures, it seems, in my opinion, that the main obstacle during the Middle Ages for a correct explanation of the formation of images and the astronomical use of the pinhole camera was the technical incompetence of most of the scholars which discussed on the subject (*e.g.* PECHAM, HESSE, BLASIUS OF PARMA) for dealing in a satisfactory way with the geometrical aspects of the problem.

Ad 20–21. Till now, the only known suggestion for using the pinhole camera to determine the solar eccentricity was that of LEVI BEN GERSON in his *Astronomy* (see *Appendix* below).

Ad 25–26. WILLIAM explains the inversion of the image from the inverted pyramid model of propagation, illustrated further on in §§ 29–31.

Ad 27. See, for example, ALHAZEN [1], lib. VI, cap. VII, pp. 214–225.

Appendix

Levi ben Gerson's Determination
of the Solar Eccentricity with a Pinhole Camera

Introduction. Although, as we have seen, in 1292 WILLIAM OF SAINT-CLOUD mentioned that by observing the apparent solar diameter at apogee and perigee the eccentricity of the Sun can be determined from the inverse proportion between the distances and the apparent diameters, we find the first known application of this procedure in the first part of chapter 56 of LEVI BEN GERSON's *Astronomy*, based on observations of the solar diameter at summer and winter solstices in 1334. The Hebrew version of this work, composed between 1328 and 1344, was translated into Latin in the last years of LEVI's life and probably with his

---

[3] STRAKER [2], p. 269.

collaboration, by PETER OF ALEXANDRIA[1], an Augustinian friar who was also the translator of LEVI's *De sinibus* and *Pronosticatio*. Chapter 56 is still unedited in both versions and I have used for the present edition of the Latin text three of the four extant copies of the work: MSS (i) Vat. Lat. 3098, (ii) Vat. Lat. 3380, and (iii) Lyon, Bibliothèque Municipale, 326[2].

1. Text

A: Vat. Lat. 3098, f. 44rb–va.
B: Vat. Lat. 3380, f. 151rb–vb.
C: Lyon, BM 326, ff. 124r–125r.

Capitulum 56

[In 56 declarabitur quantitas excentricitatis spere solis et mensura quam uidemus in sole in auge et opposito augis. Et ibi declarabitur in parte quantitas equationis solis si uerum est quod motus suus sit proportionatus centro spere sue (a).]

[1] Et possumus deuenire in quantitatis equationis centri notitiam ex quantitate excentricitatis spere solis quam per experientiam nos uidemus. [2] Et quantitatem excentricitatis inuenimus per quantitatem diametri solis in circumferentia quam describit stans circa capita Capricorni et Cancri, quia per experientias certas inuenimus quod quantitas diametri solis de circumferentia quam describit stans circa principium Cancri est 0;27° et circa 0;0,51° et stans circa

---

[1] This is a tentative explanation for the additions to the Hebrew text contained in the Latin version and whose author can only be LEVI himself (see on this subject my "The Latin translation of Levi ben Gerson's *Astronomy*", forthcoming in the *Revue d'Histoire des Sciences*).

[2] MSS Lyon, B. M. 326, and Milano, Ambrosiana, D 327, the fourth extant copy of LEVI's *Astronomy*, have a common source and I have omitted the variant readings of the latter in the critical apparatus. Only the first part of chapter 56 is edited and translated here. In the second part, LEVI finds the place where the solar equation reaches its maximum value and he adds a table for the solar anomaly. In the Latin manuscripts, however, this table (called "*tabula equationis solis ultimata*") was computed for an eccentric model where the eccentricity is about 2;23 and the maximum equation 2;17°, instead of 2;14 and 2;8°, respectively, as in the text of the chapter ("Et postquam habuimus multarum eclipsium solarium et lunarium experientias manifestas, que nos duxerunt inuenire equationis quantitatis solis notitiam in 100 capitulo huius libri, ubi declaramus quantitatem dicte equationis finalem esse 2;17°, fecimus in hoc loco tabulas equationis solis secundum dictam quantitatem inuentam", Vat. Lat. 3098, f. 45r; *cf.* also GOLDSTEIN [1], p. 95–96 and 158, and GOLDSTEIN [2], pp. 104 and 118).

(a) *A* f. 1vb | *B* f. 71vb.

1. equationis *om, B*.
2. et quantitatem] ex quantitate *AC*, et quantitate *B* | per[1] *scr. et del. A, om. B*.

principium Capricorni est circa 0;30° (b). [3] Et ad experientias istas deuenimus quia fecimus baculum unum in cuius uno capite erat una tabella faciens angulum rectum cum baculo, in qua erat una fenestrella rotunda, et in alio capite erat una alia tabella paralella prime et recipiens radium solis transeuntem per fenestrellam prime. [4] Et longitudo baculi inter primam et secundam erat quasi trium cannarum. [5] Et multum subtiliter et minutialiter pertractauimus experientiam istam et inuenimus quod in quantitate in qua distantia inter tabulam et tabulam erat 60 gr., erat quantitas illius diametri fenestrelle, que diameter erat paralella illi diametro radij recepti in secunda tabula, quam diametrum postea mensurauimus, 1;14,16,35. [6] Et inuenimus diametrum mensuratam radij recepti in secunda tabula 1;43,26,30. [7] Et ista experientia fuit anno incharnationis Yeshu Christi 1334 die 29 maij circa meridiem. [8] Continuando experientiam istam circa quantitatem predictam non inuenimus diuersitatem notoriam quousque sol principium Cancri intrauit et ultra. [9] Et ideo est notum quod quantitas diametri solis stantis in principio Cancri est de circumferentia quam tunc describit 0;27,51° (c). [10] Et circa istam quantitatem inuenimus in oppositionibus lune diametrum.

[11] Et secunda die mensis decembris recepimus habere experientias quantitatis diametri solis cum intrumento predicto. [12] Et inuenimus quod in mensura in qua distantia inter tabulam et tabulam erat 60, erat quantitas illius diametri fenestrelle, que diameter erat paralella illi diametro radij in tabula secunda recepti, quam diametrum radij mensurauimus postea, 1;14,57,29. [13] Et in ista mensura erat diameter radij in tabula secunda recepti 1;46,21,44. [14] Et continuando experientiam non inuenimus diuersitatem notoriam circa quantitatem predictam quousque sol principium Capricorni intrauit et ultra. [15] Et hinc est notum quod quantitas diametri solis stantis circa principium Capricorni est circa 0;30° de circumferentia quam describit (d). [16] Et multiplicauimus sepissime ad experientiam istam anno predicto et diametrum solis secundum proportionem augmentari continue de capite Cancri ad caput Capricorni inuenimus.

[17] Hoc autem sedato, possumus quasi in perfectam notitiam quantitatis excentricitatis spere solis hoc modo uenire. [18] Sit enim ABC circumferentia quam in sua spera describit, cuius centrum sit punctus D. [19] Et centrum terre sit in superficie istius circumferentie sub puncto D punctus E. [20] Et diameter

---

(b) *AB in mg. et C inter* circumferentia *et* quam describit *add.*: Prima diameter in se est 0;29,9,55, set arcus sibi correspondens est de 0;27,51°. Et diameter secunda in se (+ est *C*) 0;31,24,25, set arcus eius est circa 0;30°.

(c) *AB in mg. add.*: Ista diameter uisa est 0;29 et cetera, ut supra.

(d) *AB in mg. add.*: Ista diameter uisa est 0;31 et cetera, ut supra.

5. tabula] tabella *BC* | diametrum + radij *C*.
6. tabula] tabella *B*.
7. incarnationis *BC* + domini *C* | Jesu *C*.
11. recepimus] reincepimus *B*.
12. tabula] tabella *B* | 1;14,57,29] 1;14,57,59 *ABC*; *Heb. MSS*; 1;14,57,29, 1;14,58,29.
13. tabula] tabella *B*.
15. hinc] hic *C* | est *om. C*.
17. hoc modo *om. B* | uenire] deuenire *B*.

transiens per ista centra sit linea ADEC. [21] Sequitur quod punctus A est maior distantia et punctus C minor. [22] Signetur circa punctum A circumferentia FG et circa punctum C circumferentia HI et protrahantur linee FE, HE. [23] Et ponatur quod puncta F, H, sint loca in quibus linee FE, HE, contingunt circumferentias FG, HI. [24] Et protrahantur linee AF, CH. [25] Et est notum quod linea AF est equalis linee CH, quia quelibet est semidiameter solis. [26] Et est etiam notum quod angulus AFE est rectus. [27] Set angulus AEF est suppositus 0;13,55,30° et angulus CEH est suppositus 0;15°. [28] Et est notum quod in mensura in qua linea CE est 60, est linea CH 0;15,42,30, que sunt sinus anguli CEH. [29] Et in mensura in qua linea AE est 60, est linea AF 0;14,35. [30] Et quia linea CH est equalis linee AF, est notum quod in mensura in qua linea AF est 60 est linea CH 0;14,35. [31] Et quia linea CH erat 0;15,42,30 in mensura in qua linea CE est 60, necessario sequitur quod proportio linee CE ad lineam AE sit talis qualis est proportio 0;14,35 ad 0;15,42,30. [32] Et quando componimus, sequitur quod proportio linee CA ad lineam AE sit talis qualis est proportio 0;30,17,30 ad 0;15,42,30. [33] Et ista proportio est talis qualis est proportio 120 ad 62 et circa 0;14.

[34] Et ideo est necessarium quod [in] mensura in qua linea AC, que est diameter spere solis, est 120, est linea AE 62 et circa 0;14. [35] Et ideo sequitur quod quantitas linee DE, que est quantitas excentricitatis, est 2 et circa 0;14. [36] Et hoc est quod uolebamus probare. [37] Et hinc est notum quod quantitas equationis finalis solis est 2° et circa 0;8°, ut statim probabitur si ita est quod motus spere solis sit circa suum centrum, quem decet sic poni nisi constringamur aliunde aliter ponere.

2. Translation

Chapter 56

[In chapter 56 the amount of the eccentricity of the sphere of the Sun and the measure [of the diameter] of the Sun which we see at apogee and perigee will be demonstrated. And there will be demonstrated in part the equation of the Sun if truly its motion is proportionate with respect to the center of its sphere.]

[1] We can arrive at the amount of the equation of the center from the amount of the eccentricity of the sphere of the Sun which we see through experience. [2] And we found the amount of the eccentricity from the amount of the diameter of the Sun in the circumference which it describes when it is near the beginnings of Capricorn and Cancer, because by means of unquestionable trials we found that the amount of the diameter of the Sun when it is near the beginning of Cancer is approximately 0;27,51° of the circumference which it is then describing and approximately 0;30° when the Sun is located in the proximity of

---

23. *F* + et *B*.

the beginning of Capricorn (a). [3] And we made these trials with a staff at one of whose extremities there was a thin panel making a right angle with the staff, in which there was a little round window; in the other extremity there was another thin panel parallel to the first, receiving the Sun's ray passing through the little window in the first panel. [4] And the distance on the staff between the first and the second panel was almost 3 canes. [5] And we performed this observation in a very careful and precise way and we found that, in the measure where the length of the staff was 60, the amount of the diameter of the little window was 1;14,16,35, this diameter being parallel to the diameter of the ray received on the second panel, which we measured later. [6] And we found that this diameter was 1;43,26,30. [7] And this observation was made the 29 of May of the year 1334 A.D. about noon. [8] Continuing with this observation concerning the aforementioned amount, we did not find a noticeable discrepancy until the Sun reached the beginning of Cancer and further. [9] Consequently, it is known that the amount of the diameter of the Sun when it is at the beginning of Cancer is 0;27,51° of the circumference it is then describing (b). [10] And we found that the diameter of the Moon at opposition was about the same amount.

[11] The second day of December [of the same year] we began again to make observations concerning the amount of the diameter of the Sun with the aforementioned instrument. [12] And we found that in the measure where the distance between the panels was 60, the amount of the diameter of the little window was then 1;14,57,29, this diameter being parallel to the diameter of the ray received on the second panel, which we measured later. [13] And in the aforementioned measure the diameter of the ray received on the second panel was 1;46,21,44. [14] Continuing with the observation, we did not find a noticeable discrepancy concerning the aforementioned amount until the Sun reached the beginning of Capricorn and further. [15] We know from it that the amount of the diameter of the Sun when it is near the beginning of Capricorn is approximately 0;30° of the circumference it is then describing (c). [16] And we repeated this observation very often during the aforementioned year and we found the diameter of the Sun to be increased proportionally in a continuous way from the beginning of Cancer to the beginning of Capricorn.

[17] This matter once settled, we can reach an almost perfect knowledge of the amount of the eccentricity of the sphere of the Sun in the following way. [18] Let ABC be the circumference on which the Sun travels, whose center is D. [19] Let point E, in the plane of this circumference below point D, be the center of the earth. [20] And let the diameter passing through these centers be line ADEC. [21] It follows that the distance [from the center of the earth] to point A is the greatest and that to point C is the least. [22] Draw circumference FG about point A and circumference HI about point C, and draw also lines FE and HE. [23] Let us assume that lines FE and HE are tangent to circumferences FG and HI at points F and H. [24] Draw lines AF and CH. [25] It is

(a) MSS. ABC mg add: The first diameter is 0;29,9,55, but the corresponding arc is 0;27,51°. The second diameter is 0;31,24,15 but its arc is approximately 0;30°.

(b) MSS AB mg add: This apparent diameter is 0;29 *etc.*, as above.

(c) MSS AB mg add: This apparent diameter is 0;31 *etc.*, as above.

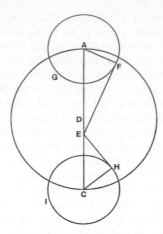

Fig. 8. *A*, 44v; *B*, 151v.

known that lines AF and CH are equal, since each of them is the radius of the Sun. [26] And it is also known that angle AFE is 90°. [27] But we have supposed that angle AEF is 0;13,55,30° and angle CEH is 0;15°. [28] And it is known that in the measure in which line CE is 60, line CH is 0;15,42,30, which is the sine of angle CEH. [29] And in the measure in which line AE is 60, line AF is 0;14,35. [30] And since line CH is equal to line AF, we know that in the measure in which line AE is 60, line CH is 0;14,35. [31] And since line CH was 0;15,42,30 in the measure in which line CE is 60, it necessarily follows that the ratio of line CE to line AE is equal to the ratio of 0;14,35 to 0;15,42,30. [32] When we compose [these ratios], it follows that the ratio between lines CA and AE is equal to the ratio between 0;30,17,30 and 0;15,42,30. [33] And this ratio is the same as that between 120 and approximately 62;14.

[34] Consequently, it is necessary that in the measure in which line AC, the diameter of the sphere of the Sun, is 120, line AE is approximately 62;14. [35] Thus, it follows that the amount of line DE, the amount of the eccentricity, is approximately 2;14. [36] And this is what we wished to prove. [37] From this it is known that the amount of the maximum of the equation of the Sun is approximately 2;8°, as we will prove later, if in fact the sphere of the Sun is moved [uniformly] around its own center, which is appropriate to suppose unless we were forced to suppose otherwise.

3. Commentary

LEVI asserts (§ 2) that we can find the solar eccentricity from observations of the apparent solar diameters when the Sun is at apogee and perigee in chapters 5, 15, and 19 of his *Astronomy* (GOLDSTEIN [3], pp. 49–50, 98, and 113; Latin version, MS Vat. Lat. 3098, ff. 6vb, 12va, and 14ra–b). In the Latin version, the same subject is also mentioned in chapter 9, in a passage which the

Hebrew version omits[1]. In both versions of chapter 19 the minimum and maximum apparent solar diameters are stated ta be 0;27,50° and 0;30°, instead of 0;27,51° and 0;30° as in the present chapter. The marginal note to § 2 in Latin MSS, probably by LEVI himself, is also omitted in the Hebrew version.

The instrument used by LEVI (§ 3), a combination of the Jacob staff and the *camera obscura*, is also described in chapter 9 of the *Astronomy* (GOLDSTEIN [3], pp. 69–70; Vat. Lat. 3098, f. 9va). In the aforementioned chapter, the length of the staff is "16 spans or more" (*i.e.*, greater than about 3 m) and the diameter of the aperture is equal to one or two units, of which the staff measures 60, being each unit 1/8 of a span (1 span = 20 or 25 cm). A very similar instrument was used by KEPLER in his observations of apparent solar diameters before and during the solar eclipse at July 1600 (KEPLER [1], pp. 335–9; STRAKER [2], p. 284). The sentence which gives the size of the instrument (§ 4) is omitted in the Hebrew version: Paris, Bibliothèque Nationale, MSS Heb. 724, f. 104b, and 725, f. 79b. It is difficult, however, to establish the exact distance between the aperture and the screen in the staff used by LEVI, insofar as the unit *canna* has a variable length in southern France and northern Italy during the fourteenth century (*cf.* ZUPKO [1], pp. 61–68). The comparison of the Hebrew and Latin texts does not help on this point because the sizes of the different pieces of the Jacob staff seem not to be always equal in both versions.

In § 9, according to the procedure explained in chapters 5 and 9 (GOLDSTEIN [3], pp. 49–59 and 70–71; Vat. Lat. 3098, ff. 6vb–7ra and 9va–b), LEVI subtracts the diameter of the aperture from the diameter of the image on the screen to find the corrected apparent solar diameter. In figure 9,

$$CF = CD - FD,$$

where FD = AB, the diameter of the aperture.

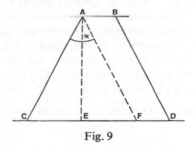

Fig. 9

Thus

$$1;43,26,30 - 1;14,16,35 = 0;29,9,55,$$

---

[1] "Hoc instrumentum quesiuimus principaliter ad inuestigandum si est aliqua spera celestis excentrica, quia cum eo possumus scire faciliter quantitatem diametri lune uisi a nobis in qualibet quattuor distantiarum suarum iuxta quod posuit Ptolomeus, et quantitatem diametri solis uisi a nobis in qualibet duarum distantiarum ..." (Vat. Lat. 3098, f. 9vb; *cf.* GOLDSTEIN [3], p. 72, § 55).

and this agrees with the value given in the margin by the Latin MSS. As AE, the distance between the aperture and the screen is 60, angle α is given by

$$CF/2 = AE \tan(\alpha/2)$$

or

$$CE/AC = \sin(\alpha/2).$$

Therefore

$$\alpha = 0;27,51,2,17\ldots°.$$

As angle α is very small, $AC = AE$, and then *corda* $CF = 2\,AE \sin(\alpha/2)$, and $\alpha = 0;27,51,3,6\ldots°$, rounded by LEVI to $0;27,51°$.

LEVI does not inform us if for his second series of observations (§§ 12–15) he was using an aperture slightly larger than the previous one or if the distance between the aperture and the screen was slightly smaller. The corrected apparent diameter of the Sun is now (figure 9)

$$CF = CD - FD = CD - AB = 1;46,21,44 - 1;14,57,29 = 0;31,24,15.$$

As above,

$$0;31,24,15 = 2\,AE \tan(\alpha/2)$$

or

$$CE/AC = \sin(\alpha/2).$$

Hence

$$\alpha = 0;29,59,18,53\ldots°.$$

If we consider $AC = AE$,

$$CF = 2\,AE \sin(\alpha/2), \quad \text{and} \quad \alpha = 0;29,59,19,54\ldots°,$$

rounded by LEVI to $0;30°$.

LEVI's procedure in §§ 17–26 is the same as that of PTOLEMY in the *Almagest* (IX, 8; X, 2) to find the eccentricity of the inferior planets from two observations of maximum elongations while the center of the epicycle lies in the apsidal line (TOOMER [1], pp. 453–6 and 470–2; NEUGEBAUER [1], pp. 152–4 and 161–3).

In figure 8 (§§ 27–36), AF and CH represent respectively the solar radius at apogee and perigee for an observer located at E. Consequently, angle AEF is half $0;27,51°$ and angle CEH is half $0;30°$. If $CE = 60$ and $AE = 60$,

$$CH = R \sin CEH = 0;15,42,28\ldots, \text{ rounded by LEVI to } 0;15,42,30$$

and

$$AF = R \sin AEF = 0;14,34,55\ldots, \text{ rounded by LEVI to } 0;14,35.$$

Now

$$CE/AE = AF/CH;$$

hence

$$AC/AE = (AF + CH)/CH.$$

If $AC = 2\,R$,
$120/AE = 0;30,17,30/0;15,42,30$, and

$$AE = 62;13,42,0,29\ldots, \text{ rounded by LEVI to } 62;14.$$

In an eccentric model, the maximum equation of center is given by

$$\sin eq_{max} = e/R\,;$$

with $e = 2;14$ and $R = 60$,

$$eq_{max} = 2;7,59,24\ldots°, \text{ rounded by } Levi \text{ to } 2;8° \,(§\,37).$$

The same value for solar eccentricity is derived by Levi in chapter 57 from three observations (24 April and 4 August 1334; the third, on 13 June 1334, is described in chapter 55) using Ptolemy's method for finding the apogee and eccentricity of the outer planets from three oppositions (Goldstein [3], p. 183; *Almagest*, X, 7; XI, 1, 5; Toomer [1], pp. 484–498, 507–519, 525–538).

References

Alhazen [1]: *Opticae Thesaurus Alhazeni Arabis libri septem … Item Vitellonis Thuringopoloni libri X, omnes instaurati … a F. Risnero*, Basileae, 1572.

Apianus [1]: P. Apianus, *Astronomicum Caesareum*, Ingolstadt, 1540.

Bīrūnī [1]: al-Bīrūnī, *The Determination of the Coordinates of Positions for the Correction of Distances between Cities*. A translation of *Kitāb Taḥdīd Nihāyāt al-Amākin Litaṣḥiḥ Masāfāt al-Masākin* by Jamil Ali. Beirut, 1967.

Brahe, [1]: *Tychonis Brahe Dani Opera Omnia*, ed. by J. L. E. Dreyer, Hauniae, 1913–1929.

Caussin [1]: Caussin de Perceval, "Le Livre de la grande Table Hakémite par Ebn Iounis", *Notices et Extraits des Manuscrits de la Bibliothèque nationale*, tome septième, Paris, an XII de la République [1804], pp. 16–240.

Curtze [1]: M. Curtze, "Die Dunkelkammer. Eine Untersuchung über die Vorgeschichte derselben", *Himmel und Erde*, 13 (1901), pp. 225–236.

Duhem [1]: P. Duhem, *Le Système du Monde*, Paris, 1913–1959.

Gemma Frisius [1]: *De radio Astronomico & Geometrico*, in: *Cosmographia, siue Descriptio universi orbis, Petri Apiani & Gemmae Frisij, … Adiecti sunt alij, tum Gemmae Frisij, tum aliorum Auctorum … Antuerpiae*, 1584.

Goldstein [1]: B. R. Goldstein, *The Astronomical Tables of Levi ben Gerson*, New Haven, 1974.

Goldstein [2]: B. R. Goldstein, "Medieval Observations of Solar and Lunar Eclipses", *Archives Internationales d'Histoire des Sciences*, 29 (1979), pp. 101–56.

Goldstein [3]: B. R. Goldstein, *The Astronomy of Levi ben Gerson (1288–1344)*, New York-Berlin, 1985.

Hesse [1]: *Preclarissimum mathematicarum opus in quo continentur perspicacissimi mathematici thome Bravardini … cum acutissimis ioannis de assia super eadem perspectiva questionibus annexis …*, Valentie, 1500.

Kepler, [1]: J. Kepler, *Ad Vitellionem Paralipomena quibus Astronomia pars optica traditur …*, Francofurti, 1604.

Kepler [2]: J. Kepler, *Gesammelte Werke*, herausg. von W. von Dyck, M. Caspar, F. Hammer & C. H. Beck, München, 1937–.

Lindberg [1]: D. C. Lindberg, "The Theory of Pinhole Images in the Fourteenth Century", *Archive for History of Exact Sciences*, 6 (1970), pp. 299–325.

Littre [1]: E. Littré, "Guillaume de Saint Cloud. Astronome", *Histoire littéraire de la France*, 25 (1869), pp. 63–74.

II

298

MANCHA [1]: J. L. Mancha, "Egidius of Baisiu's Theory of Pinhole Images", *Archive for History of Exact Sciences*, **40** (1989), pp. 1–35.

MERCIER [1]: R. Mercier, "Studies in the Medieval Conception of Precession (Part I)", *Archives Internationales d'Histoire des Sciences*, **26** (1976), pp. 197–220.

NEUGEBAUER [1]: O. Neugebauer, *A History of Ancient Mathematical Astronomy*, New York-Berlin, 1975.

OPPOLZER [1]: Th. R. von Oppolzer, *Canon der Finsternisse*, Denkschriften d. kais. Akad. d. Wiss., math.-naturwiss. Cl., **52**, Wien, 1887 (repr. New York, 1962).

POULLE [1]: E. Poulle, "William of Saint Cloud", *Dictionary of Scientific Biography*, vol. **14**, pp. 389–391, New York, 1981.

POULLE [2]: E. Poulle, "Jean de Murs et les Tables alphonsines", *Archives d'Histoire doctrinale et littéraire du Moyen Age*, **47** (1980), pp. 241–271.

REGIOMONTANUS [1]: *Ioannis Regiomontani ... de triangulis planis et sphaericis libri quinque, una cum tabulis sinuum ... in sequenti opere, quod complectitur ordinatam Astronomicorum & Geometricorum problematum descriptionem ... Omnia ... edita ... per Danielem Santbech Nouiomagum*, Basileae, 1561.

REINHOLD [1]: E. Reinhold, *Theoricae novae planetarum Georgii Purbachii Germani ab Erasmo Reinholdo Salueldensi pluribus figuris auctae, & illustratae scholiis ...*, Parisiis, 1557.

STRAKER [1]: S. Straker, *Kepler's Optics: A Study in the Foundations of 17th-Century Natural Philosophy*, Indiana University dissertation, unpublished, 1971.

STRAKER [2]: S. Straker, "Kepler, Tycho, and the 'Optical Part of Astronomy': the Genesis of Kepler's Theory of Pinhole Images", *Archive for History of Exact Sciences*, **24** (1981), pp. 267–293.

TOOMER [1]: G. J. Toomer, *Ptolemy's Almagest*. Translated and Annotated by ... London, 1984.

ZUPKO [1]: R. E. Zupko, *Italian Weights and Measures from the Middle Ages to the Nineteenth Century*, Philadelphia, 1981.

III

THE LATIN TRANSLATION OF LEVI BEN GERSON'S *ASTRONOMY*

1. INTRODUCTION

Although a medieval translation of the Hebrew text of Levi ben Gerson's *Astronomy* has been known to exist at least since Renan-Neubauer's publication (1893), it has not been the subject of a detailed study and it has been considered to be anonymous and unrelated to the translation of the treatise *De sinibus* undertaken by Petrus of Alexandria in 1342 and dedicated to Pope Clement VI. The object of the present article is to identify the author of the translation, to date it, and to provide arguments showing that Levi ben Gerson collaborated in the translation and, in fact, may even have dictated it.

2. MEDIEVAL LATIN TRANSLATIONS OF LEVI'S WORKS

There are extant medieval Latin translations of at least five scientific or philosophic works of Levi:

1) The *Book of the Correct Syllogism* (*Liber syllogismi recti*), a treatise on modal logic, preserved in the Vatican MS Ottobon. 1906 (Renan-Neubauer, 1893, p. 602; see also Manekin, 1985, pp. 87-90).

2) The *Prognostication for the Conjunction of Saturn, Jupiter and Mars in the year 1345* (*Pronosticatio super conjunctionem Saturni, Jouis et Martis anno domini 1345*), preserved in Paris, Bibliothèque Nationale, MS lat. 7378A, and Oxford, Bodleian Lib., MSS Ashmol. 192 and 393, and MS Digby 176 (Renan-Neubauer, 1893, p. 643). The Hebrew and Latin texts of this work have recently been published by Goldstein and Pingree (1990).

3) The treatise *On Harmonic Numbers* (*De numeris harmonicis*), a work on musical theory written by Levi in 1343 at the request of Philippe de Vitry, later bishop of Meaux (1351-61), to whom Nicholas of Oresme dedicated his *Algorismus proportionum* (Grant, 1965, p. 328). This text, of which the Hebrew version seems to have been lost, is preserved in the aforementioned manuscript of the Bibliothèque Nationale in Paris and in Basel, MS F.II.33. It has been published by Carlebach (1910, pp. 128-39).

4) The *Astronomy* (*Liber Bellorum Dei*, V, 1), an incomplete translation of the part conserved in Hebrew of the first section of Book

V of the *Wars of the Lord* (*Sefer Milhamot Adonai*); that is, chapters 1 through 103, 106, 109 and 110. According to the table of contents accompanying the manuscripts, however, the translation should include all three sections of Book V and not only the first. At least four copies of this text are preserved: Vat. Lat. MSS 3098 and 3380; Milan, Ambros. D 327; and Lyon, Bibliothèque Municipale, 326.

5) The treatise *On Sines, Chords and Arcs, and the Instrument "the Revealer of Secrets"* (*De sinibus, cordis et arcubus, item instrumento reuelatore secretorum*), dedicated to Pope Clement VI in 1342, is a translation of chapters four through eleven of the *Astronomy*, dealing with trigonometry and describing the Jacob Staff, an instrument for astronomical observations invented by Levi. It is preceded by a dedicatory letter and a prologue. At least four copies of this text are preserved: Paris, B.N., MS lat. 7293 (this codex was in the papal library at Avignon in 1369, *cf.* Renan-Neubauer, 1893, p. 621; the first folio of this copy, containing the dedicatory letter and most of the prologue, is lacking); Vienna, Nationalbibliothek, MS 5277; Klagenfurt, Bischöfliche Bibliothek, MS XXX.b.7; and Munich, Bayerische Staatsbibliothek, CLM 8089. The MS 2962-2978 of the Bibliothèque Royale in Brussels (ff. 194r-210r) contains a summary of this text.[1] The treatise *De sinibus* was partially edited by Curtze (1898, 1901).

The manuscript copies of *De sinibus* and the *Prognostication* tell us that Petrus of Alexandria *ordinis fratrum heremitarum sancti Augustini* was the translator of these two works. In the *colophon* of the *Prognostication*, Petrus mentions the date of Levi's death (20 April 1344) and also indicates that Solomon, Levi's brother and physician in the court of Clement VI in Avignon (Renan-Neubauer, 1893, p. 643; Goldstein and Pingree, 1990, pp. 29 and 34) collaborated in the translation of the text. There is, therefore, no doubt regarding the author of these translations or their date.

3. THE AUTHOR OF THE LATIN TRANSLATION OF THE *ASTRONOMY*

The text of the Latin translation of the *Astronomy* remains in manuscript except for the table of contents, published by Renan-Neubauer (1893, pp. 632-41), and the table for the solar equation, published by Goldstein (1974a, p. 158). Occasionally, Goldstein has used the Latin text in the critical apparatus of the edition and translation of certain chapters of the Hebrew text (Goldstein, 1979, 1985). The Latin manuscripts of the *Astronomy* do not contain any information on the date or the author of the translation and in the literature on Levi, this translation is considered to be anonymous and independent of that of the *De sinibus* (Renan-Neubauer, 1893, p. 620; Goldstein, 1974a, p. 79; 1979, p. 103). However, if the text of *De sinibus* and the corresponding passages of the *Astronomy* are compared, it becomes clear that we are dealing with a single translation: both texts are practically identical except for certain minor stylistic variations, most of which involve the substitution of impersonal verb forms or the passive voice for personal ones. Some examples of the collation follow:

De sinibus

Ad cuius probationem suppono pro nunc, quousque ex dicendis fuerit declaratum, quod si corpus radiosum fuerit indiuisibile siue puncti unius, radius terminatus ad obiectum non esset latior fenestra propter distantiam fenestre ab obiecto quamcumque possibilem in aliqua quantitate notabili sensui; quia propter mirabilem distantiam celestium corporum ad terram non est signum notabile sensui propter tam modicam distantiam, quam est illa que inter fenestram et obiectum existit (Tertium capitulum, Paris, B.N. 7293, 7vb-8ra).

Postquam de opere nostri instrumenti sum locutus quantum demonstratio superior requirebat, nunc prosequar usque ad complementum totius eius operis uel facture et usus. Et primo de eius factura; secundo de usu ipsius quoad omnes experientias perfecte habendas celestium corporum. Igitur hoc est opus seu factura ipsius. Accipiatur baculus unus rectus longitudinis 4 palmorum canne uel amplius, in quo fiat una superficies recta et plana latitudinis semidigiti per totam longitudinem baculi. Et ponatur in capite baculi una tabella quasi cornuta, cuius cornua non sint acuta in capite, sed aliqualiter sint rotunda. Et distantia unius cornu ab alio sit modicum maior quantitate unius digiti pollicis et semis, ita quod alterutrum cornu super alterutrum *mach* oculi possit sine artatione uisus locari. Et quando factus

Liber Bellorum Dei, V, 1

Supponamus pro nunc, quousque ex dicendis fuerit declaratum, quod, si corpus radiosum eset indiuisibile siue puncti unius, radius terminatus ad obiectum non esset latior propter distantiam fenestre quamcumque possibilem ab obiecto in aliqua quantitate notabili sensui; quia propter mirabilem distantiam celestium corporum usque ad terram non est signum notabile sensui propter tam modicam distantiam, que est illa que inter fenestram et obiectum existit (Capitulum quintum, Vat. Lat. 3098, 6vb-7ra).

Postquam de opere nostri intrumenti fuimus locuti quantum demonstratio supradicta requirebat, nunc prosequimur usque ad complimentum totius eius operis uel facture et usus. Et primo de eius factura; secundo de usu ipsius quoad omnes experientias perfecte habendas celestium corporum. Hoc est scilicet opus seu factura ipsius. Accipiemus enim baculum unum rectum longitudinis 4 palmorum canne uel amplius, in quo faciemus unam superficiem rectam et planam latitudinis semidigiti per totam longitudinis baculi. Et ponemus in capite baculi unam tabellam quasi cornutam, cuius cornua non sint acuta in capite, sed aliqualiter sint rotunda. Et distantia unius cornu ab alio sit modicum maior unius digiti pollicis cum dimidio quantitate, ita alterutrum cornu super alterutrum *mach* oculi sine artatione fuerit in hac forma, inuenietur centrum uisus magis infra capud quam sit tabula posita in capite baculi per 20^{am} partem palmi unius, sicut probaui per experientias multas cum maxima diligentia et labore (Quintum capitulum, *ibid.*, 10va).

Quoniam ut dixi necessarium est quod primo queratur scientia habendi longitudines et latitudines fixarum stellarum, quibus intendo me iuuare ad inueniendum loca longitudinum et latitudinum planetarum. Et uia que ad hoc ducet est ut queratur primitus locus solis, quia eius solius locus uerus potest omni tempore inueniri. Et est notum quod nullus sydus uidetur cum eo nisi cum pena, luna dumtaxat excepta; ideo cum dicto instrumento docui inueniri locum lune cum sole taliter quod sol oculos non offendat. Et cum accepta fuerit experientia ista aliquantulum ante solis occasum et statim cum post dictum occasum apparuerint alique stelle, accipiatur experientia cum luna et aliqua stella fixa, qua experientia habebitur longitudo inter eas hoc modo. Computabitur cursus lune in medio tempore inter experientias duas predictas secundum Ptolomei doctrinam, que in tam breui tempore a ueritate non notabiliter elongatur, sicut est michi per multas experientias manifestum (Octauum capitulum, *ibid.*, 15va-b).

Ne in usu predicti instrumenti error intercidat, quilibet utens eo attendat ad ista. Primo ut lumen teneatur post caput, ut supra. Secundo ut uideantur ambo sidera in angulis superioribus tabule uel prope, ita quod ambo semper sint ad superficiem superiorem equaliter prope et nunquam notabiliter longe. Tertio quod non sint uapores inter oculos et stellam uidendam, quia uapores ostendunt diuersam distantiam

uisus locari. Et quando factus fuerit in hac forma, inuenietur centrum uisus magis infra caput quam sit tabula posita in capite baculi per 20^{am} partem palmi unius, sicut probauimus per multas experientias cum maxima diligentia et labore (Capitulum septimum, *ibid.*, 7vb).

Quoniam ut diximus necessarium est quod primo queramus scientiam habendi longitudines et latitudines stellarum fixarum, quibus intendimus nos iuuare ad inueniendum loca longitudinum et latitudinum planetarum. Et uia que ad hoc nos ducet est ut queramus primitus locum solis, quia eius solius locum uerum in omni tempore possumus inuenire. Et est manifestum quod nulla stella uidetur cum eo nisi cum pena, luna dumtaxat excepta; ideo cum nostro instrumento docuimus inueniri locum lune cum sole taliter quod sol oculos non offendat. Et cum accepimus istam experientiam aliquantulum ante solis occasum [et statim cum post dictum occasum] apparuerint alique stelle, accipiemus cum luna experientiam cum aliqua stella fixa, qua experientia habebimus longitudinem inter eas hoc modo. Computabimus etenim cursum lune in medio tempore inter experientias duas predictas secundum Ptolomei doctrinam, que in tam breui tempore a ueritate non notabiliter elongatur, sicut est nobis per multas experientias manifestum (Capitulum decimum, *ibid.*, 9vb-10ra).

Ne in usu predicti instrumenti error intercidat, utens eo attendat ad ista. Primo ut lumen teneatur post caput, ut supra. Secundo uideantur ambo sidera in angulis superioribus tabule uel prope, ita quod ambo sint ad superficiem superiorem equaliter prope et nunquam notabiliter longe. Tertio quod non sint uapores inter oculos et stellam uidendam, quia uapores ostendunt diuersam distantiam inter stellam et inter stellam et stellam uel sydus et diuersam quantitatem dyametri sideris in maius et minus (Nonum capitulum, *ibid.*, 17ra).

stellam uel sydus et diuersam quantitatem dyametri sideris in maius et minus (Capitulum 11, *ibid.*, 10va).

We can thus conclude that Petrus of Alexandria is also the translator of the *Astronomy*. Since this hermit friar was working on the translation of Levi's works between 1342 and 1344, he was presumably also the author of the Latin version of the treatise *De numeris harmonicis* dated 1343. Moreover, although the text of *De sinibus* preserved in Munich CLM 8089 is considered to be an independent translation (Günther, 1890, pp. 75ff.; Renan-Neubauer, 1893, pp. 622-23; Touati, 1973, p. 55), a detailed examination proves that the text is identical to the one the other manuscripts explicitly ascribe to Petrus of Alexandria. The copyist simply changed the chapter headings and increased their number from 9 to 14,[2] adding immediately as chapters 15, 16, and 17 an incomplete copy of Peter of Saint-Omer's *Tractatus semisse*[3] without giving any indication that it is a different work.

In the manuscripts I have examined of *De sinibus* and the *Astronomy* I have not found any new information on Brother Petrus that would allow us to fill out his biography. A word added *supra lineam* in the copy of *De sinibus* preserved in the Nationalbibliothek of Vienna (f. 41r) identifies 'Alexandria' as the Italian city of the same name.[4] In any case, it is certain that all knowledge of Levi's work on astronomy among Christians in Europe from the XIVth century onward was due to the work of this translator.[5]

4. THE DATE OF THE TRANSLATION

What is the date of Petrus of Alexandria's translation of Levi's *Astronomy*? Although, as already mentioned, the manuscripts do not say it, the treatise *De sinibus* gives us an indirect way of dating. In the dedicatory letter, written in 1342, Levi says that the chapters on the Jacob Staff had not previously been translated into Latin. Adressing the Pope, Levi writes:

> And although the above-mentioned discovery has been revealed along time ago, as will be seen later, written down in Hebrew characters and divulged by hearing and in an incomplete manner together with other words from my mouth, never, however, was it translated in an orderly manner into Latin.[6]

If Levi's statement is accepted, it is evident that the translation of the *Astronomy* cannot be dated before 1342, given that the texts of *De sinibus* and of the corresponding chapters in the *Astronomy* are identical. This leads us to think that Levi dedicated some of the chapters of the *Astronomy* to the Pope under the title *De sinibus* before the complete translation of the work, begun around 1342, was finished.

There is yet another argument which points to the same date. The dedicatory letter is not the only indication that the treatise *De sinibus* was addressed to the Pope. Another piece of evidence can be found in the section dealing with trigonometry. In the *quarta dictio* of the second chapter Levi states:

> Quando igitur Sanctitas uestra de arcu noto sinum sibi ocurrentem scire uoluerit, querat in tabulis arcum notum et in eius directo sinum inueniet (Paris, B.N., MS 7293, 4va).

The corresponding passage in the Latin translation of the *Astronomy* is the following:

> Quando igitur uestra Corona ex arcu noto sinum sibi correspondentem scire uoluerit, querat in tabulis arcum notum et in eius directo sinum inueniet (Vat. Lat. MS 3098, 4ra).

Now the expression *uestra Corona* has no equivalent in the Hebrew version (H 4:142; Goldstein, 1985, p. 42) and it is probably addressed to Clement VI. This hypothesis fits in with other independent evidence. At the end of a letter which can be dated between 1340 and 1344, Isaac Qimhi (also called Petit de Nyons) writes to Levi:

> May He [i.e., God] grant that you find perpetual grace in the eyes of the Great King for the good of us all.

Touati (1973, p. 47), in his commentary on this letter, points out that the expression "the Great King" could only, at that date and place, refer to two rulers: Robert d'Anjou, King of Naples and Count of Provence, protector of Kalonymos ben Kalonymos, and Clement VI, elected Pope in 1342. Since so far nothing is known of any relationship between Robert d'Anjou and Levi, we can conclude that Levi wished to dedicate the translation of Book V of the *Wars of the Lord* to the Pope.

5. CHARACTERISTICS OF THE LATIN TRANSLATION OF THE *ASTRONOMY*

Levi's astronomical tables, published by Goldstein in 1974, provide an indirect indication of the importance of the contents of the Latin translation of the *Astronomy*. Apparently until some time between 1337 and 1339 Levi used a solar model with an eccentricity of 2;14 and a maximum equation of 2;8°. After that time, he adopted a solar model with a maximum equation of 2;17° corresponding to an eccentricity of 2;23,25,41 (Goldstein, 1974a, pp. 93-94). Now in the Hebrew manuscripts, the table for the solar correction is calculated according to the first value for the eccentricity, whereas only the Latin manuscripts contain the table based on the second value.

Goldstein's edition (1985) of the first twenty chapters of the Hebrew version of the *Astronomy* facilitates the task of comparing systematically the original and the Latin texts. This comparison immediately shows that the translation, while in general faithful to the Hebrew text in terms of content, is, in many places, not literal. The variations or accidents of translation fall, *grosso modo*, into the following categories: *substitutions*,[7] *omissions* (negative accidents) and *additions* (positive accidents). In what follows, I will not deal with omissions since, in general, they are not significant; mostly, they are due to an abbreviation of the Hebrew text or to a desire of the translator to avoid repetition.[8]

With regard to the additions, a superficial examination might suggest the possibility that some are glosses inserted by the translator, whereas others are due to different copyists. Consider some examples. The table of contents preceding the Hebrew text reads: "Chapter 15: In it we shall present instructions for finding the true position of the Sun at any time we wish" (Goldstein, 1985, pp. 21 and 306), while the corresponding Latin text says: "In chapter 15 we will show how to find with the greatest possible certainty the true solar position, its maximum equation, its apogee and the value for the obliquity of the ecliptic",[9] in exact conformity with the content of the chapter. Similarly, the end of chapter 12 of the Hebrew text reads: "As for the rest of the [instructions] needed to make an astrolabe, others have said enough about it" (H 12:54). The corresponding Latin text offers "...Ptolemy and others...".[10] In chapters 8 and 10, the Latin text specifies the materials which can be used to make certain pieces of the modified version of the Jacob Staff – "copper or bronze", "copper, bronze, or light wood"[11] – while the Hebrew version lacks these specifications (H 8:3; 10:25).

The possibility that such differences are due to different copyists can immediately be rejected: all the extant Latin copies of the *Astronomy* are identical. But also the other possibility, namely that the additions are glosses inserted by the translator, soon becomes difficult to maintain. For this hypothesis requires that we assume that the translator was thoroughly familiar with the Hebrew text even before setting out to translate. Only such complete knowledge would have enabled him (i) to

mention subjects in some passages which are discussed later on in the
Hebrew version, (ii) to comment or justify Levi's statements or
procedures, (iii) to give different geometric demonstrations on some
points, (iv) to correct errors in the Hebrew text, or (v) even provide
some parameters not found in the manuscripts of the original version.
Some examples follow.

(i) In chapter 17, after claiming that the Ptolemaic lunar model
cannot be true since the apparent diameter of the Moon at its minimum
distance is only slightly greater than at its maximum distance, the Latin
text adds a sentence lacking in the Hebrew version (H 17:2): "The same
can be demonstrated from the parallax concerning which we cannot
confirm Ptolemy's opinion",[12] a remark which anticipates the contents
of chapter 74. In the discussion of the preliminary lunar model at the
beginning of chapter 71, invoking the observation of 17 January 1333,
6;9 hours after mean noon, the Latin text indicates the mean and true
positions of the Moon (Gemini 279;6,42° and Cancer 1;10°,
respectively). However, in the Hebrew text these values appear only at
the end of the chapter when, Levi reexamines this observation for the
purpose of discussing his definitive lunar model.[13]

(ii) In the same chapter, after affirming that his observations of solar
and lunar eclipses compelled him to make the correction for the Moon
greater than 5° at apogee and less than 7;40° at perigee, the Latin texts
adds the following sentence that is absent from the Hebrew text:

> ...although we will not be able to decide this point definitely till we have
> observations of some eclipses which we expect to occur with the Sun near the
> place of its maximum equation, in such a way that we can thus reach an exact
> knowledge of this maximum equation, since from it we will know exactly the
> amount for the equation of lunar anomaly.[14]

Another difference of this kind occurs in chapter 12, where Levi
presents instructions for making an astrolabe. In the Hebrew version,
Levi suggests that to find the direction to the zenith accurately one
should take two observations of solar altitude within a short time
interval near noon (so that the solar altitude will not have changed)
turning the instrument 180° about its vertical axis between the two
observations, so that the altitude readings with the alidade can be
marked at two places on the rim symmetrically about the zenith
direction (H 12:7-10). Regarding this passage, Goldstein (1985, p. 168)
has pointed out that such a rotation would introduce a torque in the cord
which might interfere with the second observation. And if the cord were
to be detached and then attached again and time allowed for it to
become motionless, a longer time interval would elapse between the
two observations. Goldstein adds that Levi does not discuss this

problem of his observational technique. The Latin text of the same chapter, however, suggests an iterative method for avoiding the problem:

> And to prevent the error which may be produced in performing this or a similar operation, it is necessary to suspend the astrolabe and turn it as indicated above, repeating the operation two, three, four or five days.[15]

(iii) The Latin version also exhibits some significant variations concerning the geometrical proofs used. There are differences in chapter 9 in the demonstration establishing that for a correct measurement of the apparent solar diameter, it is necessary to subtract from the diameter of the image formed on the screen of the *camera obscura* an amount equal to the diameter of the opening (H 9:21-24; Vat. Lat. 3098, 9va).[16] There are also differences in chapter 10 in the description of the auxiliary triangular piece of the Jacob Staff used to determine the position of the planets with respect to the Sun (H 10:16-29; Vat. Lat. 3098, 10ra). In chapter 15, the text and the figures describing the solution of the problems arising when the angles formed by the walls of the room used to determine the solar parameters are not straight (H 15:61 ff.; Vat. Lat. 3098, 12va-b) are also different. The same holds true of those sections in chapter 20 (both text and figure) showing at which place in the eccentric the solar equation reaches its maximum value.[17]

(iv) There are also places where the Latin translation reveals no trace of the errors and faults in the text transmitted by all Hebrew manuscripts. In chapter 61, for example, where the Hebrew manuscripts use the word *'Libra'* (H 61:11; Goldstein, 1975, p. 36), the Latin manuscripts give, correctly, *'Aries'*.[18] Goldstein (*ibid.*, p. 40) has also pointed out that the passage in the same chapter in the Hebrew text, stating that the amount of the total precession in the time separating Timocharis from Levi does not reach 23°, is problematic. The Latin text, however, asserts that the total precession during the time separating Timocharis and al-Battānī does not reach 20°.[19] In chapter 80 apropos the computation of the solar eclipse of 26 June 1321, the Hebrew manuscripts give Cancer 11;39,51° for the true solar position (a value which according to Goldstein (1974a, p. 124) should be 11;40,34°), while the Latin manuscripts carry 11;40,28°, a value which is much closer to the correct one and cannot be due to an error of the translator.[20] Finally, in chapter 7, the erroneous statement "to the north" in the Hebrew text (H 7:71), which some manuscripts correct in the margin, is maintained in the Latin manuscripts because the error is avoided by changing the description of the figure given some lines above.[21]

(v) We now turn to numerical values not found in the manuscripts of the Hebrew version and provided by the Latin text. In chapter 80, for example, discussing the lunar eclipse of 23 October 1333, Levi writes that "the arc on the surface of the sphere from point G to point A is about 0;42,25° in this place at the time of the eclipse where the Sun is close to its perigee" (H 80:126; Goldstein, 1979, p. 115). The Latin version specifies the exact distance from the apogee: 125;25°.[22]

Another difference concerns the values for the mean motions of the ascending node of the Moon which Levi discusses in chapter 70 of the *Astronomy*. Goldstein (1974a, pp. 107-8) shows that the values given in the Hebrew manuscripts are not those that Levi in fact used to calculate the entries in column V of tables 13-19 and 20-21 (Goldstein, 1974a, pp. 172-81). These entries are based on a different set of values for the motion of the ascending node which have been reconstructed by Goldstein from the tables themselves. The Latin manuscripts do not carry these tables. However, instead of the values for the motion of the ascending node given in the Hebrew text, the Latin version of chapter 70 provides precisely those values in fact used to calculate the tables. The values found in the Hebrew text are:

1	day	0; 3,10,37,38,56, 2,10°
30	days	1;35,18,49,28, 1, 5°
31	days	1;38,29,27, 6,57, 7,10°
365	days	19;19,39, 1,50,53,10,50°
365,25	days	19;20,26,41,15,37,11,22,30°
366	days	19;22,49,39,29,49,13°
60	years	80;26,41,15,37,11,22,30°
1800	years	253;20,37,48,35,41,15°

The values given by the Latin version, compared with those reconstructed by Goldstein, are:[23]

		Latin text	Goldstein
1	d.	0; 3,10,37,41,14,17, 4°	0;3,10,37,41,14,24,...°
30	d.	1;35,18,50,37, 8,32°	1;35,18,50,37,...°
31	d.	1;38,29,28,18,22,49, 4°	1;38,29,28,18,...°
365	d.	19;19,39,15,51,53,49,20°	19;19,39,15,52,36,...°
365,25	d.	19;20,26,55,17,12,23,36°	19;20,26,55,17,54,...°
366	d.	19;22,49,53,33, 8, 6,24°	19;22,49,53,33,50,...°
60	y.	80;26,55,17,12,23,36°	80;26,55,17,54,...°
1800	y.	253;27,38,36,11,48°	253;27,38,57,18,...°

Finally, it must be pointed out that were we to maintain the hypothesis that all these changes are to be assigned to the translator who had an excellent knowledge of Levi's work, it would be necessary to assume that he was acquainted also with the non-astronomical sections of the *Wars of the Lord*. In the Hebrew version, for example, after having claimed that a certain philosophical principle is useful for explaining the creation of the celestial bodies, Levi adds "as will become clear in what follows, God willing" (H 2:15). In the Latin version the expression "in what follows" is made more explicit:

And this is a very useful principle for proving the creation of the celestial bodies as will be evident in Book VI of this work.[24]

All these variations, not to mention others involving extensive additions and reordering of the contents of some chapters,[25] seem quite unlikely to be the work of the translator. But there are, moreover, further differences which allow us to reject definitely this hypothesis and to prove at the same time that the Latin translation was the result of the collaboration between Levi ben Gerson and Petrus of Alexandria. These differences are additions to the original version *which cannot be derived from the Hebrew text and whose source can only be Levi himself.* I shall mention only some of them.

1) In chapter 15, Levi tells us that the room with a window facing the south, which he used for the determination of the solar parameters, was not built with right angles (H 15:63). The Latin text adds that the house was in Orange:

...as occurred to us in the room in which these observations were made in the city of Orange where we wrote this work with much effort and difficulties.[26]

2) In a passage in chapter 17 of the Hebrew version, Levi writes that the apparent diminution of the diameter of Mars, as observed on a certain occasion, was produced by the interposition of vapors caused by a comet. The Hebrew version adds that the vapor "came into being under Scorpio and it was drawn from there to somewhat below the north pole: there it burst into flame and it perished in Scorpio" (H 17:28-29). Goldstein (1972a, p. 45) has identified this comet as the one of 1337. The Latin text tells us that the comet was first observed in Taurus (which is compatible with other reports of the time as well as with modern computations, according to which the position of the comet on 23 June 1337 was Gemini 0;8°; *cf.* Thorndike, 1950, pp. 219-25) and that it remained visible for three months. It adds that Saturn was then in

conjunction with Mars which is also correct (Tuckerman, 1964, p. 686):

> And we think that the reason that the size of the apparent diameter of Mars in
> the sign of Scorpio did not increase, was a thick vapor generated between us
> and the planet because immediately afterwards a comet appeared under the
> north pole in the sign of Taurus and from there it continued to the pole and
> reached Scorpio and could be observed for more than three months. The
> influence of Saturn in conjunction then with Mars in Scorpio contributed to the
> formation of this thick vapor.[27]

3) Levi ben Gerson's reports and computations of four solar and six
lunar eclipses between 1321 and 1339, contained in chapters 80, 99 and
100 of the *Astronomy*, present in the Latin text numerous positive
accidents, the most interesting of them concerning the instruments used,
the weather conditions during observation, and the computations.

3.1) *Instruments*. Goldstein (1979, p. 105) has pointed out that in
these chapters (in the Hebrew version) no instruments are mentioned
although Levi may have used either an astrolabe or a Jacob Staff. The
Latin version, on the other hand, provides us with some information
about the subject. The most significant additions are as follows: for the
lunar eclipse of 23 October 1333, "without instrument";[28] for the solar
eclipse of 14 May 1333, "with an astrolabe";[29] for the lunar eclipse of
19 April 1334, "because according to our estimation, the Moon was
near the meridian line at the end of the eclipse although we had not
marked the meridian line in the room we were in during the
observation";[30] for the solar eclipse of 3 March 1337, "we did not have
an instrument to determine the time";[31] for the lunar eclipse of 3
October 1335, "without instrument";[32] and for the solar eclipse of 3
March 1337, "according to what we estimated from the solar ray
passing through the window without measuring, however, this ray with
the compass".[33]

3.2) *Weather conditions*. We have two passages having no parallel in
the Hebrew version. On the lunar eclipse of 23 October 1333, the Latin
text adds:

> You must know that we could not use this observation to determine the times
> of the eclipse because the sky was cloudy, although a little after the end of the
> eclipse the sky cleared and then we were able to make the above-mentioned
> observation.[34]

For the lunar eclipse of 19 April 1334, we find,

> You must know that the sky was cloudy during this eclipse. But between the
> beginning and the end of the eclipse, we could make some observations of the
> altitude of Regulus at the moment at which we knew the amount of the eclipsed

part of the Moon. These observations gave a result somewhat different from that already mentioned. However, we base our opinion more on this one than on the others because according to us the reason the one and the others did not agree was a certain thickness in the air.[35]

3.3) *Computations.* There are some significant differences between the Hebrew and the Latin texts in chapters 80 and 100 with regard to the solar eclipse of 26 June 1321 and the lunar eclipse of 23 October 1333. Here, however, I shall only mention, the differences between the two versions in the computation for the solar eclipse of 1321, found in section 5 of chapter 99, unpublished either in Hebrew or in Latin. With the exception of the values for the lunar anomaly and the true solar longitude (90;38° and Cancer 11;39°, respectively) which coincide, each version offers a completely different set of computed data, each of them being internally consistent and derived from different values for the lunar parallax in longitude (0;40,19° East in the Hebrew version and 0;40,8° East in the Latin one) and for the relative lunar velocity (0;30,20$^{o/h}$ and 0;30,31$^{o/h}$, respectively).[36]

In the report on the solar eclipse of 14 May 1333 in chapter 80, there is also a variant worthy of mention, although it is not directly involved in the computation. In the Hebrew version, Levi writes that "the part of the Sun to the north that remained visible was 0;32^d; and this agrees very well with what was perceived" (H 80:79; Goldstein, 1979, p. 113). The Latin text, however, reads:

And about 0;32^d of the solar diameter to the north remained visible which agrees very well with what was observed. We said above that this observation would demonstrate the truth of our computation because, according to us, the southern part of the Sun should be eclipsed, and thus it occurred, while, according to others, the northern part of the Sun should be eclipsed which did not occur.[37]

6. CONCLUSION

From the foregoing, we can conclude, first, that, beyond doubt, the additions to the Latin text we have examined were not made independently of Levi; and, second, that Levi's participation in the actual translation process is most probable, although it is not possible to be completely certain.

We cannot absolutely exclude the possibility that the translator used a Hebrew text different from the one we know and of which there are no preserved copies. It must be admitted, however, that this hypothesis is unlikely. The analysis of the extant Hebrew text suggests that before his death, Levi had not completed his planetary models and had not revised his lunar model as was his intention. There is also sufficient evidence to prove that Levi wrote or changed certain chapters of the *Astronomy* as late as 1339-1340. Under these circumstances, why in *ca.* 1342 would he begin to write another complete version of a text which was not yet finished?

There is a more plausible explanation of the singularities of the Latin version. It seems that Levi did not know Latin, but we know that he spoke Provençal (Touati, 1973, pp. 38-39). Thus, the Latin version might well be the result of a double translation: with the Hebrew text and his notebooks at hand, Levi would have translated the text into Provençal orally, occasionally making additions and revision as he went along, and this oral version would have then been written down in Latin by Petrus of Alexandria. This method of translation in which two persons – one knowledgeable in Hebrew or in Arabic, the other in Latin – cooperate is well documented in the Middle Ages: this is for instance the case of the translations of Avicenna and Ibn Gabirol by Dominicus Gundissalinus and the Jew Avendauth in the XIIth century, and that of the translation of al-Zarqāllu's treatise on the *saphea* by Jacob ben Makhir and Johannes Brixiensis.[38] Our suggestion is supported by Levi himself, if we take literally the words of the dedicatory letter of *De sinibus* to Pope Clement VI. After asserting that his works had not been translated until he met Petrus of Alexandria, Levi adds that Petrus,

> in order to spread the discovery of the Jacob Staff, touched my lips and, opening them, I spoke and I said to him, 'write'. And he took down all the words that I spoke on this discovery. And the words left my mouth in the following manner...[39]

The assumption that the *Astronomy* was translated in this way by Levi and Petrus until Levi's death in 1344 interrupted the collaboration, also explains why the Latin text is an incomplete translation of the Hebrew version: Petrus was unable to finish it on his own. Indeed, he needed the help of Solomon, Levi's brother, to complete the translation of the *Prognostication* after Levi's death.[40]

Leaving these conjectures aside, it is clear that the Latin translation of the *Astronomy* holds an exceptional place among medieval texts of which both the original and the translation are preserved. The two versions, one directly, the other indirectly, serve as testimonies of Levi's work. Clearly, in order to decide on the importance of the additional information contained in the Latin manuscripts for our knowledge of the astronomical work of Levi, we will have to wait for the editions of the complete texts.

NOTES

[1] *Incipit* (f. 114 r): "Extracta per me paulum ex tractatu magistri leonis judei de balneolis de baculo jacob. Prima propositio. Quadratum corde arcus minoris semicirculo est equalem ei quod fit ex sinui residui dicti arcus..."; *explicit* (f. 210r): "...Et si una fuerit ab una parte et alia ab altera parte subtrahe latitudinem stelle scite ab inuenta distantia et quod remanet erit latitudo alterius".

[2] The copyist begins (f. 83r) with the table of contents of the treatise, identical to the one contained in the other copies ("Primum capitulum continet epistolam ad dominum papam Clementem VI et prologum operas ..."; see an edition of this table in Curtze, 1898, pp. 97-8), adding at the end the date at which he started his work ("Et hic tractatus fuit translatus de Haebraeo in latinum anno Christi 1342 Pontificatus domini Clementi Papae VI anno primo. Descriptus uero huc anno 1610 15 Junij"; f. 83v). However, Levi's prologue ("Cum sapientis astronomi verba ad notitiam nostram peruenerunt ...", f. 84r) is entitled *Caput secundum. Stabiliuntur quaedam principia*, a title corresponding in the other copies to chapter 2, devoted to trigonometry and divided into five sections. In his turn, the three first sections of chapter 2 became chapter 3 (*De divisione orbium et sphaerarum; item quid dicatur arcus, quid sinus et quid sagitta*), the two remaining ones in chapter 4 (*Declaratio tabularum sinuum*), and so on, producing an enlargement of the number of chapters from 9 to 14. Leaving aside this difference, the text itself is identical to that explicitly attributed to Petrus of Alexandria (although the copy is of bad quality), as is proved by the following example of a collation:

Ad cuius probationem suppono pro nunc, quousque ex dicendis fuerit declaratum, quod, si corpus radiosum fuerit indiuisibile siue puncti unius, radius terminatus ad obiectum non esset latior fenestra propter distantiam fenestre ab obiecto quamcumque possibilem in aliqua quantitate notabili sensui; quia propter mirabilem distantiam celestium corporum ad terram non est signum notabile sensui propter tam modicam distantiam, quam est illa que inter fenestram et obiectum... (Paris, B. N. MS 7293, 7vb-8ra).	Ad cuius probationem supposito pro nunc, quousque ex dicendis fuerit declaratum, quod, si corpus radiosum esset indiuisibile siue punctum [sic] unius radij [sic] terminatus ad obiectum non esset latior fenestra per [sic] distantiam fenestre ab obiecto quamcumque possibilem in aliqua quantitate notabili sensui per numeralem seu [sic] mirabilem distantiam corporum celestium ad terram non est signum notabile sensui per tam modicam distantiam, quam est illa que intra [sic] fenestram et obiectum... (Munich, Staatsbibliothek, CLM 8089, 99v).

It is also incorrect, as Curtze (1898, p. 100) remarked, to assert (Renan-Neubauer, 1893, p. 623; Steinschneider, 1964, p. 132; Roche, 1981, p. 9) that neither Petrus nor Levi himself had called the instrument 'Jacob's Staff' (*baculus Jacob*). The name 'Jacob's Staff' appears in the dedicatory letter to the Pope ("... propter quod mihi sacramentum astronomie completum in baculo Jacob extitit reuelatum ...", Klagenfurt, B.B., MS XXX.b.7, 23r; Vienna, NB, MS 5277, 40v; Munich, CLM 8089, 84r) and the name 'revealer of secrets' in the prologue ("... et cum inueni per experientias multas cum instrumento predicto eccentricitates orbium planetarum multum diuersas ab eo quod erat conueniens ad Ptolomei sententiam, coactus fui experientias multas accipere circa uera loca cuiuslibet planetarum, ut certitudo ueritatis predictorum locorum esset in uia ad inueniendum dispositiones celorum et orbium omnium planetarum que correspondent omnibus que apparent in eis ex eccentricitate, uelocitate, retardatione, directione, retrogradatione ac statione eorum, ad quorum omnium ueritatem cum instrumento predicto perueni Deo duce. Et ideo merito instrumentum iam dictum 'reuelatorem secretorum' uocaui", Paris, B.N. MS 7293, 1ra; Klagenfurt, B.B., MS XXX.b.7, 23v; Vienna, NB, MS 5277, 41r-v; Munich, CLM 8089, 84v-85r).

[3] The *Tractatus semisse* or *semissarum* is a well-known work (*cf.* Pedersen, 1976, pp.

37-41; Poulle, 1980, pp. 206-10). In CLM 8089 of the Staatsbibliothek in Munich, the copy of *De sinibus* occupies ff. 83r to 117v, and the text of Peter of Saint-Omer, ff. 117v to 127v. The *incipit* of the latter is: "Caput decimum quartum [sic]. Instrumentum componere, quo vera loca omnium planetarum sine tabulis equationum inveniantur. Quoniam nobis non concedatur philosophiae studium nec tempus philosophandi ...", and the *explicit*: "... 6 annis mouetur 5 minutis cum dimidio. Finis huius tractatus anno 1610 22 Junij" (see an edition of this text in Pedersen, 1984, pp. 649-729).

    [4] A passage in Levi's dedicatory letter of *De sinibus* to the Pope could be useful to date his encounter with Petrus of Alexandria, but unfortunately it seems to be corrupt or difficult to understand. The invention of the Jacob Staff was motivated, according to Levi, by the discrepancies between the magnitudes of two eclipses (26 June and 9 July 1321) derived from Ptolemy's models and his own observations. Levi adds that the discovery of the instrument ramained unknown for Latin readers "during 21 days until a religious, brother Peter of Alexandria, of the order of hermit brothers of St. Augustine, arrived almost like the son of the man ..." ("sed sic permansit occultum 21 diebus donec uenit quasi similitudo hominis filij ut [sic] religiosus uir frater Petrus de Alexandria, ordinis fratrum heremitarum Sancti Augustini ...", Klagenfurt, B.B., MS XXX.b.7, 23r; Vienna, NB, MS 5277, 40v-41r; Munich, CLM 8089, 84r; without variant readings in the manuscripts, so that there is no textual basis for the tempting correction *annis* instead of *diebus*).

    The Latin text of the *Prognostication* for the conjunction of 1345 contains, however, some additional information on the relationship between Levi ben Gerson and Petrus of Alexandria. According to a passage not found in the Hebrew version, in 1339 Levi sent a prognostication on the comet which appeared during that year to Pope Benedict XII (1334-1342) in Avignon. At Christmas 1339, Petrus of Alexandria communicated to Levi his own prognostication on this comet to which he had attributed the same significance as Levi: "Et eciam signum quod postea vidimus *anno Christi 1339°* quod, ascendens de Scorpio et veniens usque Leonem contra motum signorum, *terminavit ibidem, ubi tunc Iupiter residebat*, signavit hoc idem, scilicet meridionalium conflictum, ut tunc prediximus *domino nostro, summo pontifici sancte et felicis memorie, domino Benedicto Pape xii. Michi eciam frater Petrus circa festum nativitatis Christi predixit hoc idem, per meridionales Saracenos exponens; et statim eodem anno sequutus est in Hispania eorum conflictus*" (the sentences in italics are absent from the Hebrew version; *cf.* Goldstein and Pingree, 1990, pp. 26 and 32). According to the passage quoted above, Levi and Petrus must have been in contact from at least 1339. A comparison of the Hebrew and Latin versions of the *Prognostication* shows the same sort of differences as the two versions of the *Astronomy* which will be examined in the section 5 of this article. In the case of the *Prognostication*, however, the fact that only a single copy in Hebrew has been preserved makes any conclusions based on the comparison less certain.

    Goldstein (1979, p. 105) considers it probable that the Christian mentioned by Levi in connection with his observations of the solar eclipses of 3 October 1335 and 3 March 1337 (chapters 80, sentence no. 24, and 100, sentence no. 8) is Petrus of Alexandria, a suggestion that dates the beginning of their collaboration to *ca.* 1335. The Latin text of the *Astronomy* offers no evidence in favour of this argument, however. First, sentence no. 24 of chapter 80 was omitted in the translation. Second, while *frater* Petrus mentions his name to the reader in the passage quoted above from the *Prognostication*, the Latin version of sentence no. 8 of chapter 100 is slightly different from the Hebrew and gives no name: "Set audiuimus a quodam clerico, de cuius experientia satis confidimus, quod in principio eclipsis erat solis altitudo circa 12 gradus et in fine circa 32 gradus" (Vat. Lat. 3098, 90ra). [Henceforth, the references to the original Hebrew of the *Astronomy* will be abreviated in the following way: "H 80:24" stands for "sentence no. 24 of chapter 80 of the Hebrew text or the English translation".]

    [5] It has not been possible for me to compare the partial edition and translation of the

Book of the Correct Syllogism (Manekin, 1984) with the Latin translation to check whether there are differences between both versions. Steinschneider and, after him, Renan-Neubauer remarked that the extant Hebrew manuscripts only mention the year when the work was composed (5079 A.M. = 1319 A.D.), whereas the Latin translation gives also the name of the month ("mense illud .1. augusto" according to Steinschneider, 1893, p. 71, and Renan-Neubauer, 1893, p. 602). In fact, the *explicit* of the Latin version in MS Ottobon. 1906, f. 83v, reads: "Et hic completa est intentio huius libri. Et expleuimus ipsum mense illul, id est, augusti anno creationis 5079". If Steinschneider's claim concerning the Hebrew manuscripts is correct, the addition in the Latin text is similar to those we find in the Latin version of the *Astronomy* (see later on, section 5 of this article) and, therefore, the translation could be also made by Petrus of Alexandria.

    [6] "Et licet predictum sacramentum in iamdiu fuit reuelatum, ut apparebit inferius, et annotatum hebraicis litteris, et quod alijs uerbis sine debito ordine ex ore meo prolatus [sic] forte fuerit auretenus, scilicet partialiter representatum, nunquam tamen ordinate translatum fuerat in latinum" (Klagenfurt, B.B., MS XXX.b.7, 23r; Vienna, Nationalbibliothek, MS 5277, 40v; Munich, CLM 8089, 84r).

    [7] See note 17.

    [8] There are also *mere* omissions, but in general they do not concern the astronomical content of the work. Thus, for example, the two poems by Levi on the Jacob Staff are omitted from chapter 9 (Goldstein, 1985, pp. 71-72 and 264-65; H 9:31-54).

    [9] "In 15° docebimus quomodo inueniatur solis in çodiaco uerus locus et aux eius et eius finalis equatio et distantia poli çodiaci a polo mundi ad maiorem certitudinem quam possumus optinere" (Vat. Lat. MS 3098, 1rb).

    [10] "Ideo perficitur secundum doctrinam a Ptolomeo et ab alijs traditam, quia sufficienter quoad alia est data ab eis" (Vat. Lat. 3098, 10ra).

    [11] "Et ponamus in capite baculi unam tabellam subtilem de cupro uel ere in qua sit unum paruum foramen ..." (Vat. Lat. MS 3098, 9ra); "Et in dictis fossis tabelle parue de cupro uel ere uel ligno subtili ponantur, quarum superficies interiores cadant perpendiculariter super dictis lineis ..." (Vat. Lat. 3098, 10ra).

    [12] "Hoc idem potest demonstrari ex diuersitate aspectus, in qua non inuenitur illud quod esset consequens ad sententiam Ptolomei ..." (Vat. Lat. MS 3098, 13rb).

    [13] "Que experientia habita fuit anno predicto 17 Januarij 6;9^h post equalem meridiem. Et inuenimus uerum locum lune in Cancro 1;10°, que distabat ab auge 283;18,21°. Et medius locus ipsius erat in Geminis 26;18,27° ..." (Vat. Lat. 3098, 61rb). See the corresponding passage of the Hebrew text in Goldstein (1974b, p. 285, sentence 7), and see also Goldstein (1972b, p. 280, sentence 2).

    [14] "... ponendo equationes circa augem aliquantulum maiores quam posuit Ptolomeus et aliquantulum minores circa augis oppositum, ut ex aliquibus experientijs eclipsium et alijs nobis apparet, licet adhuc non possimus de hoc iudicare perfecte quousque habuerimus experientias aliquarum eclipsium quas expectamus sole stante circa equationem maiorem ut perfecte deuenire possimus in quantitatem equationis maioris ipsius, quia ex hoc equationem prouenientem in luna pro motu diuersitatis tempore sciemus perfecte" (Vat. Lat. 3098, 62ra). See also Goldstein (1972b, p. 274, sentence 2).

    [15] "Et ut error qui in experientia ista uel simili cadere potest uitetur, tabula pluries hinc inde per duos uel tres uel quattuor uel quinque dies paretur et giretur ut supra. Et sic per experientias multas habito uero puncto sub centro astrolabij et in eo facto signo et diuisione arcus in duas partes equales, incipiendo a puncto dicte diuisionis, tota linea circumferentialis astrolabij diuidatur in quattuor partes equales" (Vat. Lat. 3098, 10va).

    [16] See the Latin text corresponding to H 9:21-26 in Mancha (1989, pp. 29-30, note 29); see also Goldstein (1985, pp. 70-71).

    [17] The passage and the figure of the Latin text corresponding to H 20:25-29 and fig.

18   THE LATIN TRANSLATON OF LEVI BEN GERSON'S *ASTRONOMY*

20.1 (Goldstein, 1985, pp. 115-16 and 228) provide a very clear example of what we call *substitution* (see Fig. 1): "Dico quod si angulus AEB non esset rectus, sinus anguli equationis esset minor linea DE. Ad cuius probationem protrahamus lineas DF, EF. Et protrahamus a puncto D columnam DG super lineam EF positam ad infinitum protractam uel protrahibilem. Et est notum quod linea DG est minor linea DE. Set linea DG est sinus anguli DFE et linea DE est sinus anguli DBE. Sequitur quod angulus DFE est minor angulo DBE. Et hoc est quod uolebamus probare" (Vat. Lat. 3098, 14va).

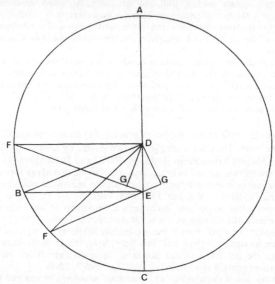

Figure 1: Vat. Lat. 3098, 14v; Vat. Lat. 3380, 102r.

[18] "... ex quo sequitur quod centrum umbre erat in eodem loco Arietis" (Vat. Lat. 3098, 48vb).

[19] "Set inter loca stellarum fixarum tempore experientiarum et Timocrati et Albeteni non sunt nisi 20° et minus" (Vat. Lat. 3098, 48va).

[20] "Et eius uerus locus erat in Cancro 11;40,28°" (Vat. Lat. 3098, 65rb).

[21] "Sit circumferentia ecliptice circulus AB, cuius centrum sit E. Et longitudo stellarum sit arcus AB. Et latitudo unius sit arcus AC et alterius arcus BD. Et C sit *meridionalis*, D septentrionalis. Et protrahamus lineas AE, BE, CD. Et est manifestum quod linea CD est corda equata. Et sinus arcus AC sit CF, et sinus arcus BD sit DG, qui sunt sciti; quia arcus eorum sunt sciti. Et sagite eorum, scilicet AF et BG, sunt scite. Sequitur quod linee FE et GE sunt scite, quia sunt residua duorum semidyametrorum super dictas sagitas. Protrahamus lineam GF, que est corda secunda, et protrahamus CF *uersus septentrionem* usque ad punctum H in quantitate linee GD. Deinde protrahamus lineam DH. Dico quod ..." (Vat. Lat. 3098, 8va).

[22] "... in loco in quo erat sol tempore eclipsis istius, in qua distabat ab auge circa 125;25°" (Vat. Lat. 3098, 66ra). In chapter 109, dealing with the observation of Saturn on 13 October 1325, the Latin text gives the values for the mean argument of center and the mean anomaly (184;58° and 127;24°, respectively), which are omitted in the Hebrew text: "Et tunc distabat ab auge 184;58° et de motu diuersitatis erat in 127;24°"

(Vat. Lat. 3098, 103rb). See the translation of the Hebrew text (H 109:35) and the values (185;0° and 127;25°) recomputed in Goldstein (1988, pp. 389 and 375, table 2, observation no. 4).

[23] "Capitulum 70. Motum autem capitis dragonis post longam inquisitionem, experientijs antiquorum eis nostras experientias adiungendo, inuenimus in una [die] retrocedendo per signa 0;3,10,37,41,14,17° et circa 4 septima; cuius motus in 30 diebus est 1;35,18,50,37,8° et circa 32 sexta. Et eius cursus in 31 diebus est 1;38,29,28,18,22,49° et circa 4 septima. Est eius motus in 365 diebus 19;19,39,15,51,53,49° et circa 20 septima. Et motus eius in 366 diebus est 19;22,49,53,33,8,6,24°. Et eius motus in 365 diebus et quarta diei est 19;20,26,55,17,12,23,36°. Et residuum sui motus super reuolutiones completas in 60 annis Christi est 80;26,55,17,12,23,36°. Et superfluum motus in 1800 annis est 253;27,38,36,11,48°. Et ordinabimus tabulas istas modo quo ordinauimus alias tabulas medij motus solis ..." (Vat. Lat. 3098, 59r).

The tables 5 and 6 in Goldstein's edition (1974, pp. 161-62) should also be different in the Latin version. While in the Hebrew text these tables have 7 and 2 columns, respectively, in the translation they would have had 13 and 3 columns since, according to the Latin text of chapter 62, they should have given not only the values of the arcs, but also the values corresponding to the sines of these arcs. See the description of these tables in the Latin version:

"Et faciemus tabulas isto modo, quia diuidemus latitudinem tabule in 7 spatia. In quorum primo scribetur numerus graduum ecliptice de uno gradu in gradum usque in 30 gradus. Et in capitibus 6 residuorum spatiorum scribentur 6 septentrionalia signa per ordinem incipiendo ab Ariete, et supra dicta 6 signa scribentur 6 alia meridionalia signa per ordinem incipiendo a Libra. Et sub signis in directo cuiuslibet gradus ecliptice in quolibet spatio, in duas partes diuiso, scribetur dictus defectus uel excessus competens gradui: *et in prima parte scribetur arcus, et sinus in secunda.* Et si gradus est signi septentrionalis, erit excessus; si meridionalis, erit defectus. Et est notum quod modo superius declarato potest sciri arcus et sinus medietatis dicti excessus uel defectus in quolibet oriçonte. Et ad hoc ordinabimus tabulas ex quibus quilibet poterit se iuuare in quolibet oriçonte latitudinis unius gradus et supra usque in 66 gradus et 27 minuta, ubi terra habitabilis terminatur, scilicet in fine Alamanie magne. Quas tabulas faciemus hoc modo. Quia diuidemus latitudinem tabule in 3 spatia, in primo scribetur numerus graduum de uno gradu usque in 66 gradus et 27 minuta, que sunt complementum 90 graduum supra maiorem latitudinem; et ideo ibi inuenietur quolibet die semel polus ecliptice in cenith, et tunc circumferentie ecliptice et oriçontis sunt una et eadem. Ex quo sequitur quod statim cum polus recedit de puncto cenith, ascendunt 6 signa. *Et in spatio secundo in directo cuiuslibet gradus scribetur arcus medietatis predicte. Et in tertio spatio scribetur sinus arcuj correspondens*" (Vat. Lat. 3098, 51ra-b).

[24] "Et istud est unum ex principijs ad probationem noue creationis istorum non modicum utile, sicut patebit in sexta parte libri istius" (Vat. Lat. 3098, 3rb).

[25] Two extensive additions worthy of mention are found, for example, in chapter 20. The first one is placed immediately after sentence 117 of the Hebrew text, in which Levi asserts that a special property of the model where the center of the mean motion lies between the center of the sphere and the center of the earth is that if the correction for 90° of apparent motion in this model is set equal to the correction for 90° of apparent motion in the simple eccentric model, "the excess of the apparent size of the planet at perigee over its amount at apogee is greater than it was in any of the previous models, for the excess of the apogee over the perigee is greater in this model, and this is self-evident to the careful reader" (Goldstein, 1985, p. 125). The Latin version goes on with a demonstration of sentence 117 by a double *reductio ad absurdum*, as follows (see fig. 2):

"Ad cuius probationem supponamus ABC circumferentiam spere. Et punctus D sit centrum spere et punctus E centrum terre et punctus F centrum motus equalis. Et protrahamus lineam EB et sit angulus AEB rectus. Et protrahamus lineam FB. Et est notum quod angulus FBE est equatio ad 90° motus uisi. Et ponatur ABC circumferentia spere excentrice equalis predicte. Et sit punctus D centrum spere et punctus E centrum terre. Et protrahamus lineam EB et sit angulus AEB rectus. Et protrahamus DB. Et est notum quod angulus DBE est equatio ad 90° motus uisi. Et ponamus quod angulus DBE in secunda figura [see fig. 3] sit equalis angulo FBE in prima. Dico quod linea DE in prima est maior quam linea DE in secunda, quod, si hoc non esset, esset equalis uel minor. Et primo dicamus quod sit equalis. Et protrahamus lineam DB in prima figura.

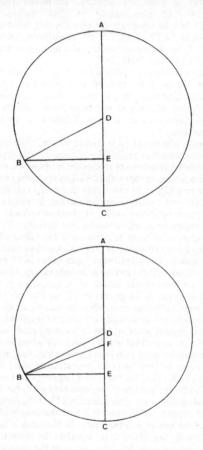

Figures 2 and 3: Vat. Lat. 3098, 15r; Vat. Lat. 3380, 103v.

Et quia linea DE in prima figura supponitur equalis linee DE in secunda, est notum quod linea EB in prima figura est equalis linee EB in secunda [*mg.*: quia sic sunt in equali distantia a centro circumferentie sue]. Et angulus DEB in prima est equalis angulo DEB in secunda, quia in utraque est rectus. Sequitur quod angulus DBE in prima sit equalis angulo DBE [*ms.*: DEB] in secunda. Set angulus FBE in prima est equalis DBE in secunda. Sequitur quod angulus FBE in prima sit equalis angulo DBE in eadem; quod est falsum, quia pars non est toti equalis. Dico etiam quod non est minor. Quia si esset minor ponamus DG in prima figura equalem linee DE in secunda. Et protrahamus lineas GH, DH, GB [see fig. 4]. Et sit angulus DGH rectus. Et erit notum, ut supra, quod angulus DBG in prima figura est equalis angulo DBE in secunda. Set angulus DBG in prima est minor angulo DHG in eadem. [*mg.*: Non uideo quomodo probetur angulum DBG esse minorem angulo DHG. Sed bene probare potest quod angulus DBE est minor angulo DHG cum auxilio 32^e primi Euclidis habito quod angulus [D]G[H] est equalis angulo [D]E[B], quia ambo sunt recti, et quod angulus DHG est maior [*ms.*: minor] angulo DBE, quia pars minor toto. Quo habito et deducto quod DHG angulus est equalis angulo FBE ex suppositione et prehabitis, manifeste concluditur totum esse minus par-

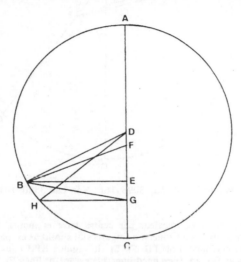

Figure 4. Vat. Lat. 3098, 15v; Vat. Lat. 3380, 103v: BE is not drawn perpendicular to AC.

te.] Sequitur quod angulus DBG est minor angulo FBE; quod est falsum, quia maius toto esse minus parte. Sequitur quod linea DE in prima est maior linea DE in secunda. Set linea DE in secunda figura est maior linea DE in alijs dispositionibus supradictis in speris excentricis. Sequitur quod linea DE in prima est maior linea DE in alijs dispositionibus supradictis in speris excentricis. Et hoc erat probandum" (Vat. Lat. 3098, 15rb-va).

The second addition occurs at the end of the chapter where Levi describes a planetary model with the equant located away from the diameter passing through the centers of the sphere and the earth. Between sentences 154 and 155 of the Hebrew text (Goldstein, 1985, pp. 129 and 218), the Latin version adds a passage explaining that a property of this model is that the equation of center is not symmetric with respect to the line which passes through the earth and the equant. The Latin text reads as follows (see fig. 5): "Item distinguitur quilibet modus istius dispositionis ab omnibus dispositionibus spere excentrice quia in quolibet modo istius dispositionis, diuisa spera per diametrum transeuntem per centra terre et motus, quelibet equatio ex una parte diuisionis est diuersa a qualibet equationi sibi similis alterius partis. Cuius demonstratio est hec. Sit

22   THE LATIN TRANSLATON OF LEVI BEN GERSON'S *ASTRONOMY*

ABCG circumferentia spere et sit punctus D centrum spere. Et punctus E sit centrum terre et punctus F centrum motus. Et AFDC sit diameter transiens per centra spere et

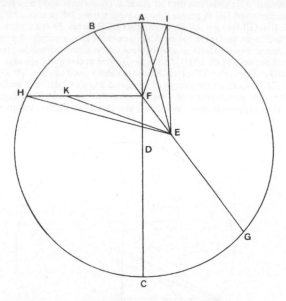

Figure 5 : Vat. Lat. 3098, 16r ; Vat. Lat. 3380, 104r.

motus. Et BFEG sit linea transiens per centra terre et motus. Dico quod omnes equationes ex parte BCG sunt diuerse a singulis sibi similibus ex parte BAG. Ad cuius probationem protrahamus lineas FH, FI. Et sit angulus BFH equalis angulo BFI. Et protrahamus lineas EH, EI. Dico quod linea FH est maior linea FI, quia angulus AFH est multo maior angulo AFI. Ideo diuidemus de linea FH maiori lineam FK equalem linee FI. Et protrahemus lineam EK. Et est notum quod angulus IFE est equalis angulo KFE. Sequitur quod triangulus IFE est equalis triangulo KFE et angulus quilibet angulo sibi simili. Sequitur quod angulus FIE est equalis angulo FKE. Set angulus FHE est minor angulo FIE. Et isto modo demonstraretur quod omnes equationes ex parte arcus BCG sunt minores omnibus sibi similibus ex parte arcus BAG. Et eodem modo demonstraretur quod in omni differentia dispositionis istius equationes unius partis sunt diuerse ab equationibus partis alterius; quod uolebamus probare" (Vat. Lat. 3098, 15vb).

Chapter 80 can be cited as an example of reordering: in the Latin text the passages corresponding to the lunar eclipses of 19 April 1334 and 3 October 1335 are in chronological order, while in the Hebrew text the latter is incorrectly called the 'fourth' lunar eclipse and its description (H 80:132-153) is placed before that of the former (H 80:154-166), called the 'fifth'. *Cf.* Goldstein, 1979, pp. 116-17.

[26] "... sicut nobis accidit in hospitio in quo experientias istas accepimus in ciuitate Auraica, in qua hunc librum cum magno labore et sudore edidimus" (Vat. Lat. 3098, 12vb).

[27] "Et hoc modo ymaginamur quod causa non crementi eius diametri in Scorpione fuit uapor grossus generatus inter nos et ipsum; quod apparet quia tunc statim apparuit

cometa sub polo septentrionali in signo Tauri, et inde processit ad polum et uenit usque ad Scorpionem, cuius processus per tres menses durauit et amplius. Ad generationem huiusmodi grossi uapores iuuit influentia Saturni, tunc in Scorpione coniuncti cum Marte" (Vat. Lat. 3098, 13va).

[28] "...post equalem meridiem circa 9ᵃ hora completa secundum nostram extimationem instrumento carendo ..." (Vat. Lat. 3098, 65rb; *cf.* H 80:17; Goldstein 1979, p. 110).

[29] "...circa 3;28ʰ post uisam meridiem ut apparuit per astrolabium ..." (Vat. Lat. 3098, 65va; *cf.* H 80:69; Goldstein, 1979, pp. 112-3).

[30] "...taliter quod nobis secundum extimationem apparuit quod luna esset circa lineam meridianam in fine eclipsis, lineam meridianam non habendo in hospitio in quo eramus experientie tempore" (Vat. Lat. 3098, 66ra; *cf.* H 80:154; Goldstein, 1979, p. 117).

[31] "Set tempore istius eclipsis non habebamus instrumentum ad capiendum horam ..." (Vat. Lat. 3098, 90ra; *cf.* H 100:7; Goldstein, 1979, p. 118).

[32] "...ei quod nobis uidebatur quadam extimationem sensibili sine aliquo instrumento ..." (Vat. Lat. 3098, 90va; *cf.* H 100:71; Goldstein, 1979, p. 122).

[33] "... eclipsatus sol fuit circa 5 digitos secundum extimationem nostram habitam per radium solis per fenestram intrantem non tamen cum circino radium mensurando ..." (Vat. Lat. 3098, 90va; *cf.* H 100:72; Goldstein, 1979, p. 122).

[34] "Et scias quod de experientia ista non possumus nos iuuare pro eclipsis temporibus quia aer erat nubilosus, licet per aliquale tempus post finem eclipsis fuit clareficus et tunc accepimus experientiam supradictam" (Vat. Lat. 3098, 65va; *cf.* H 80:112; Goldstein, 1979, p. 114).

[35] "Et scias quod aer fuit nubilosus tempore istius eclipsis. Sct in tempore intermedio inter principium et finem eclipsis accepimus experientias aliquas de altitudine Cordis Leonis, tempore quo de corpore lune erat eclipsata aliqua quantitas nobis scita, que dabant sententiam aliquam discordantem in aliquo a predicta. Tamen nos magis fundamus in hanc primam predictam quam in eas, quia oppinamur aliqua aeris spissitudo fuerit causa quod dicte experientie non concordarent cum prima" (Vat. Lat. 3098, 66ra; *cf.* H 80:155-6; Goldstein, 1979, p. 117).

[36] I am grateful to Prof. Bernard R. Goldstein for his collation of my transcription of the Latin text of this passage with the Hebrew manuscripts (Paris, Bibliothèque Nationale, 724 and 725).

[37] "Et remansit de solis diametro non eclipsata ex parte septentrionis circa 0;32 digiti, et hoc multum consentit ei quod tunc per experientiam uidimus. Nam ante dixeramus quod in experientia ista nostri computus ueritas probaretur, quia secundum computum nostrum pars [non] remanens lucida de solis diametro meridionalis esse debebat et fuit, et septentrionalis secundum alios, quod non fuit" (Vat. Lat. 3098, 65va; *cf.* Goldstein, 1979, p. 113).

[38] The words of Avendauth describing his collaboration with Dominicus Gundissalinus are as follow: "Habes ergo librum, vobis praecipiente, et me singula verba vulgariter proferente, et Dominico archidiacono singula in latinum convertente, ex arabico translatum". Those corresponding of Jacob ben Makhir are: "Profatio gentis hebreorum vulgarizante et Johanne Brixiensi in latinum reducente" (*cf.* Romano, 1977, p. 382).

[39] "... Petrus de Alexandria, ordinis fratrum heremitarum Sancti Augustini, qui ad propalandum sacramentum baculi prelibati tetigit labia mea, et aperiens in eum locutus sum et dixi ad eum qui stabat coram me: scribe. Et me sibi referente omnia uerba mysterij huius conscripsit, que exordiuntur ut sequitur..." (Klagenfurt, B.B., MS XXX.b.7, 23r; Vienna, Nationalbibliothek, MS 5277, 41r; Munich, CLM 8089, 84r).

[40] A close examination of chapter 110, the last one of the Latin translation, suggests that the work of Petrus of Alexandria was also interrupted some time after Levi's death, since there are two lexically and stylistically very different parts in this chapter. The

beginning of the second one coincides with the intervention of a second hand in the oldest copy of the text (MS Vat. Lat. 3098, 103va, lin. 49: "Et ecce apparet que sit equacio declinationis in 90° ab auge [et] in 180° motus diuersitatis ..."). In the following table, the left column contains some examples of terms and sentences repeatedly used by Petrus without variations along chapters 1 through 103, 106, 109 and the first part of chapter 110 (ff. 1ra-103va), and the right one those which replace them in the last pages of chapter 110 (ff. 103va-107ra).

Petrus' translation	*Second translator*
sed	verum
motus spere	incessus spere
latitudo diametrorum	declinatio/dilatation diametrorum
duorum	duum
medius motus	motus longitudinis medie
planeta	stella
que equatio	quenam equatio
ad duas equationes simul	ad duas equationes compositas
equatio subtrahenda, addenda	equatio ad minuendum, ad addendum
sequitur quod linea ED sit maior in sephel quam in auge...	et hoc addit super quantitatem linee ED in auge...
et est notum quod in quantitate in qua linea EG est 60, est linea DE...	et clarum est quod in mensura qua est quantitas linee EG 60, est quantitas linee...
et hoc sedato inquiremus que equatio prouenit in auge et sephel...	postquam itaque firmatum est hoc speculabimur quodnam signum peruenit in auge et in sephel...
et quia linea ND est in potentia maior ea in quantitate quadrati linee NI, sequitur quod linea...	et secundum quod quantitas linee TD addit super eam potentialiter (in potentia quemadmodum) quadratum linee BD, erit quantitas linee...
et ideo est notum quod in quantitate in qua linea EG est 47;52,10, que est sinus anguli EDG, est linea ED 35;40,7, que est sinus anguli EGD...	et ideo clarum est quod in mensura qua fuerit quantitas linee ET tanquam (quemadmodum/sicut quantitas) sinus anguli EDT, qui est 59;33,31, in ea erit quantitas linee ED tanquam sinus anguli ETD, qui est 6;53,35...
et expedit quod inquiramus quanto erit maior angulus EDH in sephel quam in auge propter quantitatem linee...	et conuenit ut inquiramus id quod addit in sephel in angulo EDC propter quantitatem linee...

REFERENCES

Carlebach, J. 1910. *Lewi ben Gerson als Mathematiker*. Berlin.

Curtze, M. 1898. "Die Abhandlungen des Levi ben Gerson über Trigonometrie und den Jacobstab", *Bibliotheca Mathematica*, 12:97-112.

Curtze, M. 1901. "Die Dunkelkammer", *Himmel und Erde*, 13:225-236.

Goldstein, B. R. 1972. "Theory and Observation in Medieval Astronomy", *Isis*, 63: 39-47.

Goldstein, B. R. 1974a. *The Astronomical Tables of Levi ben Gerson*. Hamden, Connecticut.

Goldstein, B. R. 1974b. "Levi ben Gerson's Preliminary Lunar Model", *Centaurus*, 18: 275-288.

Goldstein, B. R. 1975. "Levi ben Gerson's Analysis of Precession", *Journal for History of Astronomy*, 6:31-41.

Goldstein, B. R. 1979. "Medieval Observations of Solar and Lunar Eclipses", *Archives Internationales d'Histoire des Sciences*, 29:101-156.

Goldstein, B. R. 1985. *The Astronomy of Levi ben Gerson (1288-1344)*. Berlin-New York.

Goldstein, B. R. 1988. "A New Set of Fourteenth Century Planetary Observations", *Proceedings of the American Philosophical Society*, 132:371-399.
Goldstein, B. R. and Pingree, D. 1990. "Levi ben Gerson's Prognostication for the Conjunction of 1345", *Transactions of the American Philosophical Society*, vol. 80, part. 6. Philadelphia.
Grant, E. 1965. "Part I of Nicole Oresme's *Algorismus proportionum*", *Isis*, 56:327-341.
Günther, S. 1890. "Die erste Anwendung des Jacobsstabes zur geographischen Ortsbestimmung", *Bibliotheca Mathematica*, 4:73-80.
Mancha, J. L. 1989. "Egidius of Baisiu's Theory of Pinhole Images", *Archive for History of Exact Sciences*, 40:1-35.
Manekin, C. 1984. *The Logic of Gersonides: An Analysis of Selected Doctrines, with a Partial Edition and Translation of the "Book of the Correct Syllogism"* (unpubl. dissertation). Columbia University.
Manekin, C. 1985. "Preliminary Observations on Gersonides' Logical Writings", *Proceedings of the American Academy for Jewish Research*, 52:85-113.
Pedersen, O. 1976. "Petrus Philomena, a Problem of Identity, with a Survey of the Manuscripts", *Cahiers de l'Institut du Moyen Age grec et latin*, Université de Copenhague, 19:1-54.
Pedersen, F. S. 1984. *Petri Philomenae de Dacia et Petri de S. Audomaro Opera quadrivialia. Pars II: Opera Petri de Sancto Audomaro*. Edidit Friedericus Saaby Pedersen. Corpus Philosophorum Danicorum Medii Aevi, X.2. Hauniae.
Poulle, E. 1980. *Les instruments de la théorie des planètes selon Ptolémée. Equatoires et horlogerie planétaire du XIIIᵉ au XVIᵉ siècle*. Genève-Paris.
Renan, E. and Neubauer, A. 1893. "Les Ecrivains Juifs Français du XIVᵉ siècle", *Histoire littéraire de la France*, 31:351-789.
Roche, J. J. 1981. "The Radius Astronomicus in England", *Annals of Science*, 38: 1-32.
Romano, D. 1977. "La transmission des sciences arabes par les juifs en Languedoc", in: M. H. Vicaire and B. Blumenkranz (eds.), *Juifs et judaïsme de Languedoc*. Toulouse, 363-386.
Steinschneider, M. 1893. *Die hebräischen Übersetzungen des Mittelalters*. Berlin.
Steinschneider, M. 1964. *Mathematik bei den Juden* (2nd ed.). Hildesheim.
Thorndike, L. 1950. *Latin Treatises on Comets. A.D. 1238-1368*. Chicago.
Touati, C. 1973. *La pensée philosophique et théologique de Gersonide*. Paris.
Tuckerman, B. 1964. *Planetary, Lunar and Solar Positions at Five-day and Ten-day Intervals. I: A.D. 2 to A.D. 1649*. Philadelphia.

ADDENDA

p. 1: The Latin text of Levi's *De numeris harmonicis* has been again edited by Christian Meyer and Jean-François Wicker: "Musique et mathématique au XIVe siècle. Le *De numeris harmonicis* de Leo Hebraeus", *Archives internationales d'histoire des sciences*, vol. 50, n. 144 (2000), pp. 31–67. Although better than Carlebach's, this edition cannot be considered definitive.

pp. 5–6: See IV, esp. pp. 14–16.

p. 17, n. 5: See also C. Manekin, *The Logic of Gersonides: A Translation of "Sefer ha-Heqqesh ha-Yashar" (The Book of the Correct Syllogism), with Introduction, Commentary, and Analytical Glossary*. Dordrecht, Kluwer, 1992. The Latin text preserved in MS Ottobon. 1906 is usually considered a Renaissance translation.

IV

Levi ben Gerson's Astronomical Work: Chronology and Christian Context

Et laborauimus cum argumentis ingenij modo quo declarabitur infra in tantum quod aliquando inueniebamus dispositionem concordem cum nostris experientijs ac etiam antiquorum et cum multis alijs quas inueniebamus post illas. Et cum gaudebamus ex istis experimentis credendo ipsa punctaliter esse inuenta, extrema gaudij tristitia occupabat, quia inueniebamus experientiam unam que ualde elongabatur ab experientijs dispositionis inuente in illo planeta...

Levi ben Gerson, *Astronomy*, Latin version, Vat. Lat. 3098, 38ra.

1. Introduction

Although Levi ben Gerson (1288-1344) is best known for his exegetical and philosophical work, his contributions to science and mathematics were of considerable interest, and his *Astronomy* is one of the most original texts on the subject of the Middle Ages. Conceived as book V, part 1, of his Milḥamot 'Adonai (Wars of the Lord), but preserved on account of its length in separate manuscripts from the rest of the philosophical treatise, it contains detailed criticisms of Ptolemy and al-Biṭrūjī, reports on observations made between 1321 and 1340, the description of a new instrument invented by Levi (the cross-staff, that, combined with the camera obscura, allowed him to study the variation in apparent planetary sizes, a subject absent from the astronomical tradition), lunar and planetary theories clearly departing from Ptolemy's, and a set of new astronomical tables, mainly related to the computation of syzygies and eclipses, whose underlying parameters he derived from his own observations[1].

Chronology, a matter of intrinsic interest for historians, has an additional significance in the case of Levi's *Astronomy*, which appears to us in the preserved copies like a work in progress: incomplete in its Hebrew and Latin versions, perhaps unfinished at the time of his death, it reflects the changes in Levi's mind over time on fundamental aspects of his work — solar eccentricity, lunar model,

---

[1] A summary of Levi's astronomical contributions can be found in Goldstein, 1992; see also Goldstein, 1974a, 1979, 1985, and 1988.

precession value, and method of computation of different tables —, and informs us of his endless troubles with planetary theory, to which he devoted his last years. Nevertheless, no detailed study has been until now devoted to the matter, that is inadequately dealt with by scholars not familiar with Levi's scientific work. Levi tells us that book V, part 1, of *Milḥamot 'Adonai* was finished on 24 November 1328 (though some matters were completed with the results of observations and research after that date[2]), and that parts 2 and 3 of book V, and book VI were completed on 28 November and 5 December 1328, and 5 January 1329, respectively — a claim that, literally taken, is difficult to accept[3]. These dates are thus often interpreted as indicating the ending of a working period of months or years[4], and Levi's words in the colophon of his *Astronomy* as meaning that after 1328 only *minor additions* were incorporated to the text[5]. However, when the manuscript tradition of a work dated by Levi has been closely examined, the conclusions are quite different. It has been proved, for instance, that 1319, the year given by Levi for his *Sefer ha-Heqqesh ha-Yashar* (*Book of the Correct Syllogism*), must be considered as the completion date of a first version of the text, of which a thorough revision was made not before 1323, although in the manuscripts of this second redaction only 1319 is mentioned[6]. As for Levi's *Astronomy*, similar conclusions derive from some chapters already published, which were doubtless composed after 1335 and 1340[7].

---

[2] "The completion of part I of this book took place on the 21st day of Kislev of the year 89 of the 6th millenium [= 5089], praise be to the Creator of all things who is exalted over any blessing or prayer, who helped us according to his mercy and the greatness of his kindness, amen, amen. Levi said: do not be surprised, o reader of these words of ours, if there are introduced in this some matters that we composed after the aforementioned time to complete this research according to what is possible for us from the observations that we took after this, for so we stipulated when we composed these words, as mentioned previously" (Paris, B.N., Heb. 724, f. 257b; cf. Goldstein, 1974a, p. 74).

[3] According to the colophons of the manuscripts, paradigmatic years of Levi's productivity are also 1321 and 1323 (cf. Touati, 1973, pp. 72-3). In 1321, for instance, he finished in April his mathematical work *Sefer Maʿaseh Hoshev* (*Book of the Reckoner*), and between June and December he composed his commentaries to Averroes' *Compendium* and *Middle Commentary* to Aristotle's *Physics* (June-July), Averroes' *Compendium* of Aristotle's *De generatione et corruptione* and *De coelo* (August-September), and Averroes' *Compendium* of Aristotle's *Meteorologica* (December). Truly relying on Levi's capability, some modern scholars even claim (e.g., Shatzmiller, 1991, p. 39, and Weil-Guény, 1992, p. 357, following Renan-Neubauer, 1893, p. 615) that Levi also composed in 1321 his canons and tables for eclipses!

[4] Touati 1973, p. 51: "Il est donc bien évident que ces dates sont uniquement celles de la mise au point définitive d'une oeuvre depuis très longtemps en chantier: douze ans". Identical conclusions on Levi's *Commentary to the Pentateuch* can be found in Freyman, 1991.

[5] By way of contrast, see Goldstein, 1974a, p. 30: "Most of the observations are dated between 1330 and 1339, although the earliest was made in 1321. But they are so woven into the text that 1328 must be taken as the date of a preliminary draft, and not a complete text to which a few remarks were later appended".

[6] Manekin, 1992, esp. pp. 38-40.

[7] Chapters 46 or 80, and 100 or 122 (Goldstein, 1979 and 1988).

Moreover, the internal chronology of Levi's work could provide new light on the old and unresolved problem of his relations with Christians. In the introduction to his astronomical tables, Levi informs us that he wrote them "at the request of many great and noble Christians"[8], and so it is reasonable to think that the Christian request predated by some years their redaction. There is no doubt that the relations with personages of the Papal court in Avignon determined Levi's intellectual production during the last years of his life: in 1342 a Latin version of chapters 4-11 of the *Astronomy* (dealing with trigonometry and the *cross-staff*) made by Petrus of Alexandria, from the Hermit Brothers of St. Augustine, in collaboration with Levi, and offered to pope Clement VI, followed in 1343 by *De numeris harmonicis*, a response to a mathematical problem proposed by Philippe de Vitry, the renowned musical scholar, friend of the Pope, and in 1344 by the *Prognostication* for the planetary conjunction of 1345, the latest of Levi's preserved works[9]. Additional information confirms the diversity of these relations, for Petrus tells us that an astrological prediction was also adressed by Levi to pope Benedict XII in 1339, and Levi himself mentions a "noble Christian" and a "distinguished cleric who studied this science [i.e., astronomy] with us" in connection with his observations of the lunar and solar eclipses of 3 October 1335 and 3 March 1337, respectively[10]. Moreover, the preserved Latin translation of the whole text of Levi's *Astronomy* was also due to the collaboration between Levi and Petrus during these years[11], as if a program was undertaken in collaboration with Levi himself for making

---

[8] Goldstein, 1974a, p. 20.

[9] The Latin text of the *editio* of chapters 4-11 of the *Astronomy* that Levi dedicated to Clement VI (with the title *Tractatus instrumenti astronomie*, although the work is widely known as *De sinibus*) was partially edited by Cúrtze 1898, 1901. Philippe de Vitry (1291-1361), also a friend of Nicole Oresme (ca. 1320-1392), who dedicated to him his *Algorismus proportionum*, was elevated by Clement VI to bishop of Meaux in 1351; the Latin text of *De numeris harmonicis* was published by Carlebach 1910, pp. 128-39, and analyzed by Chemla and Pahaut 1992. The *Tractatus de coniunctione Saturni et Iovis anno Christi 1345* has been edited and translated in Goldstein-Pingree 1990, and it is very likely that it was also written for the Pope: two celebrated Christian astronomers, John of Murs and Firmin of Bellaval, who were invited by Clement VI to Avignon after Levi's death to advise him on the calendar reform, wrote for the Pope predictions based on the same conjunction.

[10] Goldstein-Pingree 1990, pp. 26 and 32; Goldstein 1979, pp. 110 (80:24) and 118 (100:8); Mancha 1992b, p. 37. The significance of Levi's relationship with Christians is also ackowledged in Hebrew sources; Isaac Qimhi (also called Petit de Nyons) in a letter to Levi that can be dated between 1340 and 1344 writes: "May the Lord of Peace [i.e., God] make magnificient your teaching and resplendent your happiness, and may He spill his gifts over you. May He grant that you find perpetual grace in the eyes of the Great King for the good of us all" (Touati 1973, p. 47, adds in his comment that the words "Great King" can only refer at that time and place to Robert d'Anjou, King of Naples and Count of Provence, or to the Pope). A similar expression is used by Levi in the Latin version of chapter 4 of the *Astronomy*: "Thus, when His Crown [*uestra Corona*] wish to know the sine that corresponds to a known arc..." (Vat. Lat. 3098, 4ra), which in the corresponding passage of the *Tractatus instrumenti* dedicated to the Pope was modified as "His Holiness" (*Sanctitas uestra*: Paris, B.N., MS 7293, 4va); cf. Mancha 1992b, pp. 26-7.

[11] On the Latin version of the *Astronomy* see Mancha 1992b.

available the results of his astronomical research to Christian readers. The recent discovery of a Provençal version of his tables and canons for syzygies and eclipses, earlier than the preserved Hebrew one[12], allows us to revise the suitability of the concepts "original text" and "translation" in referring to the different versions of his works, and confirms that the relations between Levi and the Christians already existed before the 1340's. Unfortunately, it seems as if there was a reluctance on both sides to provide us with details on the matter.

The object of this article is to establish the main stages of the redaction of the extant versions of Levi's astronomical works. The complete list of dated and datable observations reported in the *Astronomy*, which range from 1321 to 1340, shows that the period of Levi's greatest observational activity occurred between April 1333 and October 1335 (38 observations out of a total of 82, most of them of the Sun and the Moon). We are informed by Levi in chapter 98 that he completed his solar and lunar model just at the end of 1335, and we also know that Levi's work on exegetical matters was interrupted from December 1332 to some time before December 1337[13]. Thus, as the date of the latest observation on which depends the redaction of a chapter provides a *terminus post quem* for it, it can be easily proved that *at least* 85 of the 125 preserved chapters[14] of the Hebrew version of the *Astronomy* were redacted after 1335. Therefore, a plausible inference is that the apparent increase of Levi's astronomical activity after the first months of 1333 followed closely the Christian request of tables for syzygies, whose Provençal and Hebrew versions were redacted, in all likelihood, during 1336 and the first months of 1337. After some months devoted again to exegetical matters (until approximately June 1338)[15], Levi returned to his astronomical observations and, according to chapter 100[16], some of them, made in 1339, led him later to modify his solar model, to reject the precession value determined in 1335, and to examine the appropriateness of some features of his lunar model with implications for the planetary theory. These changes were reflected in the text and provide a *terminus post quem* for at least 25 of of the 85 aforementioned chapters, namely December 1339.

---

[12] These tables and canons, composed "at the request of many great and noble Christians", are preserved in at least four Hebrew manuscripts (Goldstein 1974a, 77). Although they are usually (and incorrectly) called "abridged version of Levi's tables", they constitute an independent work, composed separately, which Levi, after revision, later incorporated into the text of the *Astronomy* as chapter 99. The Hebrew text of the introduction to the canons was published in Steinschneider 1899 and translated into German in Carlebach 1910, 35-41. On the Provençal version of this work, which I recently discovered in the MSS Scaligerani 46 and 46* of the University of Leyden, see Mancha 1998b.

[13] Touati, 1973, p. 52; Weil-Guény, 1992, pp. 362-3.

[14] Levi's table of contents lists 136 chapters. The planetary theory (chapters 103 to 136) is apparently unfinished: the text of 11 of these 34 chapters is simply lacking (104, 105, 107, 108, 111, 112, 115, 116, 119, 120, and 127), whereas another 4 are incomplete (117, 118, 121, and 122). In the Latin version 31 chapters lack: after chapter 102, related to lunar theory and also incomplete in both versions, only chapters 103, 106, 109, and 110 are extant.

[15] Touati, 1973, *ibid.*; Weil-Guény, 1992, p. 364.

[16] Goldstein, 1979, esp. pp. 118-9.

2. Levi's work during 1333-1337

Levi's *Astronomy* preserves one of the most important sets of observations made by a single astronomer during Middle Ages — at least eighty-two dated (or datable) observations in all[17] throughout twenty years, the earliest 26 June 1321 (a solar eclipse) and the latest 24 July 1340 (Jupiter in conjunction with Venus — see Table I)[18]. Levi made 67 of them during three relatively brief periods: the first, from June 1327 to December 1328 (13 observations); the second, from February 1333 to October 1335 (38 observations), and the last one from September 1338 to December 1339 (16 observations). The first period precedes immediately the completion date for the *Astronomy* given by Levi; the second fit the first part of the five years (1333-37) without exegetical or philosophical production; and the third one follows very nearly the completion date for his *Commentary on Proverbs* (23 April 1338), the last work on the Bible that Levi completed before his death.

In chapter 98 of the *Astronomy*, Levi writes:

> It must be known that we have finished this book and opened [with it] a door for subsequent scholars to find a theory agreeing with all the appearances of the motions of the planets in longitude and latitude; because we have not yet made all the observations required to find the exact and true theory. In the preceding chapters, however, we described a theory still not complete, and we wrote there that, if we could make the observations required for [constructing] the exact and true theory before the publication of this book, we would work again to search in a complete way for that theory [...] Thus, do not be surprised about the date in which [according to us] we completed this book, if you find in it observations made after that date; because our report of them was written after having completed the book.
> And it must be [also] known that we worked very much on the

---

[17] Levi obviously made more observations than those recorded in the text; so, at the end of chapter 117 (Goldstein, 1988, p. 396), for instance, the Hebrew manuscripts have a blank space of more than a page where Levi, in a previous redaction of the chapter, reported observations of Mars at apogee and perigee disagreeing with his computations.

[18] Only dated and datable observations are numbered in column I. The dates listed in column III but not numbered in column I are mentioned in the text as having taken place. The number of the chapter(s) where the observations or dates are mentioned are listed in column II. If the results of a dated observation are mentioned in a chapter, but no date is stated there, the number of the chapter in column II is enclosed in square brackets. The dates of observations 25, 64, and 79 are not given in the text; see Goldstein, 1996. For observations 45 and 53, see Mancha, 1992a, pp. 289-97. The 36 planetary observations reported by Levi in chapters 113, 117, and 122, not preserved in the Latin version, have been published in Goldstein, 1988. Observations 32, 38, 72, and 81 have been published in Goldstein, 1972, pp. 279-282; Goldstein, 1974b, pp. 282-285, and Goldstein, 1979, pp. 149-151. For the eclipse observations see Goldstein, 1979, and Mancha, 1992b, pp. 32-34. On observation 65, see Goldstein, 1985, pp. 106-107, and Mancha, 1992b, p. 32. On observations 43, 46, and 49, see Mancha, 1998b, pp. 44-6. The abbreviations used in column IV are: b. = before; a. = after; n. = noon; m.n. = mean noon; mn. = midnight.

aforementioned observations, and we were unable to find a theory agreeing with them, because we computed the positions of the fixed stars according to al-Battānī's reckoning until we arrived at the moment of the lunar eclipse which occurred on 3 October of 1335 A.D., and we found then that the observed longitude of Aldebaran was 0;22° smaller than the value that follows from al-Battānī's reckoning. We then concluded that the motion of the fixed stars from al-Battānī's time to our time was smaller by the aforementioned amount than the motion that follows from al-Battānī[19], and to it must be added that, at the beginning, the center of vision was not perfectly known by us, until through a minutely careful research we arrived at its perfect knowledge at the time of the aforementioned observation. And it complemented the innovations made by us in solar and lunar theory in the month of December of 1335 A.D. Let God the Creator of everything be blessed, because He directed us to that perfection with His mercy and grace. Amen[20].

This chapter is placed at the end of the large section of the work devoted to solar, lunar, and eclipse theory (chapters 55-97)[21], and it is easy to show that all these chapters wcrc redacted after 1334-5. The lunar eclipse of 3 October 1335 closed the three-years period of astronomical observations, and it is the last of 10 solar and lunar eclipses which took place between 1321 and 1335 and were reported and analyzed by Levi in chapter 80. The eclipse played a significant role in the

---

[19] At this place Levi adds in a marginal note: "It must be known that after this observation we made another observation, exact and repeated many times, according to which α Leo is very near to the position in which it must be [situated] according to al-Battānī's reckoning, and we found the same thing for α Vir and α Sco".

[20] "Et sciendum quod hunc librum compleuimus et apperimus portam sequentibus ad inueniendum dispositionem que consentiat omnibus que in motibus cuiuslibet planetarum in longitudine latitudineque uidentur; quia nondum habuimus experientias necessarias ad inueniendum dispositionem punctalem et ueram. In superioribus autem locis posuimus dispositionem nondum completam, et tunc ibi nos scripsimus quod si haberemus experientias sufficientes ad dispositionem punctalem et ueram antequam publicaretur hic liber, rediremus ad inuestigandum querendum circa istam dispositionem perfecte [...] Et sciendum quod multum laborauimus circa experientias has predictas, nec dispositionem consentientem inuenire potuimus, quia computabamus loca stellarum fixarum secundum computum Albeteni quousque peruenimus ad eclipsim lunarem que fuit anno Christi 1335, die 3 octubris. Et tunc inuenimus Aldebaram in minori quam esset consequens ad computum Albeteni circa 0;22°. Et tunc iudicauimus quod motus stellarum fixarum a tempore Albeteni usque ad nostrum erat minor in quantitate predicta quam esset consequens ad Albeteni, cum hoc quod centrum uisus non erat nobis in principio scitum perfecte, quousque multum minutiando peruenimus ad eius perfectam notitiam predicte experientie tempore. Et istud fuit complementum cum eo quod innouauimus in sole et luna anno Christi 1335 in mense decembris. Sit Creator omnium benedictus qui ad hanc perfectionem nos duxit sua pietate et gratia. Amen" (Vat. Lat. 3098, f. 81rb-va). The marginal note reads: "Et sciendum quod post experientiam istam experientiam certam habuimus et multiplicata quam pluries quod Cor Leonis est circa illum locum in quo debet esse secundum computum Albeteni. Et similiter de Alchimec et Corde Scorpionis inuenimus".

[21] See a summary of the contents of the *Astronomy* in Table II.

Astronomy, for Levi interpreted its observational data as a definitive confirmation of his models for the Sun and the Moon, and used them to derive some fundamental parameters like the mean synodic month and the rate for precession, as well as the values for the mean solar distance and the radius of the earth's shadow[22].

In the computation of the eclipse (chapter 80) Levi used a solar model (hereafter called solar model I) with an eccentricity of 2;14 (maximum equation, 2;8o), derived (chapter 56) from observations of the apparent solar diameter made with a pinhole camera, the last of which occurred in December of 1334, and an apogee position about Cnc 3o, determined (chapter 57) from solar altitudes observed on April 24 and August 4 of that year[23]. The mean position of the Sun was also determined (according to chapter 46) in 1335[24], and Levi's value for the mean solar motion (0;59,8,20,8,44,6,3,14o/d) was derived (chapter 65) from his computation of the mean solar position at the time of the ocultation of Spica by the Moon reported by Timocharis (−293 March 9/10)[25] and the mean solar position at the summer solstice of 1334 (June 13, 14;53^h after mean noon).

The lunar position at the time of the eclipse was computed according to a lunar model described in chapter 71 (hereafter called lunar model III)[26], probably Levi's most original contribution. It is a complex eccentric model without epicycle, that avoids the unobserved variations in distance that follow from Ptolemy's model and produces maximum corrections at syzygy and quadrature departing from those of the *Almagest*, with an empirically based condition — the correction at 0o of anomaly is always 0o — which Levi extended to the planetary models[27].

---

[22] Goldstein, 1975 and 1979; Mancha, 1993, 1998a, pp. 31-4, and 1998b, pp. 307-9.

[23] Mancha, 1992a, pp. 289-97; Mancha, 1998a, pp. 44-6.

[24] Goldstein, 1988, p. 387. The corresponding passage in the Latin version reads: "Et quia dictam scientiam nundum habueramus superius coacti fuimus primo recipere ab antiquis illud quod ab eis recepimus de motibus supradictis, quia perfecte locum medij motus solis non habuimus usque ad annum incarnationis Christi 1335, ut in futuro patebit quando de sole loquimur" (Vat. Lat. 3098, f. 38rb).

[25] *Almagest*, VII.3; Toomer, 1984, pp. 335-6.

[26] In chapter 71 of the *Astronomy* Levi describes three different lunar models, of which the two first were rejected on observational grounds and the third was used to construct the lunar tables preserved in the Hebrew version of the text. The first one of these models has been analyzed in Goldstein, 1974b; the third one, in Goldstein, 1972 and 1974a.

[27] Levi argued that an epicyclic model would allow us to see both sides of the Moon, at variance with the appearances (Goldstein, 1997, p. 11), and consequently rejected the epicycle for the planetary models, "quia ex uno indiuiduo probationem accipimus ad omnia indiuidua speciei eiusdem" (chapter 43, Vat. Lat. 3098, f. 33ra-b). The mean apogee of the epicycle is determined in Ptolemy's planetary models from the equant point – or from the *prosneusis* point in the case of the Moon –, producing so what Levi calls "latitudo diametrorum motus diuersitatis" – "inclination, or oscillation, of the diameters of the motion in anomaly", translated in Goldstein, 1974b, pp. 53-63, as "reflection". According to chapter 76 Levi derived from observations that Ptolemy's values for this inclination were incorrect and that no correction due to this inclination occurred at 0° of anomaly: "Item apparet quod dispositio nostra sit uera ex eo quod de latitudine diametrorum motus diuersitatis uidemus, que nullo modo cum dispositione Ptolomei concordat, ut patet ex duabus experientijs

The latest observation mentioned in chapter 71 was made in February 1334 (Table I, no. 45), but observations related to the determination of the value accepted by Levi for lunar latitude (chapter 74) were made in May and June 1334 (Table I, nos. 44 and 48). Levi's value for the mean synodic month, $29;31,50,7,54,25,3,32^d$ (chapter 64), was derived (chapter 82) from the time elapsed between the lunar eclipse of 2/3 Choiak, Hadrian 19 (= 134 AD, Oct. 20/21), observed by Ptolemy[28], and the lunar eclipse of 1335, and in the discussion of the lunar mean motion in anomaly (chapter 85) and lunar parallax (chapter 88) the date June 15, 1334, and the year 1333, respectively, are mentioned as elapsed. Finally, the values for the solar parallax and distance, the magnitudes of the Sun and the Moon, and the radius of the earth's shadow, were also computed (chapters 90-95) from solar model I and lunar model III with data from the lunar eclipse of 1335. Obviously, for all the remaining chapters from 55 to 97, not mentioned above and depending on these models, the *terminus post quem* is also October 1335.

Until that year Levi seems to have accepted al-Battānī's value for precession[29]. At the time of the eclipse he derived (chapter 61) from his determination of the position of Aldebaran (α Tau) about 2;25h after apparent midnight, a precession

---

superius recitatis, quarum una a Ptolomeo discordat de 1;42°, alia circa 1;1°. Item ex sua dispositione propter latitudinem diametrorum in principio motus diuersitatis sequeretur equatio; quam equationem non inuenimus in una experientia quam habuimus stante luna in locis medijs inter augem et eius oppositum, quia ibi equationem non inuenimus nisi illa que ad illum locum pertinebat pro motu diuersitatis" (Vat. Lat. 3098, f. 64v; the first two observations, on 24 June 1333 and 17 January 1334, are discussed in chapter 71, but I have found no evidence for dating the last one). Using the methodological principle quoted above, Levi apparently concluded from this observation that no correction at 0° of anomaly occurs in the models for the planets; moreover, Levi experienced at this time some difficulty to construct a model (like lunar model III) producing such a correction (see in the following section the text quoted from chapter 36).

[28] *Almagest*, IV.6; Toomer, 1984, pp. 198 and 203.

[29] This assertion is derived from the following passages: "Et Albateni, qui uenit post eum [i.e., Ptolemy], inuenit istum motum in 100 annis unius gradus et 30 minutorum. Nos autem inuenimus eum in eadem mensura" (chapter 41, Vat. Lat. 3098, f. 31rb); "Et ex affectione in magno amore quem habebamus in oppinionibus antiquorum credidimus primo sententie Ptolomei, addito supplemento Albeteni de quantitate motus spere 8e et augium, et longitudinum, et motus diuersitatis in quolibet planetarum, quia credebamus hoc esse accipiendum ab eis, quia in hoc uidebatur debere ueritatem habere..." (chapter 46, Vat. Lat. 3098, f. 38vb); "Item declarabitur in futuro, quando de dispositione spere 8^e loquimur, quod Albeneti motus dicte spere inuenit gradus unius in 66 annis post 720 annos elapsos inter experientias Ptolomei et suas. Et post 480 annos elapsos inter experientias Albeteni et nostras eodem modo inuenimus sine defectu nisi tanto quantus posset errori experientie applicari. Ex quo manifeste apparet quod motus stellarum fixarum omni tempore est equalis" (chapter 58, Vat. Lat. 3098, f. 46vb). There is, obviously, something wrong in the date attributed by Levi to al-Battānī, since al-Battānī's first star catalogue in his Zīj was computed for the epoch 1st of March, 880 AD (Nallino 1899-1907, II, 269; hence 880 – 137 = 743, instead of 720); but Levi repeats the same error at least in another passage of chapter 58 (Vat. Lat. 3098, 46rb), and in chapter 61 (Vat. Lat. 3098, f. 48ra; Goldstein, 1975, p. 33).

rate of 1o in 67 Egyptian years, 123 days and 16;40 hours[30], and according to this value the positions of Regulus (α Leo), Spica (α Vir), and Antares (α Sco) for the beginning of 1336 are given as Leo 20;27o, Lib 14;21o, and Sgr 0;17o, respectively (chapter 62, section 7)[31]. Thus, the date of the eclipse is also the *terminus post quem* for the redaction of all the chapters where this precession value is used (for instance, 65, 66, 67, 81 to 85).

From the foregoing it would seem reasonable to think that, if 1335 is the *terminus post quem* for chapters 55-98, all the chapters previous to 55 were redacted before that date, maybe — according to Levi's colophon — during 1328. The extant text, however, does not support this hypothesis, for 1335 is also the *terminus post quem* for some chapters of the first part of the *Astronomy*[32]. According to chapter 98, quoted above, Levi did not solve the problem of the exact location of the center of vision until October of that year, and it implies that chapters 6 and 7, which deal with the matter, were redacted in their present form after that date. At least in two other chapters are also mentioned parameters which Levi determined with data from this eclipse: the apparent lunar radius (chapter 10, derived in chapter 92), and the radius of the moon's shadow (chapter 16, derived in chapters 80 and 94). Moreover, in chapter 41 are mentioned the values for the apparent radius of the Sun at apogee and perigee, and the position of the solar apogee (determined in chapters 56 and 57 from observations of 1334). Lastly, in chapters 130 to 135, at the end of the work, in computing the distances of the planets Levi depended on solar maximum and minimum distances derived from a solar eccentricity of 2;14 [33]; thus, the *terminus post quem* for these chapters is also 1335[34].

As for the tables for eclipses (*Luhot*), in the fifth section of the Hebrew version Levi included a worked example for the solar eclipse of 3 March 1337, where no observation is mentioned[35]. As Levi was persuaded after 1339 by the data of this eclipse and the lunar one of 26 January 1339 to increase the solar eccentricity, I assume that the calculation included in the canons was made in advance of the event, and that 3 March 1337 is the *terminus ante quem* for canons and tables. There is no doubt, moreover, that the Provençal version of these canons and tables predates the Hebrew one: its four sections are not a translation of the extant Hebrew text, it lacks the fifth section on eclipses (which, Levi tells us, he added to

---

[30] Goldstein, 1975, p. 36.

[31] Paris, B.N., MSS Heb. 724, ff. 123b-124a, and 725, f. 96b; Naples, B.N. III F.9, f. 259b; Vat. Lat. 3098, f. 55rb.

[32] Several chapters of the first part were indeed redacted after 1339, as we shall show in the next section.

[33] Goldstein, 1986; Mancha, 1998a, pp. 31-9.

[34] The calculation of the absolute distances of the planetary spheres does not involve all the parameters of the models, but only those corresponding to spheres of some thickness; i.e., the parameters given in the preliminary (and incomplete) chapters 103 (Venus), 106 (Mercury), 110 (Saturn), 114 (Jupiter), and 118 (Mars). Thus, from chapters 130-135 it is not possible to conclude that Levi's planetary theory was finished before his death.

[35] Goldstein, 1979, pp. 151-5.

the work as a gift for scholars[36]), and its variant readings for the entries in the tables show that they are also independent from those of the Hebrew version[37]. As for the *terminus post quem*, it is obvious that both versions were redacted after having derived the parameters underlying their tables, i.e., after 1334-35, as Levi himself asserts[38]. In consequence, it is most likely that both versions were composed between October 1335 and March 1337, immediately before Levi's back to his *Commentary on the Pentateuch*, interrupted in December 1332, of which he tells us that he had finished the part corresponding to *Numbers* on 16 December 1337[39].

When the Hebrew version of the canons and tables was later incorporated to the *Astronomy* as chapter 99 it was again revised, and Levi replaced seven of its tables by others based on different parameters or methods of computation, added a description of a method for domification, and substituted a new calculation of the eclipse of 26 June 1321 for the worked example of that of 3 March 1337. Levi still introduced more modifications in the Latin version of the chapter, including a description of 12 tables for finding the cusps of astrological houses for the latitude of Orange, that he composed (or intended to compose) as a gift for the Christians[40].

---

[36] "We have further composed a fifth chapter that we give as a gift for scholars in which we inform them in a simple way how to know the calculation of solar and lunar eclipses, and the true position of the Moon for any day of the days of the month, by means of some tables that we composed for this purpose" (British Library, Add. 26921, f. 13b; Carlebach, 1910, p. 40).

[37] It is clear, for instance, that the table of proportional minutes for correcting eclipses from the Hebrew version (Goldstein, 1974a, table 34), where the values for the coefficient for interpolation are given only to minutes, derives from the corresponding table (*taula 9*) in the Provençal version, which displays these values to minutes and seconds; the table for hourly lunar velocity also predates that of the Hebrew version; see Mancha, 1998b, section 5, esp. pp. 328-31.

[38] So, the text of the Provençal version reads: "...we have worked hard on these things and we have made an instrument which has shown us by experience and geometrical demonstration where the apogee of the Sun is [located], and we have found it very far from the position where it ought to be according to the computations of the Ancients [...] We found different opinions among the Ancient about the [maximum] equation of the Sun, and after hard work its amount has been revealed to us from [the observation of] lunar eclipses [...] We have also found that the equation of the moon does not agree with the amounts in the tables composed by the Ancients, and the discrepancy at some places is not small; therefore, we considered it was appropriate to investigate the [arrangement of] sphere[s] from which follows all that we see about the diversity of motions of the moon and its amounts. We have found it with the assistance of God in a way that agrees with all that we see of the Moon by experience, as we have explained in our book that we have devoted to this science" (Mancha, 1998b, p. 292, sentences 19 to 26).

[39] Touati, 1973, p. 52; Freyman, 1991, p. 129. If my conclusions on the date of composition of the Provençal and Hebrew versions of Levi's tables are correct, it is impossible to identify (as in Weil, 1991, p. 109) item 41 in the autograph list of Levi's library (*'Almana'q we-Luhot le-Heshbon mi-ketivat yadi*) with the tables requested by Christians. Note, moreover, that in item 41 does not occur Levi's expression for referring to his own works (*li 'ani Levi*).

[40] "Quoniam autem est ualde fifficile inuenire domos 12 modo predicto, ordinauimus hic supra promissum de gratia speciali 12 tabulas ad eas inueniendum facilime in oriçonte Aurayce..." (Vat. Lat. 3098, f. 82vb; Mancha, 1998b, Appendix, sentence 156). The extant Provençal version of the tables requested by the Christians represents indeed two different stages of Levi's plan of the work: its canons contains only instructions for finding mean and true syzygies, whereas the tables include,

Although the solar eclipse of 26 June 1321 is computed with the solar model I, in both versions, at the beginning of the chapter, Levi informs to the reader that he finally decided to increase the maximum solar equation and to introduce some changes in his lunar model — something that occurred, as we will see, after 1339. Thus, this year is the *terminus post quem* for the last revision of the chapter[41].

3. The breakdown of 1339

A set of observations made in 1337 and 1339 led Levi after December of that year to revise drastically his achievements from 1333-35, although *there is no evidence allowing us to establish how much time elapsed between December 1339 and the moment when these modifications were decided*. At the beginning of chapter 100, in which the latest mentioned date is 11 December 1339 (Table I, obs. 81), Levi tells us that from the observation of the solar eclipse of 3 March 1337 (as well as from the reconsideration of some data of the lunar one of October 1335) it followed that it was necessary to increase the solar eccentricity to 2;23 (maximum equation, 2;17o) — adding that a better agreement between computation and observation would be reached for the lunar eclipse of 26 January 1339 if the solar correction was computed according to this eccentricity[42]. Levi calculated a new table for the equation of center, to replace that of chapter 56, and recomputed with it in chapter 100 the eclipses already analyzed in chapter 80[43].

Also in chapter 100, Levi asserts that from the observations of the eclipses of 3 March 1337 and 26 January 1339, and the lunar observation of 24 January 1339 (as well as from reconsiderations of the eclipses of 9 July 1321 and 3 October 1335) it followed that it was probably necessary to accept some correction for the Moon at 0° of anomaly[44], whereas that of 26 January 1339 (and reconsideration of that of 3 October 1335) suggested that at 180° of anomaly the Moon was nearer to the earth than at 0° of anomaly[45] — a variation in distance which cannot be

---

as in the subsequent Hebrew version, those for computing eclipses. We do not know the final form in which the tables were finally presented to the Christians, or even if they were. It is possible that Levi finally replaced the work by the Latin version of the *Astronomy*.

[41] "Et quia postquam fuit iste computus ordinatus fuit nobis ex aliquibus eclipsibus solaribus et lunaribus reuelatum quod est necessarium addere equationi finali solis circa 0;9° et quod est necessarium ponere latitudinem aliquam in diametris sperarum motus diuersitatis in principio dicti motus, ut recitabimus in sequenti capitulo..." (Vat. Lat. 3098, f. 81va). The solar eclipse of 1321 was also computed by Levi in chapter 80, and recomputed in chapter 100 with his second solar model (Goldstein, 1979, pp. 123-5). The computation of chapter 99 is more detailed than that of chapter 80 (Mancha, 1998b, Appendix, sentences 236-271).

[42] Goldstein, 1979, pp. 118-9.

[43] The second table has been preserved only in the manuscripts of the Latin version; see Goldstein 1974a, p. 158: Table 2a.

[44] Goldstein, 1979, pp. 118-9 (100: 9, 19-23, and 29-31); Vat. Lat. 3098, f. 90ra-b.

[45] Goldstein, 1979, pp. 121 and 123 (100:61, 88); Vat. Lat. 3098, f. 90va-b.

produced by lunar model III. Levi adds that he will devote the next chapter, 101, to arrange "a model for the moon that conforms with the principles of Ptolemy [i.e., which produces the same equations]", in such a way that, whether the inclination is finally confirmed by observation, further research can easily derive the corresponding parameters[46]. Chapter 101 was completed by Levi, but the following, 102, where he intended to derive the parameters for these inclinations and to calculate the corresponding tables, is incomplete in the Hebrew and Latin versions[47]. Nevertheless, other passages suggest that Levi indeed completed his fourth lunar model[48].

The precession rate derived at the time of the lunar eclipse of 3 October 1335 was

---

[46] Goldstein, 1979, p. 120 (100:39).

[47] "Et ad scientiam quantitatum dictarum, scilicet diametrorum sperarum latitudinis et motus diuersitatis, deuenimus faciliter ex quantitate equationis quam inuenimus circa principium motus diuersitatis et circa 180° eiusdem quando experientie cuiuslibet predictorum sufficienter multiplicata extiterint [...] Et tractamus istam dispositionem complere hoc modo et etiam inquiremus modo quantitatem uisam diametri lune stantis circa augem et circa sephel et circa principium motus diuersitatis et circa 180° eiusdem. Et tunc Deo dante reuertemus ad inquirendum dispositionem lune talem que ab experientijs quas habebimus non discordet, et hoc uia argumentorum ingenij qua ad dispositionem peruenimus supradictam. Et faciemus in isto capitulo tabulas equationum lune modo quo sunt superius ordinate eis unum spatium pro motu sperarum latitudinis adiungendo" (Vat. Lat. 3098, f. 92vb).

[48] Besides the passage from chapter 99 quoted in note 40, at the end of chapter 76, after having stated that observations confirm that there is no inclination of the diameters of the spheres of the motion in anomaly at the beginning of this motion, Levi adds in a marginal note: "Nobis finaliter fuit notum ex cclipsibus solaribus et lunaribus quod *necesse est ponere aliam latitudinem diametrorum in principio motus diuersitatis*, quod potest poni per modum spere latitudinis, ut dictum est supra" (Vat. Lat. 3098, f. 64v). Another indication which suggests that Levi finally used lunar corrections other than those derived from lunar model III (Table 35 in Goldstein, 1974a) occurs in a passage of chapter 82 where Levi was reconstructing the conjunction of the Moon with Spica observed by Timocharis and reported by Ptolemy (*Almagest*, VII.3; Toomer, 1984, pp. 335-6). Levi asserts that the correction corresponding to 323;33,32° of mean anomaly is 2;49,49°, and in a marginal note adds that "as stated in the margin of chapter 65, taking into account the position of Spica that we finally found and *the lunar correction according to the values we finally found for it*, there is agreement between observation and computation with respect to the mean lunar position": "Et iam declaratum est extra in 65° capitulo quod ponendo Alcimeh in loco in quo inuenimus eum finaliter *et equationem lune ut finaliter eam inuenimus*, medius locus lune quasi cum ista conclusione concordat" (Vat. Lat. 3098, f. 70rb). The marginal note in chapter 65, mentioned in the preceding passage, reads: "Set secundum experientiam nostram de Corde Leonis, Alcimec erat tempore experientie nostre in Libra 14;40°. Sed cursus suus in tempore elapso inter istas duas experientias, ut suppositum est supra uia radicis, erat 24;36,39°; ex quo sequitur quod luna tunc fuit uisa in Virgine 20;3,21° *et equatio lune erat addenda in tempore isto secundum quod declarabitur in futuro perfecte 2;40,32°* et quod diuersitas aspectus secundum computum nostrum erat addenda 0;53° et circa 0;0,48°. Et restat quod locus medius lune tempore experientie Timocrati erat in Virgine 16;29,1°"(Vat. Lat. 3098, f. 57vb). In sum, the main change in this marginal recomputation affects the whole lunar correction (*i.e.*, equation + parallax), and instead of the values of the text (3;40,46° = 2;49,46° + 0;51°) the marginal note offers 3;34,20° = 2;40,32° + 0;53,48°. I am unable to derive from Levi's table 35 based on his lunar model III the value 2;40,32° proposed in the margin (my recomputation from table 35 gives 2;51,1° – a value very close to that in the text, 2;49,49°).

also modified some time after 1339. In the seventh section of chapter 62, a marginal note to the passage where the longitudes of α Leo, α Vir, and α Sco are given[49], informs us that "according to many observations made at the time of a lunar eclipse" (whose date is not mentioned), their true positions at the beginning of 1336 were Leo 20;50o, Lib 14;40o, and Sgr 0;40o. From these observations Levi derived a new value for precession of 1o in 66 Egyptian years, 60 days, and 3 hours[50]. At least chapters 65, 66, 67, 71, 82, and 98 of the Hebrew and Latin versions contain marginal notes for correcting previous computations based on the precession rate adopted after 3 October 1335[51]. As the only lunar eclipse observed by Levi after 1335 was the eclipse of 26 January 1339[52], this date provides a *terminus post quem* for these marginal notes and for the second change in Levi's mind about precession.

Taking into account that these modifications affect the whole of Levi's work during 1333-1335 (solar, lunar, and precession theory), and also that they were decided after the corresponding parts of the text (chapters 55 to 98) were already redacted, it is difficult not to imagine that Levi felt, at least for a moment, deeply downhearted. In any event, according to chapters 36 and 46, the changes in the lunar model opened a door for the difficult problem of the motion of the planets, whose models were still unfinished at that time[53]. In chapter 36 Levi writes:

> Thus, as some of the aforementioned models must be true, since none of the remaining ones agrees with many of the appearances that seem to follow from the motion in anomaly for the planets — as stated above —, and, however, these do not agree either with the part of the correction established by Ptolemy at the beginning of the motion in anomaly which seems to derive from the inclination of diameters, since from these models no equation agreeing with Ptolemy's follows, how can it be possible? So we were at first faced with this controversy, because in this long and clear research we were not able to find a model from which the aforementioned equation could follow, and furthermore we made some observations of the moon that, in our opinion, proved that at the beginning of the motion in anomaly there was no inclination of the diameters of this motion; thus, we then judged that no inclination occurred at this place. Consequently, we considered that Ptolemy did not derive from experience this feature of his models, but he was rather

---

[49] See the preceding section, and note 31.

[50] Paris, B.N. Heb. 724, f. 124a (text); Heb. 725, f. 96b (mg.); Naples, B.N. III F.9, f. 260a (text): "After having written this passage we found that in our time the position of Spica and the position of α Leo were at the places expected according to al-Battānī's reckoning except for 0;2° in the case of α Leo and around 0;6° for Spica. For this reason we must assume that, according to that we have found for the motion of α Leo, the motion of the fixed stars is 1° in 66 Egyptian years, 60 days, and 3 hours".

[51] See above note 19.

[52] Goldstein, 1979, pp. 118 ff.

[53] For chapter 46 see Goldstein, 1988, p. 386 (46:17-21); Vat. Lat. 3098, f. 37vb.

obliged to do so for the planets because the correction that he admitted at 180° of anomaly compelled him to accept that correction at the beginning of that motion.

This was the reason that led us to labor long and hard, trying to find a model that, without this correction, could agree with many planetary observations that we made, assuming an equation of center for the motion in longitude smaller than Ptolemy's, in order to agree with observations. So, carrying on this work and these observations we were unable to make them agree, and it was necessary to reject entirely the work already done and to begin again to research other models corresponding to these observations. For a long time our mind remained, for this reason, in great perturbation and worry, trying with great effort and hard work to find such a model, until we gathered a great number of reliable observations, which we repeated as many times as required to lead us to the truth of this research. From these observations it was known to us that is was necessary to accept that some correction at the beginning of the motion in anomaly follows from the inclination of the diameters of this motion[54].

Thus, some time after 1339 Levi's research on the models for the planets was continued (or resumed) on new grounds[55]. In fact, this year is the *terminus post*

---

[54] "Et quoniam aliquam dictarum dispositionum ultimarum est necessarium esse ueram, quia nulla aliarum correspondet multis consequentium que ex motu diuersitatis in planetis prouenire uidetur ut dictum est supra, et etiam istis ultimis non uidetur correspondere illud quod de equatione quam posuit Ptolomeus in principio motus diuersitatis prouenire uidetur ex latitudine diametrorum, quia secundum istas dispositiones in principio motus diuersitatis nulla equatio prouenit concors illi quam posuit Ptolomeus, quomodo ergo hoc fiet? [...] Equidem nobis ista controuersia in principio accidit, nam quia in ista inquisitione tam longa et clara non inuenimus aliquam dispositionem que concordaret cum equatione predicta, et etiam habuimus in luna experientias aliquas que, ut nobis uideatur, testimonium nobis dabant quod in principio motus diuersitatis latitudo nulla erat in diametris motuum; ideo protunc nulla latitudinem iudicauimus ibi esse. Et credidimus tunc quod Ptolomeus hanc radicem non fuisset per experientias consecutus, ymo quod dispositio quam constrictus est ponere in planetis propter equationem prouenientem in 180° motus diuersitatis constrinxisset eum equationem hanc ponere in principio dicti motus. Et ista fuit causa non ducens in magnos, longos et duros labores uolendo dispositionem aliquam inuenire quam concordaret cum experientijs multis quas habuimus de planetis non ponendo equationem predictam, ponendo equationem centri motus longitudinis minorem quam posuit Ptolomeus, taliter quod cum uisis experientijs concordaret. Et prosequendo hoc et multis alijs experientijs habitis inueniebamus discordiam, et totum erat necesse destruere et de nouo incipere alias dispositiones correspondentes dictis experientijs inuenire. [...] Et hac de causa longo tempore in magna mentis perturbatione et anxietate et labore multo ac sudore ad inueniendum dispositionem aptam permansimus, quousque magnam multitudinem experientiarum uerarum habuimus; quas etiam easdem multiplicauimus sepe que fuerunt necessarie ad ducendum nos in ueritatem inquisitionis istius.Et fuit nobis notum ex eis quod est necessarium in principio motus diuersitatis equationem aliquam ponere prouenientem ex latitudine diametrorum motuum" (chapter 36; Vat. Lat. 3098, f. 28rb).

[55] In chapter 117 (Goldstein, 1988, p. 396, between sentences 41 and 42 of the text; see also note 17, above) the missing report of observations of Mars disagreeing with Levi's computations was

quem for all the chapters of the *Astronomy* devoted to the planets (except for those devoted to the theory of distances). The preliminary model for Venus described in both versions of chapter 103 is constructed on the assumption that the equation of center of Venus, identical with that of the Sun, is 2;17o, which corresponds to solar model II. In chapters 109, 113, and 117, which contain observations of the superior planets and Levi's derivation of their apogees[56], the latest mentioned observations were made on 9 May 1339, 30 January 1339, and 10 January 1339, respectively (Table I, obs. 76, 78, and 82), and on 24 July 1340 in chapter 122. In fact, Levi's discussion of the models in the extant chapters on the planets — 103, 106, 110, 114, and 118 — corresponds closely in content and procedure to that of chapter 101, devoted to the preliminary work on the new lunar model. Other chapters from the first part of the *Astronomy* were redacted or at least revised after 1339: chapter 17, in which the latest observation of Mars can be dated to June-July 1339[57]; chapter 19, where the rates for the motion of the apogee of the planets are mentioned[58]; and chapters 36 to 38, where is described and discussed an arrangement of orbs producing some equation at 0o of anomaly[59].

4. The last years

No philosophical or exegetical work was finished by Levi after his *Commentary on Proverbs*, completed in April 1338, considered as the last productive year of Levi's life[60]. From the preceding remarks, however, there is no doubt that between 1339 and April 1344 Levi accomplished a great deal — the redaction at least of chapters 17, 19, 36-38, 46, 100-103, 106, 109-110, 113, 114, 117, 118, and 122 of the Hebrew version of the *Astronomy* — to which must be added his *De numeris harmonicis* (1343), the Hebrew and Latin versions of the *Prognostication* (1344), and, most of all, his collaboration with Petrus of Alexandria in preparing the

---

probably redacted when Levi still maintained that the correction for 0° of anomaly is always 0°, and deleted once he accepted an equation under these circumstances and supposed that it was possible to account also for these observations.

[56] Goldstein, 1988; Mancha, 1998a, pp. 28-30.

[57] Goldstein, 1996, p. 298, shows that the observations of the apparent size of Mars at three oppositions in Leo, Scorpio, and Capricorn, reported by Levi in this chapter (Goldstein, 1985, p. 106; Mancha, 1992b, pp. 32 and 44) were made in February 1333, April/May 1337, and June/July 1339, respectively.

[58] Goldstein, 1985, p. 113; Vat. Lat. 3098, f. 14rb.

[59] It is likely that chapter 48 (Vat. Lat. 3098, ff. 39vb-40rb) was also written after 1339.

[60] Touati, 1973, p. 52: "Cette année [1338] semble être la dernière année de production intensive de notre auteur: elle voit paraître coup sur coup ses commentaires sur *Josué*, les *Juges*, *Samuel*, les *Rois* (janvier), *Ezra*, *Néhémie*, *Daniel*, les *Chroniques* (février-mars), les *Proverbes* (avril). Pendant les six années qui lui restent à vivre, il n'écrira que peu de choses. Jusqu'en 1340 au moins, il continue à consigner ses observations astronomiques; en 1343, il compose pour Philippe de Vitry son *De numeris harmonicis*".

complete Latin version of the *Astronomy* — a laborious task[61] that, in my opinion, started *some time before the date on which the changes whose terminus post quem is 1339 were decided* and was probably interrupted by Levi's death. Chapters 4 to 11 of this version, after a slight stylistic revision[62], were offered to Clement VI in 1342 under the title *Tractatus instrumenti astronomie*.

The available evidence suggests that the preserved texts of the Hebrew and Latin versions of the *Astronomy* were composed close in time, and that, when Levi decided on the modifications derived from his observations in 1339, all the chapters whose *terminus post quem* is 1335 were already rendered into Latin. There are many arguments in favour of this assertion, but the main one is the following. At the end of the Latin version of section 7 of chapter 62, Levi includes a long passage[63], absent from the Hebrew version, which contains a list of 42 zodiacal stars (including α Leo, α Vir, and α Sco), for each of which the text provides us with the ordinal number within the corresponding constellation in

---

[61] On the characteristics of the Latin version – not translation – of Levi's *Astronomy* see Mancha, 1992b.

[62] That the texts of the *Tractatus* and the Latin *Astronomy* are practically identical was demonstrated in Mancha, 1992b, pp. 23-25. There is no doubt that the Latin *Astronomy* (as preserved in Vat. Lat. 3098) predates the *Tractatus* (extant in Paris, B.N. Lat. 7293, 1r-17v) – in contrast to what has sometimes been suggested. Consider the following example, in which only stylistic differences are involved: in a passage of chapter 6 (Goldstein, 1985, p. 53; 6:33) Levi analyzes the situation where the larger plate [CD] of the cross-staff is exactly hidden by the smaller plate [EF]; the text in MS Vat. Lat. 3098, f. 7va is: "Et sint sic ordinate quod linea EF totaliter occupet respectu oculi lineam CD" whereas MS B.N. 7293, 9vb, reads: "Et sint sic ordinate quod linea EF totaliter et punctaliter occultet lineam CD respectu oculi". It is obvious that the second text is better than the first, and it does not make any sense to propose the first if the second is previously available. When the variant readings involve conceptual differences, the text of the Parisian MS clearly deteriorates with respect to that of the Vatican MS. Consider the following example from chapter 7, for which I shall first give Goldstein's translation of the Hebrew text. A: "Proof: we consider the plane of circle ABCD to be the plane of the ecliptic, and its center is at E; it is clear that the plane of the circle of latitude passes through its poles. We consider arc DG to be part of the circle of latitude; its sine which comes from point G is line GF. We draw line EFD, and it is clear that line EFD is a single straight line, for the line from the center to the versine [*or:* half] of a chord bisects its arc" (Goldstein, 1985, p. 58; 7:35-37). B: "Demonstratio: supponamus superficiem ecliptice *circulum oriçontem* ABCD planam et eius centrum sit punctus E; manifestum est *per Theodecium* quod superficies circumferentie latitudinis transit per polos ecliptice. Ponamus arcum DG arcum latitudinis stelle, et erit sinus qui ueniet a puncto G linea GF. Et protrahemus lineam EFD, et est manifestum quod linea EFD est una linea recta, quia linea protracta a centro ad medium corde intersecat arcum per medium *est recta*" (Vat. Lat. 3098, f. 8ra; *italics:* words deleted by a thin line; **bold:** added in the margin); C: "Demonstratio: suppono superficiem ecliptice circulum oriçontem ABCD planam, cuius centrum sit punctus E; manifestum est per Theodecium quod superficies circumferentie transit per polos ecliptice. Ponatur arcus DG arcus latitudinis stelle, et sinus qui uenit a puncto G erit linea GF. Et protrahatur linea EFD, que, ut est manifestum, est linea recta, quia linea protracta a centro ad medium corde intersecans arcum per medium est recta" (Paris, B.N. 7293, f. 11vb).

[63] Corresponding to the brief passage that in the Hebrew text gives us the longitudes of α Leo, α Vir, and α Sco for 1336; see above note 31.

Ptolemy's catalogue (*Almagest*, VII.5-VIII.1), its description, ecliptical coordinates for the beginning of 1336, and magnitude[64]. As in the Hebrew version, the positions of α Leo, α Vir, and α Sco given in the text are corrected in the margin according to the precession rate determined after 26 January 1339[65]. Therefore, I conclude that this part of the Latin text was composed while Levi still accepted the value for precession determined in October 1335, and this implies that at least some parts of the Hebrew text (namely, all the chapters whose *terminus post quem* is 1339, and the marginal notes on precession) were redacted after the Latin version of chapter 62. Otherwise, it would be difficult to explain why Levi would have added to the Latin version after 1339 a list of star longitudes absent from the Hebrew text, but computed according to a superseded value for precession.

5. Conclusion

There is no chapter in the *Astronomy* for which the year 1335 can be established as a *terminus ante quem*. Nevertheless, as Levi asserts that his work was 'completed' in 1328, we must accept that a first version was finished that year — although we know that it ought to have been completely different from the extant work. Very few observations of those reported in the *Astronomy* were made between December 1328 and the first months of 1333 (none in 1329 and 1332, only 4 in 1330 and 1331), and we know that during these years Levi worked on his commentaries on the Bible (*Esther, Ruth, Genesis, Exodus*, and *Leviticus*).

Probably around 1332, a circle of "great and noble Christians" requested from Levi a calendar to establish dates and times of mean and true syzygies (thereby to know which dates satisfy the astronomical requirements of Easter) and other related astrological data. We do not know if Levi's research to fulfill this request was sponsored by the Christian circle, as seems likely, or other relevant circumstances (e.g., if Pierre Roger, the future Clement VI, was his patron)[66]. Although Levi's interest in astronomical matters is well attested as early as 1321, it is reasonable to think that the time devoted by Levi during his last years to the *Astronomy* (and also the final form of the text) are probably due to Christian interest and support. Be that as it may, it is a fact that at the beginning of 1333 Levi interrupted his work on

---

[64] Vat. Lat. 3098, ff. 55rb-56va.

[65] "Notandum quod postquam iste liber fuit completus inuenimus per experientias multas factas per nos tempore cuiusdam eclipsis lunaris quod locus Cordis Leonis in longitudine erat in Leone 20° et circa 0;50° anno incarnationis Christi 1336 in principio anni. Et Alcimech erat in Libra 14;40°. Cor autem Scorpionis inuenimus cum nostro instrumento in Sagitario circa 0;40°. Et ideo necesse est ponere motum stellarum fixarum 1° in 66 annis egiptiacis et 60 diebus et circa 3 horis, secundum illud quod inuenimus de cursu Cordis Leonis in tempore elapso inter experientias Ptolomei et nostras" (Vat. Lat. 3098, f. 55vb).

[66] It is worth mentioning that, despite this long relation with the Christian circle, there is no textual evidence suggesting that astronomical works in Latin, not rendered into Hebrew, were used by Levi.

biblical matters and devoted himself to astronomical research — an activity that occupied all his time until his death, except for a brief break of approximately a year around 1337-1338. Years 1333-1335 were devoted to observations and derivation of parameters for his new models, and years 1336-1337 to the composition of the Provençal and Hebrew versions of the requested tables.

As for the *Astronomy*, we have seen that in the form preserved in the manuscripts most of the text was composed, beyond any reasonable doubt, not before 1336[67]. There is no procedure, however, to decide which parts of the text whose terminus post quem is 1335 were redacted between 1335 and 1337, and which parts were composed after 1338. Taking into account that chapters whose terminus post quem is 1339 occur throughout Levi's text, I am inclined to think that all the chapters whose terminus post quem is 1335 were composed after 1338, and not before. At an indeterminate date before 1344, a thorough revision of the work was undertaken, and the text was modified and enlarged for the last time. It is not possible to establish when Levi changed his mind on these issues, but it seems likely that the last ones, concerning the models for the planets, took place nearer to 1344 than to 1339, and that the corresponding chapters were never finished.

Acknowledgements

I have greatly benefited from discussions on Levi's *Astronomy* with Bernard R. Goldstein, and thank him for his translations of some passages of the Hebrew version.

References

Carlebach, J. 1910. *Lewi ben Gerson als Mathematiker*. Berlin, L. Lamm.
Curtze, M. 1898. "Die Abhanlungen des Levi ben Gerson über Trigonometrie und den Jacobstab", *Bibliotheca Mathematica*, NS 12:97-112.
Curtze, M. 1901. "Die Dunkelkammer", *Himmel und Erde*, 13:225-236.
Chemla, K., and Pahaut, S. 1992. "Remarques sur les ouvrages mathématiques de Gersonide", in Freudenthal 1992, 149-191.
Dahan, G. (ed.) 1991. *Gersonide en son temps. Science et philosophie médiévales*, Louvain-Paris, E. Peeters.
Freudenthal, G. (ed.) 1992. *Studies on Gersonides - A Fourteenth-Century Jewish Philosopher-Scientist*. Leiden, E.J. Brill.
Freyman, E. 1991. "Le commentaire de Gersonide sur le Pentateuque", in Dahan 1991, 117-32.
Goldstein, B. R. 1972. "Levi ben Gerson's Lunar Model", *Centaurus*, 16:257-84.
Goldstein, B. R. 1974a. *The Astronomical Tables of Levi ben Gerson*. Hamdem, CT, Archon Books.

---

[67] Table II displays a summary of contents of Levi's *Astronomy*. I have marked with one asterisk (*) or two (**) the parts or chapters whose *termini post quos* are 1335 and 1339, respectively.

IV

Goldstein, B. R. 1974b. "Levi ben Gerson's Preliminary Lunar Model", *Centaurus*, 18:275-88.

Goldstein, B. R. 1975. "Levi ben Gerson's Analysis of Precession", *Journal for the History of Astronomy*, 6:31-41.

Goldstein, B. R. 1979. "Medieval Observations of Solar and Lunar Eclipses", *Archives Internationales d'Histoire des Sciences*, 29:101-56.

Goldstein, B. R. 1985. *The Astronomy of Levi ben Gerson (1288-1344)*. Berlin-New York, Springer.

Goldstein, B. R. 1986. "Levi ben Gerson's Theory of Planetary Distances", *Centaurus*, 29:272-313.

Goldstein, B. R. 1988. "A New Set of Fourteenth Planetary Observations", *Proceedings of the American Philosophical Society*, 132:371-99.

Goldstein, B. R. 1992. "Levi ben Gerson's Contributions to Astronomy", in Freudenthal 1992, 3-19.

Goldstein, B. R. 1996. "Levi ben Gerson and the Brightness of Mars", *Journal for the History of Astronomy*, 27:297-300.

Goldstein, B. R. 1997. "The Physical Astronomy of Levi ben Gerson", *Perspectives in Science*, 5:1-30.

Goldstein, B. R., and Pingree, D. 1990. "Levi ben Gerson's Prognostication for the Conjunction of 1345", *Transactions of the American Philosophical Society*, vol. 80, part 6.

Mancha, J. L. 1992a. "Astronomical Use of Pinhole Images in William of Saint-Cloud's *Almanach Planetarum* (1292)", *Archive for History of Exact Sciences*, 43:275-98.

Mancha, J. L. 1992b. "The Latin Translation of Levi ben Gerson's *Astronomy*", in Freudenthal 1992, 21-46.

Mancha, J. L. 1993. "La determinación de la distancia del sol en la *Astronomía* de Levi ben Gerson", *Fragmentos de Filosofía*, 3:97-127.

Mancha, J. L. 1998a. "Heuristic reasoning: approximation procedures in Levi ben Gerson's *Astronomy*", *Archive for History of Exact Sciences*, 52:13-50.

Mancha, J. L. 1998b. "The Provençal Version of Levi ben Gerson's Tables for Eclipses", *Archives Internationales d'Histoire des Sciences*, 48:269-352.

Manekin, C. H. 1992. *The Logic of Gersonides: a translation of* Sefer he-Heqqesh ha-Yashar *(The Book of the Correct Syllogism) by Rabbi Levi ben Gershom*. Dordrecht, Kluwer.

Nallino, C. A. 1899-1907. *Al-Battānī sive Albatenii Opus astronomicum*. Milano, Publicazioni del Reale Osservatorio di Brera in Milano, XL.

Renan, E., and Neubauer, A. 1893. "Les écrivains juifs français du XIVe siècle", *Histoire littéraire de la France*, XXXI: 351-789. Paris, Imprimerie Nationale.

Shatzmiller, J. 1991. "Gersonide et la société juive de son temps", in Dahan 1991, 33-43.

Steinschneider, M. 1899. "Devarim ᶜAtiqim", *Mimisrach Umimaarabh*, 4:40-43.

Touati, C. 1973. *La pensée philosophique et théologique de Gersonide*. Paris, Les Éditions de Minuit.

Toomer, G. J. 1984. *Ptolemy's Almagest*. London, Duckworth.

Weil, G. E. 1991. *La bibliothèque de Gersonide d'après son catalogue autographe*. Louvain-Paris, F. Peeters.

Weil-Guény, A. M. 1992. "Gersonide en son temps: un tableau chronologique", in Freudenthal 1992, 355-65.

Table I

DATED AND DATABLE OBSERVATIONS IN LEVI'S *ASTRONOMY*

No.	Ch.	Date and time	Observed / [Discussed]
1	80,88,99, 100	26 Jun 1321	Solar eclipse
2	80, 100	9 Jul 1321	Lunar eclipse
3	113	16 Sep 1325, 6;30^h b.n.	Jupiter with Aldebaran
4	109	16 Sep 1325, 6;00^h b.n.	Saturn with Aldebaran
5	109	27 Sep 1325, 6;30^h b.n.	Saturn with Aldebaran
6	109	13 Oct 1325, 6;00^h b.n.	Saturn
7	113	13 Oct 1325, 6;00^h b.n.	Jupiter
8	113	1 Jun 1327, 8;00^h a.n.	Jupiter
9	109	29 Dec 1327, 10;00^h a.n.	Saturn
10	113	8 Jan 1328, 5;00^h b.n.	Jupiter
11	113	3 Mar 1328, 7;00^h a.n.	Jupiter
12	113	1 May 1328, 8;00^h a.n.	Jupiter
13	117	5 May 1328, 8;00^h a.n.	Mars and Venus
14	113	16 Jun 1328, 8;00^h a.n.	Jupiter
15	122	19 Aug 1328, 8;00^h b.n.	Venus, Saturn
16	122	2 Oct 1328, 6;00^h b.n.	Venus, Jupiter
17	117	5 Nov 1328, 12;00^h a.n.	Mars with Aldebaran
18	113	7 Nov 1328, 6;00^h b.n.	Jupiter with Spica
19	117	28 Nov 1328, 6;00^h a.n.	Mars with Aldebaran
20	113	23 Dec 1328, 5;00^h b.n.	Jupiter
21	113	29 Apr 1330, 8;00^h a.n.	Jupiter
22	80	16 Jul 1330	Solar eclipse
23	122	27 May 1331, 8;00^h a.n.	Mars with Regulus
24	80, 100	14 Dec 1331	Lunar eclipse
25	17	[Feb 1333]	Mars at opposition
26	122	16 Apr 1333, 7;00^h a.n.	Mars with Regulus
27	80,88,100	14 May 1333	Solar eclipse
28	73	20 Jun 1333, 8;13^h a.m.n.	Moon with Saturn
29	73	20 Jun 1333, 10;32^h a.m.n.	Moon with Saturn
30	73	21 Jun 1333	Moon with Saturn
31	73	21 Jun 1333, 2;14^h a.obs.29	Moon with Saturn
32	71, [76]	24 Jun 1333, 8;07^h a.m.n.	Moon with α Sco
33	122	5 Jul 1333, 8;00^h a.n.	Mars, Saturn
34	71, [76]	4 Aug 1333, 8;29^h b.m.n.	Moon with μ Gem
35	80,95,100	23 Oct 1333	Lunar eclipse
36	71,80,100	23 Oct 1333, 9;30^h a.m.n.	Moon with Aldebaran
37	71	24 Oct 1333, 8;09^h a.m.n.	Moon with Aldebaran
38	71, [76]	17 Jan 1334, 6;09^h a.m.n.	Moon with Aldebaran
39	71	19 Jan 1334, 9;36^h a.m.n.	Moon with Regulus

40	71	20 Jan 1334, 10;35^h a.m.n.	Moon with Regulus
41	71	11 Feb 1334, 5;59^h a.m.n.	Moon
42	80, 100	19 Apr 1334	Lunar eclipse
43	57	24 Apr 1334	Sun
44	74, 122	7 May 1334, 3;00^h a.m.n.	Moon, Venus
45	[19],[41],56,[58]	29 May 1334, around noon	Sun
46	55,57,65	13 Jun 1334	Sun
-	86	13 Jun 1334	[lunar ascending node]
-	85	15 Jun 1334	[lunar anomaly]
47	74	24 Jun 1334, b.m.n.	Sun
48	74	24 Jun 1334, b.m.n.	Moon
49	57	4 Aug 1334	Sun
50	122	4 Oct 1334, 3^h b.sunrise	Mars, Regulus and γ Leo
51	117	4 Oct 1334, 8^h b.n.	Mars with Regulus
52	122	10 Nov 1334, 1;30^h b.sunrise	Venus with Spica
53	[19],[41],56, [58]	2 Dec 1334	Sun
-	46, 60	1335	[mean solar position]
-	86	4 Feb 1335	[lunar ascending node]
54	109	28 Mar 1335, 9;00^h a.n.	Saturn
55	117	28 Mar 1335, 9;00^h a.n.	Mars with Regulus
56	122	6 Jul 1335, 1;00^h a.sunset	Mars with Spica
57	109	20 Jul 1335, 8;00^h a.n.	Saturn
58	113	25 Jul 1335, 8;30^h b.n.	Jupiter
59	109	31 Aug 1335, 8;00^h a.n.	Saturn
60	113	4 Sep 1335, 8;00^h a.n.	Jupiter
61	61, 62	3 Oct 1335, 2;25^h a. mn.	Aldebaran
62	80,82,90,93,94,98,100	3 Oct 1335	Lunar eclipse
63	100	3 Mar 1337	Solar eclipse
64	17	[April-May 1337]	Mars at opposition
65	17	[June-August 1337]	Comet, Mars and Jupiter
66	113	22 Sep 1338, 8;00^h b.n.	Jupiter
67	117	22 Sep 1338, 7;00^h b.n.	Mars with Regulus
68	117	18 Nov 1338, 7;00^h b.n.	Mars
69	113	8 Jan 1339, 7;00^h a.n.	Jupiter
70	[19], 117	10 Jan 1339, 8;00^h b.n.	Mars
71	113	23 Jan 1339	Jupiter
72	100	24 Jan 1339, 6;01^h a.n.	Moon with Regulus
73	100	26 Jan 1339	Lunar eclipse
74	100	26 Jan 1339, 4;30^h a.mn.	Regulus
75	100	26 Jan 1339	Jupiter
76	[19], 113	30 Jan 1339	Jupiter
77	109	24 Mar 1339, 8;00^h b.n.	Saturn
78	[19], 109	9 May 1339, 12;00^h a.n.	Saturn
79	17	[June-July 1339]	Mars at opposition
80	122	28 Sep 1339, 2;00^h b.sunrise	Venus and Jupiter
81	100	11 Dec 1339, 7;14^h a.m.n.	Moon with Aldebaran
82	122	24 Jul 1340, 3;30^h a.n.	Venus and Jupiter

Table II

SUMMARY OF THE CONTENTS OF LEVI'S *ASTRONOMY*

Chapters	Subject
1-3.	Introduction; status of the astronomy.
4.	Trigonometry.
5-12.	Observation procedures: *Camera obscura,* Jacob Staff, and astrolabe (6*, 7*, 10*).
13-16.	Methods for determining the local meridian, the star positions, and the solar parameters (16*).
17-19.	Discussion of Ptolemy's models; planetary motions (17**, 19**).
20-24.	Qualitative kinematic models: motion in longitude.
25-26.	Qualitative kinematic models: motion in anomaly.
27-31.	Planetary spheres. Motions, number, and order of the orbs of a planet. Interplanetary fluid.
32-35.	Qualitative kinematic models: motion in anomaly.
36-38.	Qualitative kinematic models with equation at 0° of anomaly(**).
39-45.	Exposition and criticism of Ptolemy's and al-Biṭrūjī's astronomical doctrines (41*).
46.	Astronomy: observation and theory (**).
47-50.	Methods for determining planetary mean positions and equations. Aproximation procedures in the derivation of planetary parameters (48**).
51-54.	Cosmological matters: sphericity and circular motion of the heavens, central position and immobility of the earth, Milky Way, light of stars and planets.
55-60.	Solar theory; determination of the eccentricity and apogee; motion of the solar apogee (*).
61.	Precession (*).
62.	Spherical astronomy (*).
63-64.	Lunar theory, mean synodic month, mean motions in elongation and anomaly (*).
65-68.	Mean solar motion (*).
69-79.	Lunar theory: mean motion in longitude, ascending node, parallax, solar and lunar hourly velocity, eclipses (*).
80.	Observation and computation of eclipses (*).
81-89.	Discussion of ancient observations used in the determination of solar and lunar mean motions, motion of the solar apogee, lunar latitude, and parallax (*).
90-97.	Radius of the shadow of the earth, distances and sizes of the Sun and the Moon (*).
98.	Summary of the preceding chapters (*).
99.	Computation of eclipses (*/**)
100.	Observation and computation of eclipses (**).
101-102.	Lunar theory — incomplete(**).
103-108.	Models for the motion in longitude of the inferior planets — incomplete (**).
109-120.	Observation and models for the motion in longitude of the superior planets —

incomplete (**).
121-127.   Observations, models, and tables for the motion of the planets in latitude —
            incomplete (**).
128-129.   Discussion of the order of the planetary spheres.
130-135.   Planetary distances (*).
136.   Conclusion.

ADDENDA

p. 3, n. 9: See Addenda for III, 1.

p. 8, n. 29: This wrong date attributed by Levi to al-Battānī can explain the difference of about 0;22° between the longitude of the stars he found and those which followed from al-Battānī's reckoning, from which Levi derived a new value for precession; see J.L. Mancha, "Levi ben Gerson's Star List for 1336", *Aleph* 2 (2002), pp. 31–57, on. pp. 51–2, n. 40.

V

Heuristic reasoning:
Approximation Procedures in Levi ben Gerson's Astronomy

Communicated by J. North

1. Introduction

In his astronomical work, conceived as a part (Book V, section 1) of his great philo-sophical treatise *Wars of the Lord*, Levi ben Gerson (1288–1344) repeatedly refers to a mathematical procedure called by him *heqqesh taḥbuli* and used, for instance, in the derivation of parameters for his lunar model, in the construction of his table of sines, and in the computation of the thickness of the fluid layer between planetary spheres[1]. Al-though Levi's procedure has not been hitherto explained, Goldstein (1986, pp. 297–81), has provided many useful remarks on the meaning of the Hebrew expression *heqqesh taḥbuli* (translated as 'heuristic reasoning'), and showed that it must be associated to the category of sciences called *ᶜilm al-ḥiyal* in the *Compendium of the sciences* of al-Fārābī (d. 950), that is, a category of science that differs from a theoretical one because its rea-soning is heuristic rather than demonstrative[2], also pointing out (1985, p. 134) that Levi

---

[1] See, respectively, Goldstein, 1974a, p. 67; Goldstein, 1985, pp. 41, 134, and 138; Goldstein, 1986, pp. 290, and 300–2.

[2] Goldstein, 1986, p. 280: "the category *ᶜilm al-ḥiyal*... roughly translated... is the sci-ence of devices (*ḥiyal* is the plural of *ḥilā* that corresponds to the Hebrew *taḥbula*). In effect, al-Fārābī introduces a distinction other than the usual one between the theoretical and practical sciences, for this 'science of devices' concerns the ways of manipulating natural objects (both physical and mathematical) to make them to serve a useful purpose, and reflections on them... According to al-Fārābī, this science includes algebra (where the goal is to find a number that solves an equation); geometric proportions (where the goal is to find an unknown quantity in terms of certain given quantities); the preparation and use of mechanical devices, musical instruments, mirrors, etc." The Latin version of al-Fārābī reads: "Ex eis [i.e., ex scientiis ingeniorum] itaque sunt ingenia numerorum et sunt secundum modos plures, sicut scientia nominata apud illud nos-tri tempore *algebra et almuchabala*, et que sunt illi similia, quamuis hec scientia sit communis numero et geometrie. Et ipsa quidem comprehendit modos preparationis in inueniendo numeros, quorum uia est ut anministrantur in eis quorum radices dedit Euclides ex rationalibus et surdis, in tractatu decimo libri sui *de elementis*, et in eo quod non rememoratur ex eis in illo tractatu..." (González Palencia, 1953, p. 155); see also Rashed, 1975, p. 483.

used the same expression to describe the iterative procedure for finding the eccentricity and apsidal line of an outer planet in the *Almagest*.

Using the Latin version of Levi's *Astronomy*, composed by the Hermit friar Peter of Alexandria in collaboration with Levi himself probably between 1336 and 1344[3] (where the Hebrew *heqqesh taḥbuli* was rendered as *argumentum ingenij*), the purpose of this article is to provide an explanation of Levi's method, a description of the kind of problems for which Levi used it, and reconstructions of some of its applications in the *Astronomy*.

2. Levi's description of 'heuristic reasoning'

2.1.

Before commenting on the relevant passages on the subject in Levi's Astronomy, let us consider the equation

$$x^3 + 2x^2 + 10x = 20$$

which John of Palermo proposed ca. 1225 to Fibonacci, who proved in his *Flos* that the solution is neither an integer, nor a fraction, nor one of the irrationalities defined in Book X of Euclid's *Elements*, and presented as solution $x = 1;22,7,42,33,4,40$, without mentioning however how this result was obtained[4].

Consider now the following approximative procedure to solve it. To simplify, we write the equation in the form $f(x) = 20$. As it is clear that $2 > x > 1$, we begin with $x_1 = 1;22,30$, which yields $f(x_1) = 20;7,51,5,37,30$. Next, we consider that

$$\frac{x}{x_1} \simeq \frac{f(x)}{f(x_1)} \tag{1}$$

which leads to $x = 1;22,57,49,...,$ and $f(x) = 19;56,31,36,....$ We now make $1;22,57,49, ... = x_2$, and we assume that

$$\frac{x_1 - x}{x_1 - x_2} \simeq \frac{f(x_1) - f(x)}{f(x_1) - f(x_2)} \tag{2}$$

which leads to $x = 1;22,7,41,...,$ and $f(x) = 19;59,59,37,....$ For approximation $k + 1$ $(k = 1,2,3,...)$, we can write the previous equation in the form

$$\frac{x_{k-1} - x}{x_{k-1} - x_k} \simeq \frac{f(x_{k-1}) - f(x)}{f(x_{k-1}) - f(x_k)} \tag{3}$$

If we now make $1;22,7,41,29, = x_3$ and $19;59,59,37,.... = f(x_3)$, and compute three more approximations, we find $x_6 = 1;22,7,42,33,4,38,30,50,15,43$, which is correct to

---

[3] On the Latin version of Levi's *Astronomy* see Mancha, 1992b, and 1997.

[4] The result is too large in the sixth sexagesimal place by about $1\frac{1}{2}$; *cf.* Smith, 1925, p. 471–2. It is often repeated that Fibonacci probably used the so-called Ruffini-Horner method. This procedure was already used by As-Samaw'al in his *Treatise on Arithmetics* written in 1172 (Rashed, 1978, pp. 198–213).

the 10th sexagesimal place, and produces $f(x_6) = 19;59,59,59,59,59,59,59,59,59,54$. This procedure constitutes an example of the method called by Levi ben Gerson 'composite heuristic reasoning'[5].

2.2.

Although the expression *argumentum ingenij* is widely used in the Latin version of the *Astronomy*[6], the relevant passages to establish its meaning and to know the details of Levi's procedures occur in chapters 4, 47, and 49. At the beginning of chapter 49, devoted to explain how heuristic reasoning can be used to derive parameters for his planetary models[7], Levi writes:

> It is appropriate to know that it is not possible to do in a quick way a demonstrative research in order to show how our model must be constructed, so that from it will follow what we see by experience for any position [of the planet] –that is, a way according to which we can say: as the equation which corresponds to a certain amount of motion in anomaly near the apogee is a certain quantity, it necessarily follows that the centers [of the spheres] of these motions are distant by a certain amount–, because the computations for our models are deep and difficult, and it happens because some of these [uniform] motions do not occur around the centers of their spheres. Moreover, it is necessary to arrange this model in such a way that a maximum equation [of anomaly] at apogee and perigee equal to what we have observed will follow from it, and that an equation of the inclination of the diameters near 180° of anomaly equal to what we have observed will also follow from it.
>
> Therefore, the investigations which lead us to the truth necessarily are of the kind of heuristic reasoning, which are made from division and experience, which approach step by step to the truth until it is reached. These types of reasoning belong to the category of conditional reasoning, and there are two classes of them. One of which is taken from an excess and a defect; the second one is taken from two experiences in excess or

---

[5] The initial value (x_1) can be arbitrarily chosen, but a choice closer to the final result will reduce the number of iterations.

[6] The expression *argumentum ingenij* appears in chapters 46, 47, 48, 49, 57, 71, 88, 90, 101, 102, 103, 106, and 110, always related to the derivation of parameters for planetary models. However, in chapters 3 and 4, and in the description of chapter 49 in the table of contents, the Hebrew *heqqesh taḥbuli* is rendered as *magistrale argumentum or ars magistralis* ("In 49° [capitulo] declarabimus quod ad inueniendum radices certas in motibus planetarum non possimus tantum inniti experimentis et sensui, sed oportet habere rationes aliquas doctrinales et *magistralia argumenta*, et eas ac ea docebimus inuenire"; Vat. Lat. 3098, f. 1va).

[7] The content of this chapter (Vat. Lat. 3098, ff. 40rb-42va) is almost identical to that of chapter 110 (Vat. Lat. 3098, ff. 103ra-107ra) –the last of the chapters preserved in the Latin version and, according to the table of contents, the first of the three ones devoted to the construction of the model for Saturn–, though some of the parameters obtained by Levi in both chapters are slightly different.

from two experiences in defect. For illustrating the first class we say: if when we considered a determined first quantity, it followed an equation greater than that we have in a given second quantity, and if when we supposed a certain third quantity it followed from it an equation smaller than that we have in a given fourth quantity, it is known, according to the [rules of] proportion, that it is necessary to suppose a mean [quantity] between the first and the third, so that the ratio of the difference between the first and the mean to the difference between the first and the third is equal to the ratio of the second to the sum of the second and the fourth. To illustrate the second class, we say: if when we supposed a certain first quantity there followed from it an equation greater than that we have in a given second quantity, and if when we supposed a certain third quantity there followed from it an equation greater than that we have in a given fourth quantity, which is smaller than the second one, it is known, according to the proportion, that it is necessary to suppose a fifth quantity so that the third is the mean between the first and the fifth, and the ratio of the difference between the first and the fifth to the difference between the first and the third is equal to the ratio of the second to the difference between the second and the fourth. And one proceeds in a similar way if the second and fourth quantities are derived from equations smaller than those we have. The conditional reasoning would be simple when we say: if when we suppose a certain first quantity, there followed from it an equation of a given second quantity not equal to the third quantity which we have, it is known, according to the proportion, that it is necessary to suppose a fourth quantity in such a way that the ratio of the fourth to the first one is equal to the ratio of the third to the second[8].

In the Latin manuscripts there is at this place a long marginal note, absent from the Hebrew text, clarifying the previous passage by means of numerical examples. The text reads as follows:

Example of excess and defect: the first [quantity] is 10, that will produce 30, which is 6 [units] greater [than what we have], and this is the second; the third is 6, that will produce 15, which is 9 [units] smaller [than what we have], and this is the fourth; the fifth is 8;24, that will produce 24, what we have. Example of diverse excess: the first [quantity] is 10, that will produce 36, which is 12 [units] greater [than what we have], and this is the second; the third is 8, that will produce 30, which is 6 [units] greater [than what we have], and this is the fourth; the fifth is 6, that will produce 24, what we have. Example of diverse defects: the first [quantity] is 6, that will produce 24, which is 12 [units] smaller [than what we have], and this is the second; the third is 8, that will produce 30, which is 6 [units] smaller [than what we have], and this is the fourth; the fifth is 10, that will produce 36, what we have. The previous cases are [examples of] composite conditional reasoning; the following one is simple. The first [quantity] is 10, which produces 30, which is the second, which is not equal to 24, which is what we have, and it is the third. Therefore, 8, which is the fourth and whose ratio to 10 is equal to the ratio of 24 to 30, produces 24.[9]

---

[8] See Appendix, Text A.
[9] See Appendix, Text B.

Two aspects of Levi's description of the *argumenta ingenij* must be emphasized: firstly, they do not constitute examples of *demonstrative* research [in Aristotelian terms, *categoric* or *scientific* syllogisms], but belong to the kind of 'conditional reasoning' [*hypothetical* syllogisms]; secondly, they proceed by means of *experiment and investigation*[10], that is, making some assumptions and checking their deductive consequences, reaching the solution by successive approximations (*paulatim*), i.e., by an iteration process. Levi adds that conditional reasoning can be simple or composite.

There are two kinds of composite conditional reasoning, called by Levi (a) *from an excess and a defect*, and (b) *from two experiences in excess or from two experiences in defect*. In (a), according to Levi, if we have $f(x) = y$, where y is known, and we also have $f(x_1) = y_1$, such that $y_1 > y$, and $y_1 - y = d_1$ (the *first difference* in Levi's text), and $f(x_2) = y_2$, such that $y > y_2$, and $y - y_2 = d_2$ (the *second difference*), we can assume, in order to approximate x, that

$$\frac{x_1 - x}{x_1 - x_2} = \frac{d_1}{d_1 + d_2} \tag{A_1}$$

In (b), there are two cases: if (b_1) we have $f(x) = y$ (where x is unknown), and we also have $f(x_1) = y_1$ and $f(x_2) = y_2$, so that $y_1 > y_2 > y$, and $y_1 - y = d_1$, and $y_2 - y = d_2$, or (b_2) we have y_1 and y_2 so that $y_2 < y_1 < y$, we can assume in both cases, in order to approximate x, that

$$\frac{x_1 - x}{x_1 - x_2} = \frac{d_1}{d_1 - d_2} \tag{A_2}$$

In the case of simple conditional reasoning, if we have $f(x) = y$, where y is known, and another value also known, x_1 such that $f(x_1) = y_1$, we can consider, in order to approximate x, that

$$\frac{x}{x_1} = \frac{y}{y_1} \tag{B}$$

Thus, Levi's schemes of conditional reasoning are equivalent to assuming that there exists a proportion between four quantities, two values of a function and their respective arguments, one of which is unknown (equation B), or to assuming that linear interpolation can be applied for finding the unknown argument of a function, when three values of that function and two of the corresponding arguments are known (equations A)[11]. Other

---

[10] The Latin..*ex diuisione accepta cum experientia*.. translates the Hebrew *nisayon ve-hippus* (Goldstein, 1986, pp. 279–80; Goldstein, 1992, p. 14), where *nisayon* means *test, trial* or *experiment*, and *hippus* means *investigation or consideration*. The Latin term *diuisio* does not seem the best rendering for *nisayon*, and I have no explanation for Levi's (or Petrus') choice.

[11] Obviously, equations (A) and (B) are equivalent to the so-called *rule of two false positions* (or *rule of double false*; Latin: *regula duorum falsorum*) and *rule of false position* (Latin: *regula falsi*), already described in al-Khwārizmī's treatise on algebra, and well known in the Latin West at least since the *Liber augmenti et diminutionis* (XIIth century, attributed to an unidentified Abraham, and edited by Libri, 1838, pp. 304–76; see also Steinschneider, 1880, pp. 486 ff.), or Fibonacci's *Liber abaci* (1202; edited by Boncompagni, 1857). John of Murs in his *Quadripartitum numerorum* (1343) explains the rule of double false as follows: "Alcachaym arabico idioma sonat artem per quam ex falsis positionibus ad quos insequuntur errores proposite questionis veritas invenitur. Multiplicatur autem primus error per secundam positionem et secundus per primam

significant matters –particularly, the conditions under which equations (A) and (B) must be used– are omitted. The examples in the marginal note of the Latin text only indicate the computations to be made for obtaining the result sought (suggesting thus that Levi never intended to formulate the underlying equations), and they do not show the most interesting aspect of Levi's contribution, that is, to conceive these schemes as steps in an iteration process for problems in which a single application of these formulae does not provide the solution[12].

In Chap. 4, in constructing his table of sines, Levi's purpose is to compute the sine of $1/4°$, the desired minimum interval between consecutive entries in the column of arcs. In a way parallel to Ptolemy's in *Almagest*, I.10[13]. Levi computes the sines of arcs of 60°, 30°, 15°, 36°, 18°, 24°[= (30° + 18°)/2], 12°, 6°, 3°, and $1\frac{1}{2}°$ from Euclid, XIII.9–10 and the half-angle formula, and then asserts:

> We can also easily find the sine of $1/4°$ according to the following method. For once the sine of $(8\frac{1}{4})°$ [from the sine of $(15+1\frac{1}{2})°$] is known, we also know the sine $(4+1/8)°$, and, therefore, by successive halvings, we will know the sine of $(1/4 + 1/128)°$. And by the same procedure [we know] the sine of $(1/4 - 1/64)°$ [from the sine of 15°]. According to this procedure we find that the ratio of the sine of $(1/4+1/128)°$ to the sine of $(1/4 - 1/64)°$ is very nearly equal to the ratio of the first arc to the second one, in such a way that there is no difference between these ratios even to the fourth sexagesimal place, although they differ slightly in the fifth. Thus, according to this masterly art we conclude that the ratio of the sine of $(1/4 + 1/128)°$ to the sine of $(1/4 - 1/64)°$ is very nearly equal to the ratio of the first arc to the second. Because of it, and using the sines together in this way, i.e., according to the ratio between the arcs, we can assert that the sine of 0;15° is 0;15,42,28,37,2.[14]

---

et collectum ex utraque multiplicatione per collectum ex utroque errore dividitur, si fuerit unus errorum additus, alter vero diminutus. Sed si ambo additi vel diminuti, minor multiplicatio demitur a majori et quod superest per differentiam errorum dividitur, quod provenit est quesitum" (MS New York, Columbia University, Plimpton 188, chapter 61; ed. by L'Huillier, 1980, p. 211), that is, $x = (d_2x_1+d_1x_2)/(d_2+d_1)$, for positive and negative differences, and $x = (d_2x_1-d_1x_2)/(d_2-d_1)$, for two differences, both positive or negative. In the Middle Ages both rules were used to solve linear equations, *cf.* Smith, 1925, pp. 437 ff., and Tropfke 1980, pp. 367–74; on its application during the XVIth century to quadratic equations, see Smeur, 1978.

[12] Iteration of the rule of two false positions is a well known approximation method, usually called *method of the chord*, or *method of proportional parts*, or also, wrongly, *regula falsi*. The first formulation of this method that I have been able to locate occurs in chapter XXX of Cardan's *Ars magna* (1545), where it is called *regula aurea* and used to solve some equations of fourth, third, and second degree. The conditions for which this method is valid seem, however, to have been misunderstood by Cardan, who solely enunciates equation (A_2), adding that x_1 and x_2 must be chosen so that, if $x_1 = n$, $x_2 = n + 1$: "..nota, quòd inuentum primum semper differt vnitate ab inuento secundo, aliter non rectè est operatus" (Cardan, 1663, pp. 274a).

[13] Toomer, 1984, pp. 48–56.

[14] See Appendix, text C, and also Goldstein, 1985, pp. 41 and 138. Note that Levi computes the sines for R = 60. Hereafter I will abbreviate "60 · sin α" as "Sin α". Accurately computed, Sin 0;15° = 0;15,42,28,29,18.

Faced with the same problem that Ptolemy had, namely the impossibility of finding "by geometrical methods"[15] the chord corresponding to an arc which is one-third of the arc of a given chord, Levi computes by successive halvings the sines of arcs each time smaller until, according to him, can be asserted that

$$\frac{(1/4 + 1/128)^\circ}{(1/4 - 1/64)^\circ} \simeq \frac{\sin(1/4 + 1/128)^\circ}{\sin(1/4 - 1/64)^\circ}$$

or, in other words, that up to four sexagesimal places, the previous expression is an identity[16]. Thus, the *argumentum ingenij* is here, after having halved the angles, to use equation (B) because, though literally false, it can provide the value sought for the sine to the required number of sexagesimal places.

The passage from chapter 47 is especially significant because Levi identifies a problem which was solved in the *Almagest* by means of *argumenta ingenij*. While discussing in a general way the method for determining the apogees of the superior planets, Levi writes:

And we can easily find almost exactly the position of the apogee of these planets, and the amount of their equations of center, and the position of their motions in anomaly in the following way: we made three observations of each of these planets, in such a way that in each of these observations the distance between the planet and the mean sun is 180°; [the positions corresponding to] those observations are called by Ptolemy 'extreme points of the night'. In the *Almagest* the way has been declared by which we can reach [these parameters] from these observations, and [the procedure] belongs to the kind of heuristic reasoning.[17]

Ptolemy's derivation of the eccentricity and apogee position of the outer planets has been repeatedly described, and the problem, as formulated in *Almagest*, X.7 for Mars, constitutes a special case of the problem of deriving the radius of the lunar epicycle

---

[15] 'By geometrical methods' is the rendering in Toomer (1984, p. 54; 1977, p. 138) of Ptolemy's διά τῶν γραμμῶν (Neugebauer, 1975, p. 24: 'by rigurous methods'), and it alludes to ruler and compass constructions, according to Euclid's postulates 1–3; with this restriction only 'plane' constructions are possible, that is, problems whose solution involves algebraic equations of degree not higher than the second. The expression is also found in Hipparchus' *Commentary to Aratus*, and a Latin equivalent in Pliny's *Naturalis Historia* (ii.63) is: *ratione circini semper indubitata* (Neugebauer, 1975, pp. 771–2). The problem of finding crd α from a given crd 3α (or sin α from sin 3α) requires the solution of the cubic equation $4 \sin^3 \alpha + \sin 3\alpha = 3 \sin \alpha$. On Ptolemy's solution see also Aaboe, 1964, pp. 121–5, and Heath, 1921, ii pp. 281–2. The Greeks knew other geometrical methods to trisect an angle (Heath, 1921, i, pp. 235–44), but they require curves which cannot be constructed with ruler and compass.

[16] In fact, these ratios differ by 0;0,0,0,8,.... In *De revolutionibus*, I.12, Copernicus proceeded as Levi did, that is, halving until reaching the required accuracy degree, the ratio of arcs is equal to the ratio of the corresponding sines: once established that, for R = 100000, $3°/1\frac{1}{2}° >$ sin $3°/$ sin $1\frac{1}{2}° = 5235/2618$, halving again he found that sin$(1/2 + 1/4)° = 1309$, and that $1\frac{1}{2}°/(1/2 + 1/4)° = 2618/1309$; consequently, $1\frac{1}{2}°/1° = $ sin $1\frac{1}{2}°/$ sin $1°$, whence sin $1° = 2618/1 1/2 = 1745$.

[17] See Appendix, text D.

from three lunar eclipse observations, solved by Ptolemy in *Almagest*, IV.6[18]. However, Ptolemy clearly points out that there is a significant difference between these two solutions: for the Moon, the epicyclic radius can be "demonstrated", whereas for the superior planets a "rigorous demonstration" cannot be carried out[19]. Ptolemy's solution of the problem is a very ingenious, though lengthy, approximation, which requires three iterations for Mars, and two for Jupiter and Saturn[20]. Levi's words in this passage are, moreover, interesting because they show that the schemes described in chapter 49 (equations A and B) do not constitute a complete description of the 'heuristic reasoning'[21].

In any event, we can now establish the meaning attributed by Levi to the expression *heqqesh taḥbuli*, and also explain his apparent lack of interest in a general treatement of the subject. The two basic rules of the method, and the criteria for their application, constitute more a practical art than a part of a theoretical science. For Levi, whose ideal of *mathematical procedure* was without doubt Euclid's work, a mathematical or geometrical problem must be solved by means of 'demonstrative reasoning', that is, its solution must be deduced from theorems previously demonstrated (or directly from axioms). A problem, however, is solved by means of 'heuristic reasoning' when we assume (without proof) that a particular relation exists between the known quantities and the unknown that we seek; thus, when we formulate that relation, it cannot be properly expressed as a mathematical identity (using the sign '='), since it involves an indeterminate degree of approximation (requiring therefore the sign '≃'). Consequently, when we use such expressions, we cannot consider the value obtained as the solution unless we verify that it indeed satisfies the requirements of the problem. Thus, the characteristics of an *argumentum ingenij* are (a) the *unjustified assumptions* used as premises for obtaining a value for the unknown quantity, (b) the *checking* of that value, and (c) *iteration*, i.e., further use of the *assumptions* when the checking demonstrates that the value obtained is not correct. The reasoning which begins by making these

---

[18] On *Almagest*, X.7 (Toomer, 1984, pp. 484–98), see Neugebauer, 1975, pp. 172–7, and Pedersen, 1974, pp. 273–83. On *Almagest*, IV.6 (Toomer, 1984, pp. 190–203), see Neugebauer, 1975, pp. 73–80, and Pedersen, 1974, pp. 172–7. A detailed analysis of Copernicus' use in *De revolutionibus* of Ptolemy's procedure in *Almagest*, X.7, can be found in Swerdlow-Neugebauer, 1984, pp. 307–69.

[19] Toomer, 1984, p. 486.

[20] Hill (1900, p. 34b), demonstrated that Ptolemy's problem requires the solution of an equation of the eight degree. Ptolemy's solution was criticized by Jābir Ibn Aflaḥ (12th century) in his *Islāḥ al-Majistī* (cf. Swerdlow, 1987), and also by F. Viète in his *Apollonius Gallus* (*App. II*), who extended his criticism to Copernicus and Regiomontanus. Kepler (*Astronomia nova*, II.16) agreed with Viète on the difficulty of the method, but added that *si difficilis captu est methodus, multo difficilior investigatu res est sine methodo* (Kepler, 1609, p. 95). It is worthy of mention that Delambre, in his notes to Halma's edition of the *Almagest*, wrote that Ptolemy reached his solution *par de fausses positions ou des rectifications successives* (Halma, 1813–6, II, p. 10). In the Hebrew version of Averroes' *Epitome of the Almagest*, Ptolemy's method is called a *mofet taḥbuli* (heuristic demonstration), which combines *science* (*hokma*) and *practice* (*ma'aseh*); cf. Lay, 1991, III, p. 113).

[21] As we shall see (section 3.6) Levi's determination of the solar apogee was made by means of an 'heuristic reasoning' for which equations (A) and (B) cannot be used.

assumptions truly ends when the checking provides, in fact, a *verification*, though the verification, however, does not constitute a proof of the procedure used. Its unique justification lies therefore on its usefulness, and, once the correct values have been reached, the description of these *obscure and difficult arguments* can be omitted, for, as Levi writes, *they have no benefit except perhaps for seeking honor* [22].

3. 'Heuristic reasoning' in the Astronomy

In this section I shall analyse six cases of 'heuristic reasoning' in order to show the variety of problems solved by this method. The first two are examples of a single

---

[22] "Ex hijs enim intelliget prudens lector per qualia argumenta et uias et modos ad ea que scribimus in hoc libro peruenimus; ex hijs etiam positionem nostram uerificabimus manifeste et ostendemus quod nullo modo alio poterunt saluari motus planetarum et orbes, hijs que de motibus eorundem in longitudine et latitudine apparent saluatis et principijs naturalibus custoditis. Et licet pro nunc illa subtilia argumenta, quia nisi ante assecutam notitiam ueritatis modicum utilitatis afferent, et ideo nobis seruire cognoscentibus ueritatem non possent nisi ad honorem forsitam consequendum" (*Astronomy*, chapter 3, Vat. Lat. 3098, f. 3vb; see also Goldstein, 1985, p. 30). It has been pointed out that in Levi's mathematical treatise *Sefer Ma^{c}aseh Hoshev* there is no trace either of the 'analytical' procedures used: "L'influence d'Euclide est manifeste dans la structure de l'ouvrage, mais aussi dans le style... les démonstrations sont effectuées de manière *synthétique*, sans que soit révélée l'analyse qui y a mené. En particulier, l'on ne trouve aucune trace de cet outil possible de l'analyse qu'est l'algèbre, même pour des problèmes qui en relèveraient" (Chemla-Pahaut, 1992, p. 152); thus, Levi's attitude with regard to his heuristic procedures seems to be similar to Ptolemy's towards the computational methods called by Pedersen (1974, pp. 78–93) "the mathematics implicit in the *Almagest*". The difficult problem of Levi's sources (a matter for further research) deserves special attention. I have avoided using the expressions *regula falsi* and *regula duorum falsorum* for referring to what Levi called *simple* and *composite heuristic reasoning*, because we have no evidence that equations (A) and (B) were taken by Levi from some treatise on algebra or practical arithmetics where they were used –as it was usual in the Latin West at least since the beginning of the XIIIth century– to solve different kinds of problems involving linear equations. This possibility cannot be excluded: some of Levi's words in chapter 49 are strongly reminiscent of the usual language used to describe the rule of two false positions (*..ex pluri et pauciori..*), and the same occurs in his description of the denominators of the second term of equations (A), where one would expect that Levi, who was quite familiar with linear interpolation, simply used $(y_1 - y_2)$ instead of $(d_1 + d_2)$ and $(d_1 - d_2)$. [In fact, it occurs so in chapter 4, when Levi explains how to proceed when a sine is sought for which we cannot find the corresponding arc in the table; he asserts that, if α and β are two consecutive entries in the column of arcs, and $\alpha < \gamma < \beta$, the sine of γ must be found from $(\gamma - \alpha):(\beta - \alpha) = (\sin \gamma - \sin \alpha):(\sin \beta - \sin \alpha)$: "Et si ille arcus non inuenitur in tabulis, arcus proximus sibi queratur et differentia eorum notetur; et secundum proportionem differentie arcus non inuenti in tabulis ad differentiam duorum sibi proximorum in eis accipiatur de differentia sinus" (Vat. Lat. 3098, f. 5r)]. But, be that as it may, it is evident that Levi clearly understood his rules as applications of the theory of proportions (*...est notum ex proportione quod necesse est poni mediam unam [quantitatem] inter primam et tertiam...*). Moreover, the *regula falsi* as well as the *regula duorum falsorum* were only used to solve linear equations, for which they directly give the exact solution, and thus iteration of both rules seems to be Levi's original contribution.

application of equation (B) to avoid longer computations. The third, fourth, and fifth cases are examples of Levi's use of equations (A). In the third, a single application of equation (A$_1$) is presented, although the preliminary value (x_1) with which the computation begins must have been obtained by Levi with a previous application of equation (A$_1$) or (A$_2$). In the fourth case, Levi made two successive uses of equation (B) to obtain the preliminary values (x_1 and x_2), and then applied equation (A$_1$). The text is silent on the procedure used in the fifth case, and I will present two possible reconstructions, the first by iteration of equations (A), the second like that was used by Levi in the fourth case. The sixth case constitutes a problem similar to Ptolemy's in *Almagest*, X.7, but less complicated: without using equations (A) and (B), Levi solved it by an iterative method, though he only computed the first approximation.

3.1. Derivation of parameters of the preliminary model for Saturn (Chap. 49)

Immediately after the long passage quoted in Sect. 2, where simple and composite 'heuristic reasoning' are described, Levi begins – as an example of the procedure – the construction of a preliminary model for Saturn without an epicycle and without 'inclination of diameters' at 0° of anomaly[23]. As initial conditions, the model must

---

[23] That is, an eccentric model for the second anomaly as those mentioned by Ptolemy in *Almagest*, XII.1 (Toomer, 1984, pp. 555–6) and described by Regiomontanus in *Epitome*, XII.1–2, but with a modified equation of center. The equation due to the 'inclination of diameters' is the correction to Ptolemy's equation of anomaly which follows from the equation of center [see below, 3.1 (b)]. Levi's expression must be considered in the context of a physical interpretation of Ptolemaic planetary models, and refers to Ptolemy's requirement that, in order to compute the mean anomaly of the planet when the center of the epicycle is not in the apsidal line, the epicyclic mean apogee must be determined by the line joining the center of the epicycle and the center of the equant (instead of by the line joining the center of the epicycle and the center of the deferent sphere). This consequence of Ptolemy's equant was already criticized and rejected by Ibn al-Haytham in his *Doubts concerning Ptolemy* (cf. Sabra, 1978), since it requires an oscillatory motion of the epicycle which cannot be physically justified, and repeatedly mentioned in Arabic (Ragep, 1993, pp. 48–51 and 427–9) and Latin texts (Mancha, 1990, pp. 73–8). Levi devoted a section of chapter 43 of his *Astronomy* to this argument: "Tertia demonstratio geometricalis impossibilitatem epiciclorum ostendens es ista: quia in latitudine diametrorum epiciclorum, que est necessaria poni secundum dispositionem Ptolomei ut ex ea sequitur illud quod nos uidemus de motibus planetarum errantium, impossibilitas inuenitur; quia latitudinem istorum diametrorum semper esse ad centrum deferentis sine positione alterius motus necessarium est omnino, quia diameter terminans maiorem et minorem distantiam epicicli semper in directo centri deferentis, ut patet cuilibet intuenti. Et hoc est superius clare ostensum. Set Ptolomeus fuit constrictus ex eo quod de experientia inuenit de motibus planetarum errantium quod in omnibus poneret dictam diametrum dirigi ad alium punctum quam ad centrum predictum. Et non declarauit modum quo posset poni in diametro talis motus, nec mirum, quia impossibile est ipsum ullius nature ymaginationi subesse. Nam et aliqui naturales cognouerunt inconuenientia motus istius quem posuit in predictis diametris, et per rationem naturalem quod est impossibile probauerunt, quia motus diametrorum istorum per totum girum non circuit, set partim circuit et retrocedit, et motus qui per totum girum non circuit set partim circuit et retrocedit describit punctum in actu in quo necesse est

produce (a) Ptolemy's equations of anomaly when the center of the epicycle is on the apsidal line, for instance, 5;53° and 6;34° for $\bar{\alpha} = 90°$ at apogee and perigee, respectively, and (b) the equation derived from the 'inclination of the diameters' in Ptolemy's tables for $\bar{\kappa} \simeq 90°$ and $\bar{\alpha} = 180°$, i.e., 0;48,37°. In this preliminary model such inclination occurs as a consequence of the disposition of the centers of the spheres of mean motion and anomaly, without assuming additional spheres for that purpose.

(a) Figures 1 and 2 show Levi's model at the positions which produce Ptolemy's equations of anomaly at apogee and perigee of the eccentric, respectively[24]. The observer is placed at D (Earth). B and E are, respectively, the centers of the concavity and the convexity of the sphere of mean motion (sphere I), which is moved to the east around point B carrying the remaining orbs placed above it. The mean position of the planet is determined by the farthest point from the Earth of this convexity, in the direction of F. The lowest sphere of anomaly (sphere II) is moved around E and carries clockwise point G, and the highest one (sphere III) is moved around D carrying counterclockwise point

---

esse quietem, ut in 8° Physicorum Aristotiles declarauit, quam corpori celesti inesse impossibilis declaratur in scientia naturali" (Vat. Lat. 3098, ff. 34rb–va). Levi's reluctance to admit some equation due to the 'inclination' of the diameter of the epicycle at 0° of anomaly (which was revised during the last period of his life; see below note 55), as well as his rejection of Ptolemy's values for that 'inclination' in the case of the Moon, were based on observational arguments: "Item apparet quod dispositio nostra sit uera ex eo quod de latitudine diametrorum motus diuersitatis uidemus, que nullo modo cum dispositione Ptolomei concordat, ut patet ex duabus experientijs superius recitatis; quarum una [on June 24, 1333] a Ptolomeo discordat de 1;42°, alia [on January 17, 1334] circa 1;1°. Item ex sua dispositione propter latitudinem diametrorum in principio motus diuersitatis sequeretur equatio; quam equationem non inuenimus in una experientia quam habuimus stante luna in locis medijs inter augem et eius oppositum, quia ibi equationem non inuenimus nisi illa que ad illum locum pertinebat pro motu diuersitatis" (Chap. 76, Vat. Lat. 3098, f. 64v). The first two observations are discussed in Chap. 71, but I have found no evidence which allows us to date the last one, from which by extension Levi seems to have concluded, apparently until 1339, that no equation due to Ptolemy's 'inclination' at 0° of anomaly must be produced in the planetary models (cf. Chap. 46; Goldstein, 1988, pp. 386–7).

[24] Note that in Levi's cosmology an orb is carried by the orb placed immediately below it: "Ideo sequitur quod inferior spera alicuius planete mouet superiorem eiusdem in equalitate et similitudine sui motus et non econtrario, quia spere inferioris conuexum est uia et spatium quod partim deciderio pertransit superior spera motu. Et non est possibile quod uia et spatium motu mobilis moueatur, ymo uia et spatium est firmum et hoc motu immobile, ut per se satis est notum. Secunda ratio: uidemus enim quod natura semper nobiliori superiorem et nobiliorem dat locum, et quia spera circundans ut sic est nobilior circundata, et spera in qua est luna firmata uel alius planeta quicumque est nobilior alijs que sibi in suo motu subseruiunt, ideo sequitur quod luna uel alius planeta quicumque sic in spera circundante et nobiliori locetur. Et ex hoc sequitur quod spera inferior inter speras alicuius planete sic subseruiens superiorem moueat eiusdem planete ut nobilem, et secunda tertiam, et sic de singulis quousque omnes motus earum recipiantur in ultima in qua planeta firmatur, sicut in nobilissima earum, ad quam et ad cuius motus omnes alie ordinantur et motus earum" (chapter 29, Vat. Lat. 3098, f. 21rb); see also Goldstein, 1986, p. 273. In the figures, the orb of the diurnal motion, which must be placed inside the concavity of the sphere of the mean motion, has been omitted.

H, the planet. Point A is called the 'apogee' (*aux*), and point C 'perigee' (*oppositum augis*), though, in fact, when the planet is at C is farther from the Earth than at A[25].

For $\bar{\kappa} = 0°$ and $\bar{\alpha} = 90°$ (Fig. 1), the lowest sphere of anomaly moves point G around E through $\angle FEG = 90°$, and then the highest one moves the planet H in the opposite direction around point D through $\angle GDH = 90°$. $\angle FDH$ is thus the equation of anomaly, to be added to the mean position of the planet, F. When $\bar{\kappa} = 180°$ (Fig. 2), sphere I has revolved through this angle around the center of its concavity, B, and the center of its convexity, E, is now situated below it. Point G has been moved through by the lowest sphere of anomaly from the direction of F until the position showed in the figure is reached, and H, the planet, has been moved from G in the opposite direction also through $\bar{\alpha}$. $\angle FDH$ is again the equation of anomaly. Levi's first step is to compute the amount of line DE on which depend the different values of $\angle FDH$ at 'apogee' and 'perigee' (hereafter, ζ_a and ζ_p, respectively).

At 'apogee', $DE = BE - BD$, and, at 'perigee', $DE = BE + BD$. If $BE = e_1$, and $BD = e_2$, the values of ζ_a and ζ_p can be obtained from the following equations (not stated in the text)

$$\tan \zeta_a = \frac{(e_1 - e_2) \sin \bar{\alpha}}{60 + (e_1 - e_2) \cos \bar{\alpha}}$$

and

$$\tan \zeta_p = \frac{(e_1 + e_2) \sin \bar{\alpha}}{60 + (e_1 + e_2) \cos \bar{\alpha}}$$

from which, for $\alpha = 90°$, $e_1 - e_2 = 60 \cdot \tan \zeta_a$, and $e_1 + e_2 = 60 \cdot \tan \zeta_p$.

Levi asserts that, since $\angle EGD$ and $\angle FDH$ are equal, and $EG = 60$, in Fig. 1, DE = $60 \cdot \tan 5;53° = 6;10,57$, and in Fig. 2, DE = $60 \cdot \tan 6;34° = 6;54,25$. Thus, $6;54,25 - 6;10,57 = 0;43,28 = 2$ BD, and BD = $0;21,44$. Consequently, BE = $6;10,57 + 0;21,44 = 6;54,25 - 0;21,44 = 6;32,41$.

(b) Levi's next step is to check if the resulting amount of line BD produces the aforementioned $0;48,37°$ when the planet is at 90° of apparent motion and $\bar{\alpha} = 180°$. These $0;48,37°$ derive from Ptolemy's tables in the following way. In Levi's model, since BD is very small, the difference between 90° of apparent motion (measured from D, the Earth) and 90° of mean motion (measured from B, the center of the concavity of the sphere of mean motion) can be neglected. At this position, the equation of anomaly for $\bar{\alpha} = 180°$ would be 0° without the correction due to the equant, and the true longitude of the planet would be equal to its mean longitude. But in Ptolemy's tables (Toomer, 1984, p. 549) for $\bar{\kappa} = 90°$ the equation of center is $6;31°$. Thus, the corresponding equation

---

[25] The Latin translation also uses the word *sephel*, transliterated from the Hebrew, instead of *oppositum augis* for similar models. The word is defined in Chap. 103 as the place where the equations of anomaly are greatest, though in fact this place is farther from the center of the Earth than the apogee ("...et C est *sephel*, nostro uocabulo, licet in rei ueritate magis distet a centro terre; set ipsum *sephel* uocamus utentes uocabulo antiquorum quia ipse est locus ubi sunt equationes prouenientes pro motu diuersitatis maiores"; Vat. Lat. 3098, f. 96rb). In Ptolemy's models the maximum equations of anomaly occur when the center of the epicycle is at the perigee of the deferent, i.e., at the nearest distance from the Earth.

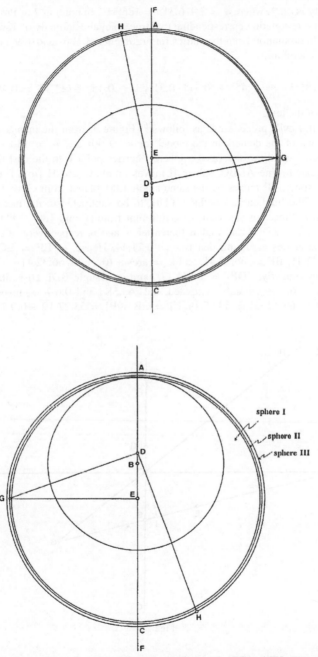

Figs. 1, 2. It is assumed that the radii of the spheres of anomaly, EG and EH, are equal, though, in fact, the thickness of sphere III is equal to the diameter of the planet

of anomaly is $\zeta(\alpha)$, where $\alpha = \bar{\alpha} + 6;31° = 186;31°$. In column 7 of Ptolemy's tables for Saturn, the equation corresponding to a true epicyclic anomaly of 186;31° must be found by interpolation from the entries for arguments of 186° and 189°, namely 0;45° and 1;6°. Therefore,

$$\zeta(186;31°) = 0;45° + (0;21° \cdot 0;31°/3) = 0;45° + 0;3,37° = 0;48,37°,$$

the value of the text.

From it, Levi's procedure is as follows[26]. Figure 3, from the manuscript[27], shows the positions of the centers in the model for $\bar{\kappa} \simeq 90°$ and $\bar{\alpha} = 180°$ (the distance BD has been greatly exaggerated). Sphere I carries point F to the east till line FD is perpendicular to line ADBC, sphere II moves points G and H from F to G through 180°, and sphere III moves by the same amount the planet from G to H. The correction due to the 'inclination' is thus ∠FDH, to be subtracted to the mean position of the planet, F. Since for any value of $\bar{\kappa}$ different from 0° and 180°, ∠FDH (hereafter, ξ) is the sum of ∠DFB and ∠DGB (hereafter, ε and η, respectively), Levi first computes these angles assuming that BD = 0;21,44. Hence, as EF = EG = 60, and BE = 6;32,41, BF = 66;32,41, and ε = arcsin (0;21,44/66;32,41) = 0;18,43° = ∠BDI. Consequently, ∠DBF = 89;41,17°, and in triangle BDI, DI is almost 0;21,44 and BI = 0;0,7 (accurately computed, 0;21,43,58.. and 0;0,7,5.., respectively). As GB = 120 − 66;32, 31 = 53;27,19, GI = GB + BI = 53;27,19 + 0;0,7 = 53;27,26.

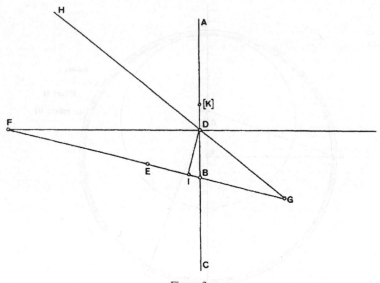

Figure 3

---

[26] Vat. Lat. 3098, f. 40va-b.
[27] Vat. Lat. 3098, f. 40v.

Since $GD^2 = GI^2 + DI^2$, $GD = 53;27,30$, and $\eta = \arcsin(0;21,44/53;27,30) = 0;23,18°$. So, $\xi = 0;18,43° + 0;23,18° = 0;42,1°$, which differs by $0;6,36°$ from the correction derived from Ptolemy's tables, i.e., $0;48,37°$, and Levi's conclusion is that line BD must be greater than $0;21,44$.

In the model, the values of ε and η can be derived from the equations (not stated in the text):

$$\tan \varepsilon = \frac{e_2 \sin \bar{\kappa}}{(60 + e_1) - e_2 \sin(90 - \bar{\kappa})}$$

and

$$\tan \eta = \frac{e_2 \sin(\bar{\alpha} - \bar{\kappa})}{(60 - e_1) - e_2 \cos(\bar{\alpha} - \bar{\kappa})}$$

As ∠FDH in Fig. 3 represents the maximum value of the correction which depends on e_2, this eccentricity can be derived from

$$\sin \zeta_{max} = \sin 0;48,37° = \sin \varepsilon_{max} + \sin \eta_{max} = \frac{e_2}{60 + e_1} + \frac{e_2}{60 - e_1}$$

whence, if $e_1 = 6;32,41$, $e_2 = 0;25,9,6,..$

Levi, however, approximates the required value with equation (B) by computing

$$\frac{x}{0;21,44} \simeq \frac{0;48,37°}{0;42,1°}$$

whence $x \simeq 0;25,8,49,... \approx 0;25,9$.

Since BD is now $0;25,9$, the amount of line DE will be $6;32,41 - 0;25,9 = 6;7,32$ at 'apogee', and $6;32,41 + 0;25,9 = 6;57,50$ at 'perigee'. Levi correctly asserts that from these values will follow equations of anomaly smaller and greater, respectively, than required, the difference between them being greater than that previously obtained with $BD = 0;21,44$, namely $0;41° = 6;34° - 5;53°$.

To obtain that difference, instead of computing accurately $\zeta_a = \arctan(6;7,32/60) = 5;49,45°$, and $\zeta_p = \arctan(6;57,50/60) = 6;37,13°$, whence $6;37,13° - 5;49,45° = 0;47,28°$, Levi again approximates it with equation (B):

$$\frac{0;21,44}{0;25,9} \simeq \frac{0;41°}{x}$$

whence $x \simeq 0;47,27°$.

From it, Levi finally concludes (Vat. Lat. 3098, f. 41ra) that to obtain $5;53°$ and $6;34°$ at 'apogee' and 'perigee' with the eccentricities $BE = 6;32,41$ and $BD = 0;25,9$, the Earth must be placed above point D at point K, and derives that the distance DK must be about $0;36,11$, which, according to him, satisfies very well the initial conditions[28].

---

[28] If $DK = 0;36,11$, computing accurately, $\zeta_a = \arctan[6;7,32/(60 - 0;36,11)] = 5;53,17°$, and $\zeta_p = \arctan[6;57,50/(60 - 0;36,11)] = 6;33,18°$. I have simplified this final step of Levi's calculation: most of the intermediary values computed to verify that $0;36,11$ satisfies Ptolemy's equations are different in the Hebrew and Latin versions, and some of them seem to be corrupt. Levi does not explain in this section how ∠FKH can be computed, asserting only that, for $\bar{\kappa} \simeq 90°$ and $\bar{\alpha} = 180°$, there is no perceptible difference between ∠FDH and ∠FKH ("Et est

3.2. Correction to al-Battānī's position for Saturn's apogee (chapter 109) [29]

Levi's purpose in Chap. 109 is to test al-Battānī's values for Saturn's mean motion and apogee, by comparing nine observations of the planet made by him between 1325 and 1339 with predicted values derived from al-Battānī's model, from which he concludes that Saturn's apogee has progressed 27;16° since Ptolemy's time (instead of 18;23° that follows from al-Battānī's rate of precession), and that its mean position at that time is 1;8° greater than al-Battānī assumed [30].

Levi's proceeds in the following way[31]. After asserting that there are noticeable discrepancies between the observed positions of Saturn and those derived from Ptolemy's model (using al-Battānī's parameters for precession and for the mean motion of the planet), Levi claims that they cannot be attributed only to some error in Saturn's mean motion in longitude. Rather, they must also be due to an error in the position of Saturn's apogee, whose motion must be greater than al-Bāttānī took it to be, for these discrepancies are of unequal amounts at different places on the eccentric, and the greater the distances

---

notum intelligenti hunc librum quod stante planeta in 90° medij motus in longitudine et in 180° motus diuersitatis non est differentia notabilis inter angulum FKH et angulum FDH propter istam eccentricitatem tam modicam; quod quilibet potest inuenire faciliter"; Vat. Lat. 3098, *ibid.*). Obviously, the distance DK in this kind of models depends on the values for the eccentricity and the epicyclic radius in Ptolemy's models, and the greater these parameters, the more DK increases: so, in a model of this kind for Mars $e_1 = 39;53,56$, $e_2 = 4;43,4$, and DK $= 1;15,45$.

[29] The table of contents of the Latin version of the *Astronomy* gives the following description of this chapter: "In 109 inquiremus locum augis Saturni et ipsius Saturni locum in medio motu in tempore nostro" (Vat. Lat. 3098, f. 2rb).

[30] An English translation of the Hebrew text of chapters 109, 113, and 117 is given in Goldstein, 1988. In the last two, Levi also compares his own observations of Jupiter and Mars with computations made using al-Battānī's parameters, concluding in the first case that the position of Jupiter's apogee is 2° greater than what follows from al-Battānī's reckoning and that its mean longitude must be diminished about 2;5°, from which he derives that the motion of its apogee has been 20;11° since Ptolemy's time, i.e., 1° in about 60 Julian years (Goldstein, 1988, ch. 113:29, p. 392). In the case of Mars, Levi concludes that al-Battānī's mean longitude must be diminished by about 0;45°, and that the eccentricity of the planet must be corrected, though no new value is given (the chapter is incomplete in the Hebrew version, and the lack of text in 117:41–42 (Goldstein, 1988, p. 396) is probably related to Levi's final decision to admit, as Ptolemy did, some correction at 0° of anomaly. [In Goldstein, 1988, p. 376, Levi's values of $\bar{\kappa}$ given in col. 4 for observations 9 and 16 represent the values of $\bar{\kappa}$ derived from al-Battānī's before Levi made the aforementioned corrections to its mean longitude and apogee, whereas the remaining values of $\bar{\kappa}$ take the correction into account (the persisting difference for observations 31 and 33 is perhaps due to Levi, if he omitted the correction in Jupiter's mean longitude). In the same way, Levi's values of $\bar{\kappa}$ given in col. 4 of Table 4 (*loc. cit.*, p. 377) for observations 17 and 28 are those derived from al-Battānī's reckoning before subtracting the aforementioned 0;45° from its mean longitude, and the remaining ones after subtracting them.] In these three chapters, Levi derives his corrections to al-Battānī's parameters in three parallel passages (namely 109:13–27, 113:12–30, and 117:12–19; *loc. cit.* pp. 387–8, 391–2, and 395), where an identical procedure is used, although the expression *heqqesh taḥbuli* is used only in chapter 113. Since chapters 113 and 117 are not preserved in the Latin version, I have preferred to reconstruct Levi's reasoning in Chap. 109.

[31] Vat. Lat. 3098, f. 103ra–b.

from the eccentric perigee, the more its observed positions (λ_o) surpass the computed ones (λ_c). As Ptolemy based his determination of Saturn's apogee on observations of the planet at opposition, Levi first considers two observations made on

(a) December 29, 1327, 10^h after noon: λ_o = Cnc 23;36°, λ_c = Cnc 22;48,31° ($\bar{\kappa}$ = 216;45°, $\bar{\alpha}$ = 178;35°), and
(b) March 28, 1335, 9^h after noon: λ_o = Lib 24;21°, λ_c = Lib 22;46° ($\bar{\kappa}$ = 305;15°, $\bar{\alpha}$ = 176;34°).

Since Levi accepts al-Battānī's equations, the problem is to find a position for Saturn's apogee able to produce agreement between computation derived from it and observation. Levi's reasoning is that the different equations of center corresponding to the different values of $\bar{\kappa}$ account for the discrepancy in the difference $\lambda_o - \lambda_c$ in both observations,

(a) $23;36° - 22;48,31° = 0;47,29°$,
(b) $24;21° - 22;46° = 1;35°$,

which is

$$1;35° - 0;47,29° = 0;47,31°.$$

Consequently, his first step is to determine how much the position of the apogee must be increased to eliminate that discrepancy. Instead of computing accurately, Levi asserts that, according to Ptolemy's tables, an error of 1° in the position of the apogee will produce a defect of 0;6° in λ_c in (a), and a defect of 0;4° in λ_c in (b). Indeed, in al-Battānī's tables (Nallino, 1907, pp. 112 and 109) the equations of center for arguments of 216° and 215° are, respectively, 4° and 3;54°, and those corresponding to 305° and 304°, respectively, 5;9° and 5;13°; thus, the differences between them are 0;6° and 0;4°, as in Levi's text[32]. Therefore, since $0;6° + 0;4° = 0;10°$, and the sum of the parts of the difference ($\lambda_o - \lambda_c$) due in both cases to the equation of center is 0;47,31°, Levi, using equation (B), considers that

$$\frac{1°}{0;10°} \approx \frac{x}{0;47,31°}$$

whence,

$$x \simeq 4;45° \text{ [accurately computed, 4;45,6°]}.$$

If we increase the longitude of Saturn's apogee by this amount, we have, for observation (a), $\bar{\kappa}$ = $216;45° - 4;45° = 212°$, and, for observation (b), $\bar{\kappa}$ = $305;15° - 4;45° = 300;30°$. Now, to compute the effect of these new values of $\bar{\kappa}$ on the equation of center, Levi again applies equation (B), which leads, in (a) to

$$\frac{1°}{0;6°} \simeq \frac{4;45°}{x}$$

whence,

$$x \simeq 0;28,30°,$$

---

[32] Computing accurately, i.e., taking into account the equation of center for ($\bar{\kappa} - 1°$) and the corresponding equation of anomaly, the differences are 0;6,9° and 0;4,14°.

to be subtracted from the previously computed position, so that λ_c = Cnc 22;48, 31° − 0;28, 30° = Cnc 22;20,1°. In (b), equation (B) leads to

$$\frac{1°}{0;4°} \simeq \frac{4;45°}{x}$$

whence,

$$x \simeq 0;19°,$$

to be added in this case to the previously computed position, so that λ_c = Lib 22;46° + 0;19° = Lib 23;5°.[33]

Once these corrections have been made, for observation (a), $\lambda_o - \lambda_c$ = 23;36° − 22;20,1° = 1;15,59°, and for observation (b), $\lambda_o - \lambda_c$ = 24;21° − 23;5° = 1;16°; that is, the differences between λ_o and λ_c are now practically identical. Finally, Levi concludes that complete agreement between observation and computation can be reached simply increasing by about 1;8° the mean longitude of the planet, and (since the previously obtained equations of center must remain unchanged) increasing also by the same amount the apogee position, that is, adding 5;53°[= 4;45° + 1;8°] to the initially supposed position, from which Levi obtained $\bar{\kappa}$ = 216;45° for observation (a). According to Levi, the difference between 1;16° and 1;8°, namely 0;8°, results from the different equation of anomaly that derives from the new mean motion in longitude: as the mean anomaly is given by

$$\bar{\alpha} = \bar{\mu}_\Theta - \bar{\mu},$$

where $\bar{\mu}_\Theta$ and $\bar{\mu}$ are, respectively, the solar mean motion and Saturn's mean motion, if $\bar{\mu}$ is increased by some amount, $\bar{\alpha}$ must be therefore decreased by the same amount, producing in both cases an increase of the equation of anomaly. Levi's 0;8° were probably taken, in a rough way, from al-Battānī's tables, where the difference between consecutive entries of the equation of anomaly for arguments near 180° is 0;7° (Nallino, 1907, p. 113)[34].

---

[33] Again, computing accurately, the differences that correspond to an increase of 4;45° in the position of the apogee are 0;30,2° and 0;19,55°, instead of Levi's approximate values 0;28,30° and 0;19°.

[34] After increasing by 1;8° the mean logitude and the position of the apogee, I find, computing accurately, that the equation of anomaly increases in (a) by 0;6,55°, and in (b) by 0;6,23°, instead of Levi's 0;8°. In the remaining part of the chapter (Goldstein, 1988, pp. 389–90: 109:27–50), Levi compares other observations of Saturn, made when the center of the epicycle is near apogee and perigee of the eccentric, with the positions derived from al-Battānī to which has been added 5;53° to the apogee and 1;8° to the mean motion. For these observations, the difference $\lambda_o - \lambda_c$ is positive at apogee, and negative at perigee (at perigee, for instance, with 184° < $\bar{\kappa}$ < 185°, the minimum and maximum differences are −0;17,40° for observation dated September 16, 1325, and −0;24,33° for observation dated September 27, 1325; at apogee, +0;11,9° on August 31, 1335, with $\bar{\kappa}$ = 305;44°, and +0;29,9° on May 9, 1339, with $\bar{\kappa}$ = 350;59°). From these discrepancies, Levi claims that another 3° must be added to the position of Saturn's apogee –probably using the same approximation procedures explained below. [Therefore, the unique values of $\bar{\kappa}$ given by Levi in chapter 109 which are computed, without correction, according to al-Battānī, are those that correspond to observations (a) and (b); the remaining ones (109:278–46) are those that derive

3.3. Solar parallax and distance at the time of the lunar eclipse of 3 October 1335 (Chap. 90)

The purpose of Chap. 90 is to find, at the time of the aforementioned eclipse, the true radius of the shadow of the Earth at the lunar distance, the distance from the Earth to the apex of its shadow cone, and the distances of the Moon and the Sun from the Earth[35]. In Fig. 4 (from the manuscripts[36]) C, F, and G are, respectively, the centers of the Sun, the Earth, and the Moon at the time of the eclipse, and CBE, CAI, and FDI are right angles; consequently, lines CA and FD are parallel. GH, according to Levi, is almost parallel to CA and FD. At the beginning of the chapter, from the ratio of the relative maximum and minimum solar distances, 62;14/57;46, the relative solar distance at the time of the eclipse, 59;25,5, and the apparent solar diameter at perigee, 0;30°, Levi derives that $\angle CEB$, the apparent radius of the Sun at that time, is 0;14,34,59,24,12°. He also states that 0;54,57,32,47° is a preliminary value for $\pi_0 + \pi_s$, the sum of the total horizontal parallax and solar parallax (= $\angle FGD = 0;53,20° + 0;1,37,32,47°$)[37], and that 0;42° is

---

from al-Bāttānī plus 4;45°, and it explains the difference of 3° in Goldstein, 1988, pp. 382–3, since for observations nos. 2, 3, 4, 30, 32, 42, and 43, $\bar{\kappa}$ is given in Levi's text before adding these 3° to Saturn's apogee. On this point the Latin text is more explicit than the Hebrew; in 109:29 it reads: "set secundum illud quod concordaueramus superius, ante addictionem 3° computando secundum dispositionem Ptolomei, Saturnus debebat esse in Geminis 28;15,40°"; Vat. Lat. 3098, f. 103rb.] Nevertheless, Levi's reasoning seems here to be wrong, insofar as this second correction to Saturn's apogee will disrupt the previously obtained agreement between prediction and observation for observations (a) and (b). Be that as it may, according to Levi, the position of Saturn's apogee ca. 1335 is 260;16°, which results from the addition of 8;53°[= 4;45° + 1;8° + 3°] to the position derived from al-Bāttānī's rate of precession (about 251;23°). As Ptolemy's value, about 1200 years before Levi's date, was 233° (*Almagest*, XI.7; Toomer, 1984, p. 541), Levi concludes that it has progressed 27;16° since Ptolemy's time (and that its motion is 1° in 44 Julian years, as stated in chapter 19 of the *Astronomy*; cf. Goldstein, 1985, p. 113).

[35] Chapters 90, 91, and 92 of the Latin text of Levi have been edited in Mancha, 1993.

[36] Vat. Lat. 3098, f. 77r.

[37] In Chap. 72 Levi asserts that the lunar parallax is about 0;55° at syzygies, and 0;57° at quadratures ("Nam inuenimus in oppositionibus quantitatem diuersitatis aspectus predicte circa 0;55° et in quartis circa 0;57°. Et experientia hoc modo fuit a nobis multotiens replicata, et ideo uidetur quod experientia non sit a ueritate remota, quia, si hoc esset, in tot experientijs tanta concordia non fuisset..."; Vat. Lat. 3098, f. 63va). In chapter 88 (Vat. Lat. 3098, f. 72ra), the value for the total horizontal parallax (or the difference between lunar parallax and solar parallax), 0;53,20°, is associated with the analysis of solar eclipses of 26 June 1321 and 14 May 1333. Although no derivation of the preliminary value for solar parallax, 0;1,37,32,47°, is given, Levi claims that it was obtained using 'heuristic reasoning', and that the matter will be discussed in a "more precise" way in chapter 90 ("...quando in hoc inquisiuimus tam perfecte quod non contradicit ista positio sibi ipsi, fuit nobis manifestum quod expedit ponere quantitatem diuersitatis [aspectus] lune sespectu spere 8^e circa 0;54,57,32,47°. Et de hoc in capitulo 90 perfectius inquiremus"; Vat. Lat. 3098, f. 72ra). Starting, for instance, with 0;2,51°, the Ptolemaic value for solar parallax, the procedure decribed below yields CF = 8635;53,59 t.r. and $\pi_s = 0;0,24°$, and a second attempt with 0;2° yields CF = 2761;36,30 t.r. and $\pi_s = 0;1,15°$; equation (A$_2$) leads then to 0;1,37,30°.

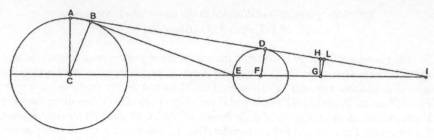

Figure 4

the angular radius of the shadow (i.e., ∠HFG), both in the measure where FG = 60[38]. The problem is to find, from these data, the solar distance, CF, at the time of the eclipse. As the exact value of π_s is not known, we can write

$$\pi_o + \pi_s = 0;53,20° + x$$

Since

$$\sin \pi_s = 1/CF,$$

the problem is thus to solve the following two equations with two unknowns

$$y = \frac{60[(0;11,20 + x)^{\cdot} \sin(0;53,20 + x)] + 60(0;53,20 + x)}{0;15,16,14,25,49(0;53,20 + x) - 60(0;11,20 + x)^{\cdot} \sin(0;53,20 + x)} \qquad (4)$$

and

$$\sin x = \frac{1}{y} \qquad (4a)$$

where 0;11,20 is the difference between π_o and the radius of the shadow, 0;15,16,14,25,49 is Levi's value for Sin∠CEB, and x and y are, respectively, the solar parallax (∠DCF) and distance (CF) at the time of the lunar eclipse of 3 October 1335.

---

[38] This value is derived in chapter 80 (Goldstein, 1979, 80:143–153; pp. 116–7,). According to Chap. 90 (Mancha, 1993, pp. 101 and 113; Vat. Lat. 3098, f. 77ra), at the time of the eclipse of 3 October 1335 the solar argument was 106;8°. At the time of the lunar eclipse of 23 October 1333, "when the distance of the Sun from its apogee was about 125;25°" (Vat. Lat. 3098, Chap. 80, f. 66ra), the radius of the shadow was 0;42,25° [both values were computed using 0;13,44° as apparent radius of the Moon as measured from the center of the Earth: "...ex quantitate diametri lune quam uidemus per radium lune per fenestram instrumenti intrantem, quam inuenimus 0;27,51° de circumferentia quam tunc describebat; sequitur quod eius uera quantitas de circumferentia quam describit circa centrum terre sit minor quantitate predicta circa 0;0,24°, ut declarabitur in futuro" (Chap. 74, Vat. Lat. 3098, f. 64ra); in Chap. 92, however, the lunar radius derived from the distance between the centers of the Earth and the Moon, 62;33,33,4,... t.r., is 0;13,42,8,33°; cf. Mancha, 1993, 92:15, p. 119]. From them Levi derived by interpolation in chapter 94 (Vat. Lat. 3098, ff. 78va-79ra) that the radii of the shadow with the Sun at apogee and perigee are, respectively, 0;42,35° and 0;41,37°.

Levi's procedure is as follows. In the first approximation, $x_1 = 0;1,37,32,47°$. As the angles are small, he assumes that FD = 0;54,57,32,47 and GH = 0;42; in consequence, if FD = 1, GH = 0;45,51,8,11,8,52. Since GH is almost identical to the perpendicular from point G to line ADI[39],

$$\frac{IF}{IG} \simeq \frac{FD}{GH}$$

and

$$\frac{IG}{GF} \simeq \frac{GH}{FD - GH}$$

So,

$$\frac{GH}{FD - GH} \simeq \frac{0;45,51,8,11,8,52}{0;14,8,51,48,51,8} = 3;14,27,28,15,51,33$$

If FD = 1,

$$FG = \frac{60}{0;57,33,15,50} = 62;32,58,16,24,8,29^{40}$$

Consequently,
GI = FG · 3;14,27,28,15,51,33 = 202;43,14,43,10,37 t.r., and
FI = FG + GI = 265;16,12,59,34,45,29 t.r.
If FI = 60, FD = 0;13,34,15,49,55. Therefore,
$\angle DIF = \angle AIC = arcSin0;13,34,15,49,51 = 0;12,57,32,47,2°$ [acc. 0;12,57,33,59,36°].
But, according to the text, at the time of the eclipse, $\angle CEB = 0;14,34,59,24,12°$; thus, if EC = 60, CB = Sin$\angle$CEB = 0;15,16,14,25,49 [acc. 0;15,16,17,4,54], and if CI = 60, since CA = Sin$\angle$AIC, CA = 0;13,34,15,49,55. Now, CA and CB are solar radii; thus,

$$\frac{CE}{CI} = \frac{CA}{CB} = \frac{0;13,34,15,49,55}{0;15,16,14,25,49}$$

whence

$$\frac{CE}{EI} = \frac{CE}{CI - CE} = \frac{0;13,34,15,49,55}{0;1,41,58,35,54} = 7;59,5,18,29,39.$$

As EF = 1, EI = EF + FI = 266;16,12,59,34,45,29 t.r., CE = 7;59,5,18,29,39 · EI = 2126;7,1,9,10,15,42 t.r. [acc. 2126;7,1,9,10,15,41,58; the reading 2126;7, 1,9,12,15,42 is probably a copyist's error], and CF = CE + 1 = 2127;7,1,9,10,15,43 t.r. is the distance between the centers of the Sun and the Earth at the time of the eclipse. If CF = 60, FD = 0;1,41,32,45,13,30; therefore, $\angle DCF = arcSinFD = 0;1,36,57,58°$ [acc. 0;1,36,58,9°], which does not coincide with x_1. So, the difference between x_1 and the value obtained for π_s is 0;1,37,32,47 − 0;1,36,57,58 = 0;0,0,34,49.

---

[39] At the end of the chapter (Mancha, 1993, pp. 108–9 and 117) Levi proves that if GH = 0;42, the perpendicular from G to ADI, called GL in the corresponding figure, measures 0;41,59,53.

[40] Throughout chapter 90 Levi assumes, according to Chap. 88 (Vat. Lat. 3098, f. 77ra), that Sin 0;54,57,32,47° = 0;57,33,15,50 (accurately computed, 0;57,33,2,7,28,..).

In the second approximation, $x_2 = 0;1,37°$. Levi repeats the computations and finds that CE/EI = 7;55,45,40,15,17,5, and EI = 266;28,36,19,31,27 t.r. Thus, CE = 2113;59,17,48,47,25 t.r [acc. 2112;59,17,48,45,51. However, this wrong value was used later to compute FD], and CF = 2114;59,17,48,47,25 t.r. If CF = 60, FD = 0;1,42,17,4,37[41], and ∠DCF = 0;1,37,31,19,36°. Levi then writes (Mancha, 1993, pp. 116 and 122):

> It follows that the maximum solar parallax is 0;1,37,31,19,36°, which is 0;0,0,31,
> 19,36° greater than what was supposed [i.e., 0;1,37°], and when we supposed
> 0;1,37,32,47° as the maximum solar parallax, the consequence was 0;0,0,34,49°
> smaller than what was supposed; it follows from it, by means of heuristic reasoning,
> that we must take from these 0;0,0,32,47° according to the ratio, in such a way that
> we obtain an amount which does not contradict itself. When we had done so, we
> found that it was necessary to suppose that the maximum solar parallax is 0;1,37,17°,
> because in this case the consequence agrees very well with the supposition.

That is, using equation (A₁) Levi computes

$$\frac{0;1,37,32,47 - x}{0;1,37,32,47 - 0;1,37} \simeq \frac{0;0,0,34,49}{0;0,0,34,49 + 0;0,0,31,19,36}$$

whence $x = 0;1,37,17,15,23$, in agreement with Levi's value in the text. Finally, Levi verifies that $x = 0;1,37,17°$ satisfies equations (4) and (4a): since CE/EI is now 7;57,17,45,12,53 [acc. 7;57,17,45,11,42], and EI = 266;22,47,36,21,49,44 t.r., CF = 2120;2,1,39,52,24,30 t.r. So, if CF = 60, FD = 0;1,41,53,6,36. Therefore, ∠DCF = 0;1,37,17,22° [acc. 0;1,37,17,35°], which coincides to the third sexagesimal place with x, and Levi is satisfied with this approximation[42].

3.4. Determination of the thickness of the fluid layer between planetary spheres (Chap. 131)

Levi's theory of cosmic distances[43] depends on the following assumptions:

---

[41] 0;1,42,17,4,37 is probably a copyist's error; read: 0;1,42,7,41,37. The first figure, computing accurately, leads to ∠DCF = 0;1,37,31,30,59°, whereas 0;1,42,17,4,37 leads to 0;1,37,40,28°.

[42] From these relative (59;25,5; see above, at the beginning of this section) and true solar distances (2120;2 t.r.) at the time of the eclipse of 1335, using his first solar model with an eccentricity of 2;14 (derived from observations of the solar apparent diameter made in 1334 with a combination of the Jacob Staff and the camera obscura; Mancha, 1992, pp. 289–97), Levi computes in chapter 91 the true maximum and minimum distances between the centers of the Earth and the Sun as 2220;28,59,3,37,39 and 2061;4,25,40,7,55 t.r., respectively (Mancha, 1993, pp. 117–8). Levi was aware that equation (4) is extremely sensitive to small changes in its parameters. So, in chapter 93 (Vat. Lat. 3098, f. 78ra–va) he computed that with a parallax of 0;55,4° (instead of about 0;54,58°) and a radius of the shadow of 0;41,54° (instead of 0;42°) the solar distance increases to 2437;4,41,14,47 t.r., and with 0;54,51° and 0;42,6° the result is 1883;23,58 t.r. If we use the standard procedure to compute the solar distance (Almagest, V.15) with Levi's parameters (i.e., 0;13,55° for the angular radii of the Sun at apogee and the Moon at syzygies, and 0;42,35° for the radius of the shadow), we obtain 2220;53,1 t.r.

[43] Cf. Goldstein, 1986.

(1) The observed motion of each planet results from the combined motions of the orbs that constitute its sphere. The lowest and concentric orb (the concavity of the sphere) is moved with the daily rotation ($360^{o/d}$); the greatest daily motion to the west of the highest orb is $360^{o/d}$ less the slowest daily motion to the east, for the Sun and the Moon, which do not retrograde, or $360^{o/d}$ plus the greatest daily motion to the west, i.e., the greatest daily retrograde motion, for the planets.

(2) Between the planetary spheres there is a fluid – called by Levi 'the body which does not preserve its shape' – whose role is to allow the motion of a sphere to take place around an immobile thing, as Aristotelian physics required[44], and also to avoid any effect of the motion of the highest orb of a planet on the motion of the lowest orb of the planet placed above it.

(3) The distance of the apogee of the convexity of the sphere of the Moon is 62;48,42, 36 terrestrial radii (*Astronomy*, Chaps. 90 and 92; Mancha, 1993)[45].

(4) The distance of the concavity of the sphere of the Sun, derived from the lunar eclipse of 3 October 1335, is 2052;4,47 t.r.[46]

---

[44] This was Levi's answer to one of Maimonides' objections to Ptolemaic cosmology in the *Guide of Perplexed* (II.24). Maimonides wrote, for instance, that "the point around which Mars revolves, I mean to say the center of its eccentric sphere, is outside the concavity of the sphere of Mercury and beneath the convexity of the sphere of Venus" (Pines, 1963, pp. 323–4), and that the same occurs with the Sun, Jupiter, and Saturn, whose eccentric spheres revolve around centers located above the sphere of the Moon. For Levi, Maimonides' objection does not imply that all the orbs of a planet must be moved around the center of the world; it is sufficient that the motion of each orb of a planet takes place around the convexity of the orb situated immediately below it, and the motion of the lowest orb of that planet around an immobile thing: "Et ex quo est sic, est notum quod illud quod sequitur ex uerbis istis [i.e., from Aristotle's doctrine] est quod omne corpus celeste mouetur circa conuexum alterius corporis infra ipsum, quia impossibile est quod moueatur circa se ipsum; ideo ex hoc sequitur quod spatium super quod transit suo motu sit firmum, ita quod non moueatur hoc motu. Et si ista sphera moueatur alio motu sequitur illud in illa quod sequebatur in prima, scilicet quousque terminetur in fine ad aliud corpus quod circa centrum ultimi moti de sua natura quiescat. Et quando bene considerauimus ista uerba ex parte illa ex qua hoc conclusit Philosophus, est notum quod ex eis non sequitur quod quelibet sperarum moueatur circa centrum terre; ymo est necessarium quod quelibet sperarum moueatur circa conuexum illius corporis quod est infra ipsam, et ultimum moueatur circa aliud de sua natura circa centrum suum quiescens... [quod] est corpus non seruans suam figuram medians inter speras unius planete et speras alterius" (*Astronomy*, Chap. 43; Vat. Lat. 3098, f. 34vb).

[45] According to chapters 90 and 92 (Mancha, 1993, pp. 107, 112, and 114–9), the true maximum distance between the centers of the Earth and the Moon is 62;33,33,4 t.r. and the true radius of the Moon 0;15,9,32 t.r.; consequently, the distance between the center of the Earth and the apogee of the convexity of the sphere of the Moon is 62;33,33,4 + 0;15,9,32 = 62;48,42,36 t.r. From 62;33,33,4 and the ratio between relative maximum and minimum lunar distances, 65;16 and 62;44, we can obtain the true minimum lunar distance, 60;7,51,23 t.r.

[46] According to Chap. 92, the true solar radius is 8;59,39,16,45 t.r. Thus, the true distance of the apogee of the convexity of the solar sphere is 2229;29 t.r. ($\approx$ 2220;28,59 + 8;59), and the true distance of the apogee of the concavity of the solar sphere is 2052;4,47 t.r. ($\approx$ 2061;4,25 – 8;59,39).

(5) The relative maximum and minimum distances derived from his planetary models are in the same ratio as the true maximum and minimum distances[47].

In Chap. 131 Levi computed the thickness of the planetary spheres according to the Ptolemaic order of the planets, i.e., Mercury and Venus between the Moon and the Sun (Chaps. 134 and 135 contain the same calculations considering, respectively, Jābir ibn Aflaḥ's ordering with the Sun below both Mercury and Venus, and al-Biṭrūjī's ordering with the Sun between Mercury and Venus). The problem which Levi solved by means of 'heuristic reasoning' can be enunciated in the following way. In Fig. 5[48], the part of the fluid layer contiguous to the convexity SS' of the sphere of the planet P_i is moved at the same rate as that surface; as the radial distance from it increases, the motion of the fluid diminishes uniformly until it reaches zero at VV', a motionless part assumed to be imperceptibly small[49]; the same occurs with the fluid moved by the concavity WW' of the sphere of the planet immediately above P_i, namely P_{i+1}. As the distance SV depends on the strength of the impulse that the fluid receives from the motion of the convexity SS', and the distance VW on the strength of the impulse received from the concavity WW', and the motions of SS' and WW' differ, the distances SV and VW also differ. However, SV and VW can be expressed by the same amount, x, in units of which TS and TW, respectively, are 60. We assume that x is the relative thickness of the fluid when the motion of the surface of the sphere is $360^{o/d}$. As SS' is moved with the greatest daily motion of the planet P_i to the west, that is, v, we have

$$\frac{v}{360^{o/d}} = \frac{sv}{x}$$

where sv is the thickness of SV if $TS = 60$; therefore, in the same units, $TV = 60 + sv$. As, in units of which $TW = 60$, $TV = 60 - x$, we can find the distance tw (i.e., TW in units of which $TS = 60$) from the ratio

$$\frac{60 + sv}{60 - x} = \frac{tw}{60}$$

---

[47] In computing cosmic distances Levi uses, as Ptolemy did, the ratios between relative maximum and minimum distances from his models. For Ptolemy, these ratios are identical to the ratios between maximum and minimum distances of the planets, and also to the ratios between the distances of the convexity and the concavity of their spheres; but for Levi, the (true or relative) minimum distance of a planet does not coincide with the (true or relative) distance of the concavity of its sphere – and Levi's motivation for doing so is to avoid the unobserved variations of planetary apparent diameters which follow from Ptolemy's models. Thus, for instance, in Table 1 of Goldstein, 1986, p. 275, the ratios 104;45/15;15 and 69;18,52/50;41,8 for Venus and Saturn are, respectively, the ratios between the relative distances of the convexity and the concavity of their spheres; the ratios between the relative maximum and minimum distances of these planets are 104;45/101;39 and 69;18,52/63;44,22, respectively (cf. Chapt. 103 and 110; Vat. Lat. 3098, ff. 96ra-97va, and 103rb-107ra).

[48] For ease of reference, this figure reproduces with slight variants Fig. 2 in Goldstein, 1986, p. 281.

[49] "... it is sufficient for this thickness to be one span [= 0.25 m] or less... for it has no function except to prevent the motions from disturbing each other, and a small amount is sufficient for this" 49; (Chap. 130:41; Goldstein, 1986, p. 288).

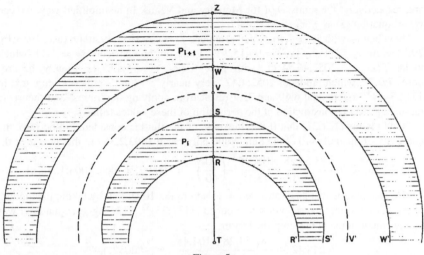

Figure 5

If we know the distance TS in terrestrial radii, we can obtain from tw the true distance TW of the concavity of the sphere of P_{i+1} from

$$\frac{TS}{60} = \frac{TW}{tw}$$

As we know the ratio $d/D(= tz/tw)$ from the model, once known TW, and according to assumption (5), we can also find the true distance of the apogee of the convexity of the sphere of P_{i+1} from

$$\frac{d}{D} = \frac{TW}{TZ}$$

If we know TZ for P_{i+1}, we can use the same procedure to find the true distances of the concavity and the convexity of the sphere of P_{i+2}, and so on. For the planets, v^o/d and the ratio d/D are known, and the true distances of the convexity and the concavity, respectively, of the spheres of the Moon and the Sun are also known from assumptions (3) and (4). Thus, as sw and tw depend on x, Levi's problem is to find a value for x which, starting from the distance of the convexity of the sphere of the Moon, yields the distance assumed for the concavity of the sphere of the Sun, namely 2052;4,47 t.r., or, in other words, to solve the cubic equation

$$2052;4,47 =$$

$$\frac{62;48,42,36 \left(60 + \frac{347;52,35\,x}{360}\right) \cdot 83;54,28 \left(60 + \frac{360;45,16\,x}{360}\right) \cdot 104;45 \left(60 + \frac{360;44,27\,x}{360}\right)}{36;5,32 \cdot 15;15 \cdot (60 - x)^3}$$

$$(5)$$

where 347;52,35, 360;45,16, and 360;47,27 are the greatest daily velocities (v) to the west of the Moon, Mercury and Venus, respectively, and 83;54,28/36;5,32, and 104;45/15;15

are, respectively, the ratios D/d for Mercury and Venus. In a simplified way, we can write equation (5) as $f(x) = 2052;4,47$.

Levi begins with $x_1 = 6;40$, which produces $f(x_1) = 1952;49,40$ t.r. (acc. 1953;15, 29); therefore, $f(x) - f(x_1) = 99;15,7$ t.r. Later on, he considers $x_2 = 7;11$, which yields $f(x_2) = 2057;46,54$ t.r., with $f(x_2) - f(x) = 5;42,7$ t.r. Finally, he proposes 7;9,20, which produces 2052;4,47 "very nearly"[50]. Levi's rather obscure text in 131:23–24[51] asserts that 6;40 and 7;11 result from applying the 'ratios' and 'heuristic reasoning'. My reconstruction of the derivation of these values follows.

Levi claims (131:21) that if the thickness of the fluid is considered to be zero, the concavity of the daily orb of the Sun would only reach 1003;2,36 t.r.; that is, if $x = 0$,

$$62;48,42,36 \cdot \frac{83;54,28}{36;5,32} \cdot \frac{104;45}{15;15} \cdot \left(\frac{60}{60}\right)^3 = 1003;2,36 \text{ [acc. } 1003;1,50,...]$$

He also asserts that "the ratios led us to add to the 1003;2,36 by taking the amount of the fluid as about the ratio of 4 to 5 cubed" (131:23)[52]. This suggests that Levi was taking equation (5) in the simplified form

$$62;48,42,36 \cdot \frac{83;54,28}{36;5,32} \cdot \frac{104;45}{15;15} \cdot y^3 \approx 2052;4,47$$

where

$$y^3 = \frac{60 + \frac{347;52,35\,x}{360}}{60 - x} + \frac{60 + \frac{360;45,16\,x}{360}}{60 - x} + \frac{60 + \frac{360;44,27\,x}{360}}{60 - x}$$

Then, using equation (B), Levi computes

$$\frac{1^3}{y^3} \simeq \frac{1003;2,36}{2052;4,47}$$

whence $y^3 = 2.0458... \approx 2$. Consequently, $y = \sqrt[3]{2} \approx 1.25 = 5/4 = 75/60$. Since

$$\frac{75}{60} = \frac{60 + x}{60 - x}$$

we have $x = 6;40$, the value of the text.

As for 7;11, Levi's reasoning was probably as follows (using again 5/4 as approximation to the cubic root of 2): if an increment of 1 in the value of the cubed ratio

---

[50] Chapter 131 :3–20 (for the value 6;40) and 25–41 (for 7;11); Goldstein, 1986, pp. 289–92.

[51] "[23] It is clear that the ratios led us to add to the 1003;2,36 by taking the amount of the fluid as about the ratio of 4 to 5 cubed, because for each one of them the ratio of the distance of the convexity of the sphere of Mercury from the center of the Earth to the distance of the concavity of the daily sphere above it, is equal to the ratio of 60 to 75, and so it is appropriate that we take the ratio cubed in the product that increases the 1003;2,36 over the amount produced by this ratio, the 99;15. [24] When we examined this in the way, we found by heuristic reasoning that it is appropriate to consider the amount of the fluid adjacent to it that the concavity of the daily sphere moves with its motion to be 7;11 in the measure where the radius of the concavity of that sphere is 60" (Goldstein, 1986, pp. 290–1).

[52] The ratio 4:5 stands, in a rough way, for the ratio between the distance of the convexity of the sphere of a planet and the distance of the concavity of the daily orb of the planet situated above it.

produces a distance of 1952 t.r. –i.e., if $(1/1)^3 = 1$ produces a distance of 1003;2,36 t.r., and $(5/4)^3 [\approx 2]$ produces 1952;49,40 t.r.–, what is the value for the cubed ratio which will produce 2052;4,47 t.r.? Using again a variant of the simple conditional reasoning,

$$\frac{(5/4)^3 - 1^3}{y^3 - 1} \simeq \frac{2 - 1}{y^3 - 1} \simeq \frac{1952}{2052}$$

whence $y^3 - 1 = 1.05122... \approx 1.06$, and $y^3 = 2.06$. Consequently, $y = \sqrt[3]{2.06} = 1.272 = 76.32/60$. Now,

$$\frac{76.32}{60} = \frac{60 + x}{60 - x}$$

and $x = 7;10,59 \approx 7;11$, the value of the text[53].

As for Levi's final value, 7;9,20, the relevant passage for reconstructing its derivation from 6;40 and 7;11 occurs in 131:42:

"Since the addition of 31 minutes to the amount of the fluid that moves according to the daily motion added 104;18 [read 104;58] to the distance of the concavity of the daily sphere of the Sun, it is clear according to the ratio, using heuristic reasoning, that a diminution of 0;1,40 from the 0;31 will diminish the distance of the concavity of the daily sphere of the Sun by these 5;47 [read 5;42,7], i.e. there remains 7;9,20 for the amount of the fluid that moves according to the motion of the concavity of the daily sphere."[54]

where $0;31 = 7;11 - 6;40$, the 104;58 are a rounding of $104;57,14 = 2057;46,54 - 1952;49,40$, and $5;42,7 = 2057;46,54 - 2052;4,47$. Levi asserts that $z(= 7;11 - x)$ can be obtained from

$$\frac{z}{0;31} \simeq \frac{5;42,7}{104;58}$$

whence $z = 0;1,40$. It is obvious that this equation derives from Levi's rule for positive and negative differences, because, if we apply equation (A1), we have

$$\frac{7;11 - x}{7;11 - 6;40} \simeq \frac{2057;46,54 - 2052;4,47}{(2057;46,54 - 2052;4,47) + (2042;4,47 - 1952;49,40)}$$

$$\approx \frac{5;42,7}{104;58}$$

that, accurately computed, leads to $x = 7;9,18,57,... \approx 7;9,19$, which yields $d_{min} = 2052;2,49$ (using 104;18, the textual reading, $x = 7;9,18,18,...$). The value of the text, 7;9,20, produces 2052;4,27, and it is not possible to decide whether it comes from a computational error or from a rounding of a further approximation using $x_2 = 7;11$ and $x_3 = 7;9,19$ (which yields $x = 7;9,19,34,...$, and $d_{min} = 2052;4,49$).

---

[53] Apparently, Levi did not realize that with 6;40 and 1952;49,40, and using equation (B), we have $(75/60)^3/y^3 \simeq 1952;49,40/2052;4,47$, whence, computing accurately, $x = 7;9,21$, very close to his final value 7;9,20.

[54] Goldstein, 1986, p. 292. My emendations are enclosed in square brackets.

40

3.5. Derivation of parameters for lunar model IV (Chap. 101)

Chapter 101 of the *Astronomy* is devoted to the preliminary steps for the construction of a new lunar model[55], which, as those described in the equivalent chapters for the planets preserved in the Latin version (103, 106, and 110, for Venus, Mercury and Saturn, respectively), must be able to produce Ptolemy's corrections for $\bar{\alpha} = 90°$ at syzygies and quadratures. According to Levi, these corrections are, respectively, 5;0,2°

---

[55] In Chap. 71 of his *Astronomy* Levi described three lunar models, of which the first and the third ones have been analyzed by Goldstein (1974b; 1974a, pp. 53–74, respectively). Lunar model III was elaborated *ca.* 1335, and Levi's tables for lunar corrections (Goldstein, 1974a; Mancha, 1997) were computed according to it. In chapter 100 of the Hebrew and Latin versions of the *Astronomy* Levi asserts that the observations made on 3 March 1337, 26 January and 11 December 1339 (solar and lunar eclipses, and a lunar observation, respectively) suggested that it was perhaps necessary to accept (a) some correction for the Moon at 0° of anomaly, and (b) that at 180° of anomaly the Moon is nearer to the Earth than at 0° of anomaly (on the Hebrew text of chapter 100, *cf.* Goldstein, 1979, pp. 118, 119, and 123). This variation in distance cannot be produced from lunar model III. At the end of chapter 100 the Latin text reads as follows: "Et ex experientia quinte eclipsis lunaris supradicte [= 26 Jan 1339] uidetur quod tempus eius a medio usque ad finem fuit maius in aliquo quam sequeretur ex ista doctrina; *quod si est uerum necesse esset ponere propinquiorem lunam nobis in* 180° *motus diuersitatis quam in eius principio.* Et istud nos mouet quod circa huius inuestigationem intendamus proposse, quod et nunc facimus. Et ad dirigendum in uiam ueram dispositionem lune inueniendi faciliter cum omnia proposita inquirenda de luna inquisiuerimus sciuerimusque perfecte, talem dispositionem lune proponimus ordinare cuius equationes consentant radicibus Ptolomei; quo facto facile erit deuenire in dispositione ista fienda in quantitates latitudinum diametrorum quas per experientias inueniemus... et hoc statim fiet in sequenti capitulo" (Vat. Lat. 3098, 90v). Although in both versions of Chap. 100 the problem seems still undecided, other chapters provide enough evidence to think that Levi finally decided to modify his third lunar model. Thus, at the beginning of Chap. 99 we read: "Et quia ad requisitionem multorum christianorum nobilium laborauimus ad ordinandum computum coniunctionum et oppositionum mediarumque uerarum facili modo ualde, in tantum quod etiam infans paruum possit practicari faciliter, nobis apparuit quod hic esset iste computus conscribendus, quia in eo est profactus mirabilis ad inueniendum faciliter tempus coniunctionis et oppositionis cuiuslibet tam in preterito quam in presenti quam etiam in futuro. Et quia postquam fuit iste computus ordinatus fuit nobis ex aliquibus eclipsibus solaribus et lunaribus reuelatum quod est necessarium addere equationi solis finalis circa 0;9° et quod est necessarium *ponere latitudinem aliquam in diametris sperarum motus diuersitatis in principio dicti motus, ut recitabimus in sequenti capitulo,* si Deo placuerit, computum coniunctionum et oppositionum secundum hanc equationem solis sic correctam ordinabimus et latitudinis antedicte" (Vat. Lat. 3098, 81va). At the end of chapter 76, after having stated that observations confirm that there is no correction due to 'inclination of diameters' at 0° of anomaly, a marginal note adds: "Nobis finaliter fuit notum ex eclipsibus solaribus et lunaribus quod necesse est *ponere aliam latitudinem diametrorum in principio motus diuersitatis*, que potest poni per modum spere [*sic*] latitudinis, ut dictum est supra" (Vat. Lat. 3098, 64v). Levi devoted Chap. 101 and 102 (this last incomplete in both versions) to preliminary work on this new model.

and 7;37,2°, although, in fact, these values derive from al-Bāttānī's tables[56]. Figures 6 and 7 reproduce Levi's figures from the manuscripts[57] showing angles and positions of the centers at syzygy and quadrature, respectively. A is the apogee, C is the *sephel*. The lower orb of anomaly carries point G from east to west through 90° around point D, the center of the Earth, and the higher orb moves through the same angle in the opposite direction around point E carrying point H, the Moon. Thus, the combined motion of these two spheres must produce values of 5;0,2° and 7;37,2° for ∠IDH at syzygies and quadratures, respectively.

Levi asserts: "...we have investigated using heuristic reasoning and we have found that it is necessary to assume that the amount of line ED is 5;43,36 when the Moon is at syzygy ... and 9;9,12 at quadrature...". Consequently, BD = 1/2(9;9,12 − 5;43,36) = 1;42,48, and BE = ED + BD = 7;26,24. Levi does not indicate how these values were obtained, but he demonstrates that they produce the aforementioned corrections. The proof is as follows[58]:

In Fig. 6, ∠ADH = 5;0,2°, and ∠ADG = ∠GEH = 90°. If EG = 60, and ED = 5;43,36, then ∠EGD = arcSin ED = 5;28,37°. We have that ∠EHI = 90° − ∠HEI = 84;31,23°. Since EH = 60, HI = 5;43,36, and EI = Sin ∠HEI = 59;43,25

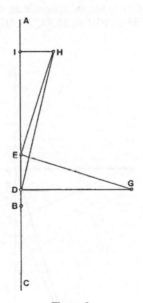

Figure 6

---

[56] Ptolemy's corrections for 90° of anomaly at syzygy and quadrature are, respectively, 4;59° and 7;34° (Toomer, 1984, p. 238). In al-Bāttānī's tables they are 5;0,2° and 7;37,2° [= 5;0,2° + 2;37°]; cf. Nallino, 1899–1907, ii, p. 80.

[57] Vat. Lat. 3098, f. 91r.

[58] Vat. Lat. 3098, ff. 99vb-91ra.

[acc. 59;43,33]. Thus, ID = ED + EI = 65;27,1, and HD = 65;42,1 [Euclid, I.47]. If HD = 60, HI = 5;13,47. Therefore, ∠ADH = arcSin 5;13,47 = 5;0,2° [acc. 5;0,1°. Note that 5;0,2° requires HI = 5;13,48].

In Fig. 7, ∠CDH = 7;37,2°, and ∠EDG = ∠GEH = 90°. If FG = 60, and ED = 9;9,12, then ∠EGD = 8;46,32° [acc. 8;46,30°]. Now ∠EHI = 90° − ∠HEI = 81;13,28°. Since EH = 60, HI = 9;9,12, and EI = 59;17,46 [acc. 59;17,51]. Thus, ID = 68;26,58, and HD = 69;3,31. If HD = 60, HI = 7;57,10. Therefore, ∠CDH = arcSin7;57,10 = 7;37,2° [acc. 7;37,0°. Note that 7;37,2° requires HI = 7;57,12].

In this case, Levi's problem is to solve the equation

$$\sin u = \frac{x}{\sqrt{\left[(60 \cdot \sin\left\{90 - \arcsin\frac{x}{60}\right\} + x)^2 + x^2\right]}} \tag{6}$$

where x = ED at syzygy (when u = 5;0,2°), or x = ED at quadrature (and then u = 7;37,2°).

It is easy to reconstruct Levi's procedure. At syzygies (Fig. 6), since Sin 5;0,2° = 5;13,48 (with R = DH), and ∠HDI < ∠HEI, we can start with an arbitrary value for x_1 greater than 5;13,48. So, let x_1 = 6;35,29. Thus ∠HEI = 6;18,25°, ∠EHI = 83;41,35°, HI = 6;35,29, EI = Sin ∠EHI = 59;38,13, and DH = 66;33,20. If DH = 60, then HI = 5;56,32, and ∠ADH = arcSin HI = 5;41,1°. Thus, the difference is 5;0,2° − 5;41,1° = −0;40,59°. In a second approximation, for instance, let x_2 = 5;30. Consequently, ∠HEI = 5;15,34°, ∠EHI = 84;44,26°, HI = 5;30, EI = 59;44,51, and

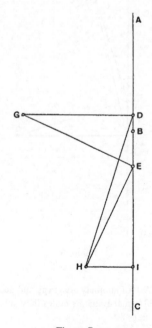

Figure 7

DH = 65;28,44. If DH = 60, HI = 5;2,23, ∠ADH = 4;49,6°. The difference is now 5;0,2° − 4;49,6° = 0;10,56°. Thus with equation (A$_1$)

$$\frac{6;35,29 - x}{6;35,29 - 5;30} \simeq \frac{0;40,59}{0;40,59 + 0;10,56}$$

and x_3 = 5;43,47. Now, if ED = 5;43,47, ∠HEI = 5;28,47°, ∠EHI = 84;31,13°, EI = 59;43,33, HI = 5;43,49, and DH = 65;42,21. If DH = 60, HI = 5;13,56, ∠ADH = 5;0,10°. Therefore, 5;0,2° − 5;0,10° = −0;0,8°. Using again equation (A$_1$) with x_2 = 5;30, and x_3 = 5;43,47 we have

$$\frac{5;43,47 - x}{5;43,47 - 5;30} \simeq \frac{0;0,8}{0;10,56 + 0;0,8}$$

and x_4 = 5;43,37. If ED = 5;43,37, ∠HEI = 5;28,38°, ∠EHI = 84;31,22°, HI = 5;43,37, EI = 59;43,34, and DH = 65;42,11. If DH = 60, HI = 5;13,48, ∠ADH = 5;0,2°. No further approximation is required[59].

To find the amount of line ED at quadratures (Fig. 7), as Sin 7;37,2° = 7;57,12, and ∠HDI < ∠HEI, we choose for x_1 a value greater than 7;57,12, for instance, x_1 = 8;52,10. Thus, ∠HEI = 8;30,3°, ∠EHI = 81;29,57°, EI = 59;20,27, and DH = 68;47,4. If DH = 60, HI = 7;44,12, and ∠CDH = 7;24,31°. The difference is 0;12,31°. If x_2 = 8;59, ∠HEI = 8;36,39°, ∠EHI = 81;23,21°, EI = 59;19,25, and DH = 68;53,42. If DH = 60, HI = 7;49,25, and ∠CDH = 7;29,32°. The difference is now 0;7,30°. With equation (A$_2$),

$$\frac{8;59 - x}{8;59 - 8;52,10} \simeq \frac{0;7,30}{0;5,1}$$

whence x_3 = 9;9,13. If ED = 9;9,13, ∠HEI = 8;46,31°, ∠EHI = 81;13,29°, EI = 59;17,52, and DH = 69;3,39. If DH = 60, HI = 7;57,10, and ∠CDH = 7;37°. Difference: 0;0,2°. Using again equation (A$_2$) with x_2 and x_3,

$$\frac{9;9,13}{9;9,13 - 8;59} \simeq \frac{0;0,2}{0;7,28}$$

whence x_4 = 9;9,16. If ED = 9;9,16, ∠HEI = 8;46,34°, ∠EHI = 81;13,26°, EI = 59;17,51, and DH = 69;3,41. If DH = 60, HI = 7;57,12, and ∠CDH = 7;37,2°. No further approximation is required[60].

---

[59] If we use, as in Chap. 131, equations (A) and (B), we can start, for instance, with x_1 = 5;13,48, which produces 4;36°, and a difference of 0;24,2°. With equation (B), 5;13,48/x ≃ 4;36/5;0,2, whence x_2 = 5;41,7, which leads to 4;58,1°, and a difference of 0;2,1°. With equation (A2), (5;13,48 − x)/(5;13,48 − 5;41,7) ≃ 0;24,2/0;22,1, and x_3 = 5;43,37.

[60] With equations (A) and (B), if we start, for instance, with x_1 = 7;57,12, we obtain 6;43,39° and the difference is 0;53,24°. With equation (B), 7;57,12/x ≃ 6;43,39/7;37,2, whence x_2 = 9;0,19, which produces 7;30,30°, and a difference of 0;6,32°. Equation (A$_2$) leads to x_3 = 9;9,7, which produces 7;36,55°, and one more approximation gives x_4 = 9;9,16.

3.6. Determination of the solar apogee (Chap. 57)

In Levi's *Astronomy*, the 'heuristic reasoning' closest to Ptolemy's procedure in *Almagest*, X.7, occurs in Chap. 57, devoted to the determination of the solar apogee. Levi measured the times of two identical solar altitudes, h_1 and h_2 (not given), symmetric with respect to the summer solstice, occurred on April 24 and August 4 of 1334, the latter however 5;42^h before the former. The Sun moved from h_1 to Cnc 0° in 50 days, 20;18^h, from Cnc 0° to h_2 in 50 days, 22;8^h, and thus from h_1 to h_2 in 101 days, 18;26^h. Using his own value for the eccentricity, 2;14, and al-Battānī's value for the mean solar motion[61], Levi's problem is to find the direction of the apsidal line (λ_a) that reproduces these data.

Rounding al-Battānī's value for the mean motion to 0;59,8,21$^{o/d}$, Levi computes the arc between altitudes h_1 to h_2 as 100;18,28°, and he asserts that, since the Sun goes through the second arc more slowly than through the first, λ_a is greater than Cnc 0° by an amount proportional to the excess of the equation corresponding to the first arc over the equation corresponding to the second, which is equivalent to 0;4,31°, the mean solar motion in the excess of the second time interval over the first, namely 1;50^h. Levi concludes that, according to his own solar equation table, 0;4,31° correspond approximately to 3°, and the apogee must be placed "about the end of the third degree of the sign of Cancer"[62]. He adds that it can be proved by means of an *argumentum ingenij*[63].

In Fig. 8[64], let B, D, and C, be respectively h_1, h_2, and the beginning of Cancer. E is the Earth, I the center of the eccentric, and M the apogee. Levi draws lines BI, DI, BE, CE, and DE, prolonged until it cuts the eccentric at A, and then AB, AC, and BC. Lines AP, BH, and AG are drawn perpendicular to BE, to AC, and to the prolongation of CE, respectively. Levi's procedure starts with the known value for arc h_1h_2 (i.e., ∠BID) and a provisional value for the total correction ∠EBI + ∠EDI, from which he computes successively the amounts of AC, AB, ∠AIB, ∠AID, and ∠CEM. With ∠CEM, he computes again ∠MID, ∠MIB, and finally ∠BID.

Levi begins by asserting that if we assume that λ_a is Cnc 3°, it follows that the correction for arc h_1h_2 is 3;11,46°. We know that ∠BIC = 50;6,58° and ∠CID =

---

[61] The mean solar motion derived from al-Battānī's tables is 0;59,8,20,46,56,14$^{o/d}$ (Kennedy, 1956, p. 156), and Levi's own value 0;59,8,20,8,44,6,3,14$^{o/d}$ (Goldstein, 1974a, p. 106). Levi derived this value in chapter 65 (Vat. Lat. 3098, ff. 57va-58ra) from his computation of the mean solar position at the time of the ocultation of Spica by the Moon reported by Timocharis (*Almagest*, VII.3; Toomer, 1984, pp. 335–6: -293 Mar. 9/10) and the mean solar position at the summer solstice of 1334 (June 15, 14;53^h after mean noon) computed from his solar model I. The mean solar motion in this interval was computed as 1627 complete revolutions and 108;37,29°, and the time elapsed between these two dates as 594358;22,2,30^d (which, computing accurately, yield 0;59,8,20,8,44,6,3,12,29,...$^{o/d}$).

[62] In Levi's solar correction table (Goldstein, 1974a, p. 156) the entries for arguments of 2° and 3° are, respectively, 0;4,20° and 0;6,29°.

[63] Vat. Lat. 3098, f. 45va-b.

[64] I have combined in a single figure the two in Levi's text (Vat. Lat. 3098, f. 45v).

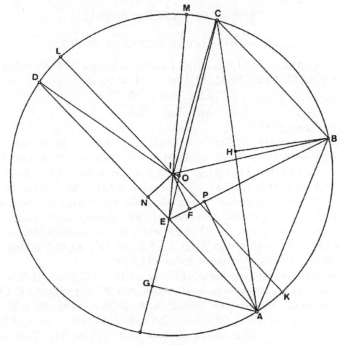

Figure 8

50;11,30°; therefore, ∠BID = 100;18,28°, and ∠BED = 97;6,42°[65]. Thus, ∠CEB = ∠CED = ∠AEG = 48;33,21°, and AG = Sin ∠AEG = 44;58,27 [acc. 44;58,33]. We also know that ∠DAC = 1/2∠DIC = 25;5,45°. Now, in triangle ACE,

$$\angle AEC = 180° \quad \angle CED = 131;26,39°,$$

and

$$\angle ACE = 180° - (\angle AEC + \angle CAE) = 23;27,36°.$$

Then, AG = Sin ∠ACG = 23;53,8 [acc. 23;53,11], and, taking AE = 60, as

$$\frac{AE}{AC} = \frac{\sin \angle ACG}{\sin \angle AEG}$$

therefore AC = 112;58,27. Now, ∠BAD = 1/2∠BID = 50;9,14°. Consequently, in triangle ABE,

---

[65] That is, 100;18,28° − 3;11,46° = 97;6,42°. If λ_a = Cnc 3°, the distance on the eccentric from point M to point D is 47;11,30°, and from point M to point B 306;53,2°, and, by interpolation in Levi's table, the corresponding equations are 1;31,39° and 1;40,14° (computing accurately, 1;31,32° and 1;40,5°). This yields 3;11,53° (or 3;11,37°) as the total correction, instead of the textual 3;11,46°.

$$\angle BEA = 180° - \angle BED = 82;53,18°,$$

$$\angle ABE = 180° - (\angle BAD + \angle BEA) = 46;57,28°,$$

and $AF = \text{Sin} \angle ABE = 43;51,2$ [acc. 43;51,4]. But in triangle AEF, $AF = \text{Sin} \angle AEF = \text{Sin} \angle AEB = 59;32,16$ [acc. 59;32,18]. Thus, if $AE = 60$, as

$$\frac{AE}{AB} = \frac{\text{Sin} \angle AEB}{\sin \angle ABE}$$

therefore $AB = 81;27,53$.

We know that $\angle BAH = \angle BAC = 1/2\angle BIC = 25;3,29°$. Consequently, $\angle ABH = 90° - 25;3,29° = 64;56,31°$. If $AB = 60$, $BH = \text{Sin} \angle BAH = 25;24,42$ [acc. 25;24,44], and $AH = \text{Sin} \angle ABH = 54;21,8$ [acc. 54;21,10]. But, in fact, $AB = 81;27,53$; thus, $BH = 34;30,9$, and $AH = 73;47,47$. We have also that $AH = AC - AH = 112;58,27 - 73;47,47 = 39;10,40$. Therefore, in triangle ABH, by Pythagoras' theorem, $BC = 52;12,25$ [acc. 52;12,17]. But BC is the chord of $\angle BIC$, namely $50;49,24$ [acc. 50;49,28]. So, when $AB = 81;27,53, BC = 52;12,25$, and if $BC = 50;49,24$, $AB = 79;18,20$. Whence $\angle AIB = 2 \cdot \text{arcSin} (79;18,20/2) = 82;44,26°$ [acc. 82;44,2°]. Thus, $\angle AID = \angle AIB + \angle BID = 100;18,28° + 82;44,26° = 183;2,54°$.

Now we draw line KIL parallel to AED. As $\angle AID = 183;2,54°$, $\angle DIL = \angle AIK = 1;31,27°$. We draw also lines IN perpendicular to AD, IO perpendicular to CE, and IP perpendicular to BE. We know that $IE = 2;14$. As $\angle IDN = \angle DIL$, $IN = \text{Sin} \angle IDN = 1;35,46$. So, if we take $IE = 60$, $IN = 42;52,39$, and $\angle IEN = \text{arcSin} 42;52,39 = 45;36,53°$. But $\angle IEN = \angle MED$, and $\angle CED = 48;33,21°$. Therefore, $\angle CEM = \angle CED - \angle MED = 48;33,21° - 45;36,53° = 2;56,28°$. Now we can compute $\angle MID = \angle IDE + \angle MED = 1;31,27° + 45;36,53° = 47;8,20°$, and $\angle MEB = \angle BED - \angle MED = 97;6,42° - 45;36,53° = 51;29,49°$. When $IE = 60$, $IP = \text{Sin} \angle IEP = \text{Sin} \angle MEB = 46;57,16$; but as, in fact, $IE = 2;14$, $IP = 1;44,52$. Thus, $\angle IBE = \text{arcSin} IP = 1;40,10°$ [acc. 1;40,9°]. Whence $\angle MIB = \angle IBE + \angle MEB = 1;40,9° + 51;29,49° = 53;9,59°$, and $\angle BID = \angle MIB + \angle MID = 53;9,59° + 47;8,20° = 100;18,19°$. Since the computed and observed angles differ only by $0;0,9°$, Levi considered the result satisfactory and assumed that $\angle CEM = 2;56,28° \approx 3°$, i.e., that $\lambda_a = \text{Cnc } 3°$.

Although Levi does not add further details on the procedure, it is evident that he understood it as an iteration process: if the resulting value for $\angle CEM$ does not produce the observed $\angle BID$ with the required accuracy, we can repeat the calculation starting from the total correction [= $\angle EBI + \angle EDI$] corresponding to the value for $\angle CEM$ previously found until the resulting $\angle BID$ coincides with the observed $\angle BID$. Levi probably believed that no significant improvement would follow from computing a second approximation starting with $\angle CEM$ $2;56,28°$, although indeed it produces better agreement with the observational data ($\angle CEM = 3;0,6°$, $\angle BID = 100;18,25°$, with a difference of $0;0,3°$ between the computed and observed $\angle BID$). A third approximation produces the same values again.

Acknowledgments.   I am grateful to B. R. Goldstein for his comments on a draft of this paper.

Appendix

The following texts are given according to my unpublished edition of the Latin version of Levi's *Astronomy*, based on the four preserved copies: MSS Vat. Lat. 3098 and 3380, Lyon, Bibl. Municipale, 326, and Milan, Ambros. D 327. I shall omit the critical apparatus, and locate the edited passages in the oldest copy, Vat. Lat. 3098.

Text A: chapter 49.

Et expedit sciri quod non potest demonstratiua inquisitio subito fieri ad ostendendum qualiter nostra dispositio debeat ordinari ut ex ea sequatur illud quod per experientiam nos uidemus in quolibet loco –ut possimus sic dicere: quia equatio est tanta in tanta quantitate motus diuersitatis circa augem, necesse est quod centra istorum motuum in tanta quantitate distent ab inuicem–, quia computationes in ista nostra dispositione sunt ualde dificiles et profunde, et hoc ideo quia aliqui motuum sunt non circa centra suarum sperarum. Et cum hoc opportet istam dispositionem taliter ordinari quod ex ea sequetur maior equatio in auge et in eius opposito tanta quantam uidemus et quod ex ea sequetur equatio latitudinis diametrorum circa 180° motus diuersitatis tanta quantam uidemus. Et ideo necesse est quod inquisitiones que ad inueniendum ueritatem nos dirigunt sint de genere argumentorum ingenij, que sunt ex diuisione accepta cum experientia que paulatim ad ueritatem accedit quousque perueniatur ad eam. Et ista met argumenta sunt de genere argumentorum condicionalium, et horum due sunt species. Una quarum est ex pluri et pauciori accepta. Secunda est accepta ex duabus experientijs de pluri diucrso uel ex duabus de pauciori diuerso. Verbi gratia de specie prima quia dicemus: si quando ponebamus quantitatem primam tantam sequebatur equatio maior quam uidemus in quantitate secunda tanta, et si quando ponebamus quantitatem tertiam tantam seque-batur equatio minor quam uideamus in quantitate quarta tanta, est notum ex proportione quod necesse est poni mediam unam inter primam et tertiam cuius differentia a prima se habeat ad differentiam inter primam et tertium sicut se habet secunda ad resultans ex secunda et quarta simul. Verbi gratia de secunda specic quia dicemus: si quando ponebamus quantitatem primam tantam sequebatur equatio maior quam uideamus in quantitate secunda tanta, et si quando ponebamus quantitatem tertiam tantam sequebatur equatio maior quam uideamus in quantitate quarta tanta, que est minor secunda, est notum ex proportione quod necesse est poni quantitatem quintam taliter quod tertia sit media inter primam et ipsam et eius differentia a primase habeat ad differentiam inter primam et tertiam sicut se habet quantitas secunda ad differentiam inter secundam et quartam. Et eodem modo esset sciendum si quantitas secunda et quarta essent [note ex equationibus que sunt minores quam uideamus]. Argumentum conditionale esset simplex quia dicemus: si quando ponebamus quantitatem primam tantam et sequebatur equatio in quantitate secunda tanta non concordi quantitati tertie quam uidemus, est notum ex proportione quod necesse est poni quantitatem quartam que ad primam se habeat sicut se habet tertia ad secundam (Vat. Lat. 3098, f. 40rb-va).

Text B: chapter 49, marginal note.

Exemplum de pluri et pauciori: prima 10, ualent 30, quod est plus 6, secunda; tertia 6, ualent 15, quod est minus 9, [quarta]; quinta 8;24, ualent 24, quod uidemus. Exemplum de pluri diuerso: prima 10, ualent 36, quod est plus 12, secunda; tertia 8, ualent 30, quod est plus 6, quarta; quinta 6, ualent 24, quod uidemus. De pauciori diuerso: prima 6, ualent 24, quod est minus 12, [secunda]; tertia 8, ualent 30, quod est minus 6, quarta; quinta 10, ualent 36, quod uidemus. Argumenta superiora sunt conditionalia composita; sequens uero est simplex. Prima 10, ualent 30, que sunt quantitas secunda que non concordat cum 24 que uidemus, que sunt quantitas tertia; ergo 8, que sunt quantitas quarta et se habet ad 10 sicut 24 ad 30, ualent 24 (Vat. Lat. 3098, f. 40va, *mg.*).

Text C: chapter 4.

Possumus etiam habere faciliter scientiam sinus arcus quarte partis gradus unius hac arte secundum quam procedam. Nam scito sinu arcus 8;15° scitur sinus 4;7,30°, et sic continue descendendo scietur sinus arcus quarte partis unius gradus et 128^e partis gradus alterius. Et per uiam eandem sinus arcus quarte gradus unius dempta 64^a parte gradus eiusdem. Et secundum hunc modum inuenimus quod proportio sinus arcus quarte partis gradus unius et 128^e partis gardus alterius ad sinum arcus quarte partis gradus unius dempta 64^a parte gradus eiusdem est quasi proportio arcus primi ad arcum secundum, in tantum quod in quartis minutiarum uel fractionum proportionis diuersitas non apparet, licet modicum apparet in quintis. Et sic secundum hanc artem magistralem est conclusum quod quod proportio sinus arcus quarte partis gradus unius et 128^e partis gradus alterius ad sinum arcus quarte partis gradus unius [dempta 64^a parte gradus eiusdem] est quasi proportio arcus primi ad arcum secundum. Propter quod ex sinibus iunctis procedendo hoc modo, scilicet secundum proportionem arcus ad arcum, asserere possumus quod sinus arcus 0;15° est 0;15,42,28,32,7 (Vat. Lat. 3098, 4vb).

Text D: chapter 47.

Et possumus faciliter quasi ad punctum quantitatis loci augis istorum planetarum pertingere, et quantitatis equationis centri et loci eorum motus diuersitatis hoc modo. Accipiatur de quolibet istorum planetarum tres experientie et in qualibet experientia distet planeta a medio loco solis per 180°. Et has experientias Ptolomeus 'extremitates noctis' appellat, et in Almagesti est declarata uia per quam ad istas experientias deuenimus, et pertinet ad genus argumentorum ingenij (Vat. Lat. 3098, f. 39vb). In my opinion, although there are no variant readings for this passage, the Latin text must be corrected to read somewhat as: "...uia per quam ad illud ex istis experientijs deuenimus...". 'Extremitates noctis' is a rendering of ἀκρώννκτοί σχηματισμοί (*Almagest*, X.6; Toomer, 1984, p. 484: "literally 'configurations [at which the planet rises and sets] at the beginning and end of night").

References

Aaboe, A. 1964. *Episodes from the Early History of Mathematics*. Washington.
Boncompagni, B. 1857. (ed.) *Scritti di Leonardo Pisano*, vol. I. Roma.
Cardan, H. 1663. *Opera omnia*. Lyon, v. 4 (repr. Stuttgart, *1964*).
Chemla, K.-Pahaut, S. 1992. "Remarques sur les ouvrages mathématiques de Gersonide", in Freudenthal *1992*, pp. 149–91.
Freudenthal, G. 1992. (ed.) *Studies on Gersonides – A Fourteenth-Century Jewish Philosopher-Scientist*. Leiden.

Goldstein, B. R. 1974a. *The Astronomical Tables of Levi ben Gerson*. Hamden, CT.

Goldstein, B. R. 1974b. "Levi ben Gerson Preliminary Lunar Model", *Centaurus*, 18:275–88.

Goldstein, B. R. 1979. "Medieval Observations of Solar and Lunar Eclipses", *Archives Internationales d'Histoire des Sciences*, 29: 101–56.

Goldstein, B. R. 1985. *The Astronomy of Levi ben Gerson* (1288–1344). *A Critical Edition of Chapters 1–20 with Translation and Commentary*. Berlin, New York.

Goldstein, B. R. 1986. "Levi ben Gerson's Theory of Planetary Distances", *Centaurus*, 29: 272–313.

Goldstein, B. R. 1988. "A New Set of Fourteenth Century Planetary Observations", *Proceedings of the American Philosophical Society*, 132: 371–99.

Goldstein, B. R. 1992. "Levi ben Gerson's Contributions to Astronomy", in Freudenthal, 1992, pp. 3–19.

González Palencia, A. 1953. *Al-Fārābī. Catálogo de las Ciencias*. Madrid-Granada, 2ª ed.

Heath, T. A. 1921. *A History of Greek Mathematics*. 2 vols. Oxford (repr. New York, 1981).

Halma. 1813–6. *Composition mathématique de Claude Ptolémée, traduite...par M. Halma*; *et suivie de notes par M. Delambre*. Paris.

Hill, G. W. 1900. "Ptolemy's Problem". *The Astronomical Journal*, No. 485, Vol. XXI, November 14, pp. 33–5.

Kennedy, E. S. 1956. "A Survey of Islamic Astronomical Tables", *Transactions of the American Philosophical Society*, NS 46.

Kepler, J. 1609. *Astronomia Nova Αιτιολογητοζ seu Physica Coelestis*, Pragae.

Lay, J. 1991. *L'abrégé de l'Almageste, attribué à Averroés, dans la version hébraïque*. Ecole Pratique des Hautes Etudes, Section des Sciences Religieuses. Thèse de doctorat (nouveau régime) soutenue le 6 Mars.

L'Huillier, G. 1980. "Regiomontanus et le *Quadripartitum numerorum* de Jean de Murs", *Revue d'Histoire des Sciences*, 33: 193–214.

Libri, G. 1838. *Histoire des sciences mathématiques en Italie*, vol. I. Paris.

Mancha, J. L. 1990. "Ibn al-Haytham's Homocentric Epicycles in Latin Astronomical Texts of the XIVth and XVth Centuries", *Centaurus* 33: 70–89.

Mancha, J. L. 1992a. "Astronomical Use of Pinhole Images in William of Saint-Cloud's *Almanach Planetarum*", *Archive for History of Exact Sciences* 43: 275–98.

Mancha, J. L. 1992b. "The Latin translation of Levi ben Gerson's *Astronomy*", in Freudenthal 1992, pp. 21–46.

Mancha, J. L. 1993. "La determinación de la distancia del sol en la *Astronomía* de Levi ben Gerson", *Fragmentos de Filosofía*, 3: 97–127.

Mancha, J. L. 1997. "The Provençal Version of Levi ben Gerson's Tables for Eclipses", *Archives Internationales d'Histoire des Sciences*, in press.

Nallino, C. A. 1899–1907. *Al-Battānī sive Albatenii Opus astronomicum*. Milano.

Neugebauer, O. 1975. *A History of Ancient Mathematical Astronomy*. Berlin, New York.

Pedersen, O. 1974. *A Survey of the Almagest*. Odense.

Pines, S. 1963. Trans. of Maimonides, *The Guide of the Perplexed*. Chicago.

Ragep, F. J. 1993. *Nasīr al-Dīn al-Tūsī's Memoir on Astronomy*, Berlin, New York.

Rashed, R. 1975. Comments on G. Beaujouan, "Reflexions sur les rapports entre théorie et pratique au moyen âge", in *Boston Studies in the Philosophy of Science*, 26: 437–84.

Rashed, R. 1978. "L'extraction de la Racine nᵢᵉᵐᵉ et l'Invention des Fractions Décimales (XIᵉ-XIIᵉ Siècles)", *Archive for History of Exact Sciences*, 18: 191–243.

Sabra, A. I. 1978. "An Eleventh-Century Refutation of Ptolemy's Planetary Theory", *Studia Copernicana*, XVI, pp. 117–31. Warsaw.

Smeur, A. J. E. M. 1978. "The *Rule of False* applied to the quadratic equation, in three Sixteenth Century Arithmetics", *Archives Internationales d'Histoire des Sciences*, 28: 66–101.

V

50

Smith, D. E. 1925. *History of Mathematics*. Boston, vol. II (repr. New York, 1958).

Steinschneider, M. 1880. "Abraham ibn Esra". *Supplement zur Zeitschrift f'ur Mathemathik und Physik* XXV, pp. 59–128 (repr. in *Gesammelte Schriften*, Berlin, 1925, pp. 407–98).

Swerdlow, N. M. 1987. "Jābir ibn Aflaḥ's Interesting Method for Finding the Eccentricities and Direction of the Apsidal Line of a Superior Planet", in D. A. King and G. Saliba, eds., *From Deferent to Equant. A Volume of Studies in the History of Science in the Ancient and Medieval Near East in Honor of E. S. Kennedy*. Annals of the New York Academy of Sciences, vol. 500. New York, pp. 501–11.

Swerdlow, N. M. -Neugebauer, O. 1984. *Mathematical Astronomy in Copernicus's De Revolutionibus*. Berlin, New York.

Toomer, G. J. 1977. Review of Olaf Pedersen, *A Survey of the Almagest. Archives Internationals d'Histoire des Sciences*, 27: 137-50.

Toomer, G. J. 1984. *Ptolemy's Almagest*. London.

Tropfke, J. 1980. *Geschichte der Elementarmathematik. Bd. 1. Arithmetik und Algebra*. Berlin, New York, 4. Aufl.

VI

THE PROVENÇAL VERSION OF LEVI BEN GERSON'S TABLES FOR ECLIPSES

SUMMARY. – The object of this paper is to present a hitherto unknown Provençal version of the tables and canons for eclipses which Levi ben Gerson (1288-1344) composed *ca.* 1336 at the request of "many great and noble Christians". This Provençal version represents a redaction of the work earlier than the preserved Hebrew one and provides new light on Levi's contributions to astronomy. Sections 4 and 5 contain a transcription of the Provençal text, with translation and commentary, and a description of the tables. A critical edition of the enlarged version of the canons for eclipses as it appears in chapter 99 of the Latin version of the *Astronomy* is presented in the Appendix, in order to facilitate its comparison with the Provençal text and with the corresponding (still unpublished) Hebrew versions.

CONTENTS

1. Introduction
2. On the manuscripts
3. The Provençal Canons and Tables
4. Canons
   4.1. Transcription
   4.2. Translation
   4.3. Commentary
5. Description of the Tables
Appendix. Critical edition of chapter 99 of the Latin version of Levi's *Astronomy*

1. *Introduction*

The relationship of Levi ben Gerson with Christians constitutes an old and unresolved problem. It concerns the possible influence of Scholastic authors on some aspects of Levi's philosophy as well as the possibility that the papal court at Avignon was involved in supporting his astronomical research in the context of the calendar reform [1]. Related to this problem, I drew attention for the first time to

---

[1] Pines (1967, 31-37) was the first who pointed out the resemblance between Gerson's, Duns Scotus', and Thomas Aquinas' doctrine of divine attributes; for further references, see Freudenthal (1992), xiv. Clement VI (1342-1352) was interested in the problems of the calendar, and in September 1344 invited John of Murs and Firmin of Bellaval (both of whom,

the significance and singularities of the Latin version of Levi's *Astronomy* in my contribution to the Peyresq Colloquium on Levi (June 1988), pointing out that the main characteristic of this Latin text is that it contains additions to the original version that could not have been derived from the Hebrew text and whose source can only be Levi himself, a fact that I interpreted as proof of Levi's participation in the process of translation. I also proved that chapters 4 to 11 of this version are identical, except for minor stylistic variations, with those of the independent *editio* of these chapters, widely known as *De sinibus* [2] – dedicated in 1342 to Pope Clement VI and attributed in the manuscripts themselves to Petrus de Alexandria –, concluding then that this hermit friar was also the translator of the *Astronomy* and guessing that the Latin version might well be the result of a double translation: Levi would have translated the text into Provençal orally, occasionally making additions and revisions as he went along, and this oral version would have then been written down in Latin by Petrus [3]. It was previously known that the treatise *De numeris harmonicis* [4] – a mathematical work related to musical theory of which the Hebrew version, if it really existed, seems to have been lost – was written by Levi in 1343 at the request of Philippe de Vitry [5] and, although the preserved

as well as Levi, wrote for the Pope predictions based on the planetary conjunction of 1345) to Avignon to consider and advise on its correction (Deprez [1889]; North [1983], 84-86); the connection between the papal court and Levi's astronomical activity was suggested by Goldstein (1974a), 20.

[2] That the text of *De sinibus is* only a slightly different version of the corresponding passages of the Latin version of the *Astronomy*, and not an independent translation, was already stated in Mancha (1989), 30. According to the colophon in the copy of *De sinibus* preserved in Ms. Paris, B. N. 7293, 17^{rb-va} (this codex was in the papal library at Avignon in 1369; cf. Renan-Neubauer [1893], 621), the title of this *editio* was *Tractatus instrumenti astronomie* (Treatise on the astronomical instrument): *Explicit tractatus instrumenti astronomie magistri Leonis iudei de Balneolis habitatoris Aurayce ad summum pontificem dominum Clementem VI translatus de hebreo in latinum anno Incarnationis Christi 1342 et pontificatus dicti domini Clementis anno primo.* I assume that this was the title intended by Levi. The treatise was partially edited by Curtze (1898, 1901).

[3] The method is well attested for texts translated between the XIIth and the XVIth centuries; cf. d'Alverny (1989). The collaboration between Levi and Petrus de Alexandria was also pointed out in Mancha (1992a). An enlarged version of my contribution to the Peyresq Colloquium was published in Mancha (1992b), I also suggested the year 1342 for the beginning of the translation, but this hypothesis needs to be revised: the available evidence suggests that both versions were composed close in time, and that most of the Latin text was ready earlier than the latest parts of the Hebrew version. (On the chronology of Levi's astronomical work see Mancha, 1997.)

[4] For the Latin text, see Carlebach (1910), 128-139; see also Chemla, Pahaut (1992), 169-181, and 183-191.

[5] Philippe de Vitry (1291-1361) was a renowned musical scholar, author of the *Ars nova* and friend of Petrarch. His motet *Petre clemens* celebrates the election of Pierre Roger to the papacy as Clement VI, and it was Clement who would elevate Vitry to bishop of Meaux in 1351. Nicole Oresme (*ca.* 1320-1382) dedicated to him his *Algorismus proportionum.*

copies do not indicate the name of the translator, I suggested that it was also made by Petrus. Moreover, Petrus is identified in the colophon of the Latin manuscripts of Levi's *Prognostication for the Conjunction of 1345* as the translator of that work. In 1990, Goldstein and Pingree's edition of the Hebrew and Latin versions of the *Prognostication* showed differences between them that are very similar to those of both versions of the *Astronomy*, and a passage in the Latin text suggests that Levi's relation with the papal court (and perhaps the collaboration between Levi and Petrus) went back to 1339.[6] Levi also mentions a "noble Christian" and a "distinguished cleric who studied this science with us" in connection with his observations of eclipses dated 1335 and 1337, and it has been suggested that Levi's assistant was Petrus himself[7].

In the introduction to the Hebrew version of Levi's tables (*Luḥot*) for finding mean and true syzygies and for computing the circumstances of eclipses – which survive in manuscripts other than those that preserve his *Astronomy*[8] – Levi informs us that the set was composed "at the request of many great and noble Christians", an assertion repeated in the Hebrew and Latin versions of chapter 99

---

[6] The comet observed in 1339 was interpreted by Levi as sign of a conflict between southerners and northerners in a prediction adressed to pope Benedict XII (1334-1342) and Brother Peter predicted to Levi the same thing at Christmas of that year (Goldstein, Pingree [1990], 26 and 32).

[7] Goldstein (1979), p. 105, 80:24, and 100:8; Mancha (1992b), 37.

[8] These tables and canons are preserved in at least four Hebrew manuscripts (British Library, Add. 26921, and Or. 10725; Munich 314, and Vat. 391,3) The last manuscript contains only the canons (Goldstein, [1974a], 77); hereafter I will refer to the first three as *A*, *B*, and *M*, respectively. The Hebrew text of the introduction to the canons, corresponding approximately to sentences [1]-[40] of the Provençal version, was published in Steinschneider (1899), according to Ms. *M*, and translated into German in Carlebach (1910), 35-41. The Hebrew manuscripts containing Levi's *Astronomy* will be denoted as *P* (Paris, B.N. 724), *Q* (Paris, B.N. 725), and *N* (Naples, B.N. III F.9). Although they are usually called "abridged version of Levi's tables" (Goldstein [1974a], 12 and 77; Sirat [1991], 325), they constitute an independent work, composed separately, though obviously related to his main astronomical work. When the tables and canons for eclipses were later incorporated into the *Astronomy* as chapter 99, the canons were enlarged and, as a consequence of Levi's decision to change their presentation, the method of computation, or the parameters involved in them, some of the tables were replaced by others (for instance, for the motion of the mean ascending node, column V of tables numbered 13-19 by Goldstein, whose epoch is 1302, replaces tables 20-21, whose epoch is 1321; for lunar velocity, table 22 IV replaces 22 IVa, and, for corrections to the time of the mean syzygy and the true position of the sun for each day of 1321, tables 40-44 replace 40a-44a); another 12 tables for astrological houses, entirely new, were composed by Levi (or at least that was his intention) for the Latin version of the same chapter (cf. Appendix, 99: 156-163). The epoch of the tables (1321) has sometimes been wrongly taken as the date of their composition (see, for instance, Renan-Neubauer [1893], 615; Thorndike [1934], 309; Shatzmiller [1991], 39; Weil-Guény [1991], 357).

of the *Astronomy* [9]. It is not difficult to establish the date of composition of the set. Its tables for mean syzygies are computed using a value for the mean synodic month ($29;31,50,7,54,35,3,32^d$) derived by Levi from the time between the lunar eclipses of 20 October 134 (observed by Ptolemy) and 3 October 1335 (observed by Levi) [10], and this date is the *terminus post quem*. The tables in the set that depend on the solar equation of center are based on Levi's model with an eccentricity of 2;14, derived in chapter 56 of the *Astronomy* from observations of the variations of the solar apparent diameter made in May and December 1334 using a combination of the Jacob staff and the *camera obscura* [11]. In chapter 80, computations of three solar and five lunar eclipses between 26 June 1321 and 3 October 1335 using this solar model are compared with observations made by Levi himself, and he concluded that the model provided acceptable agreement with the phenomena [12]. As the Hebrew version of the introduction to Levi's tables for syzygies includes a worked example for the solar eclipse of 3 March 1337 computed with the same solar model [13], and Levi was persuaded by the observational data of this eclipse and the lunar one of 26 January 1339 to increase the solar eccentricity [14], we can conclude that tables and canons requested by noble Christians were composed between October 1335 and March 1337 (*terminus ante quem*). This allows us to assert that from some time before 1335 until Levi's death a Christian circle probably connected with the papal court at Avignon (or, in any event, with cardinal Pierre Roger, later Clement VI) was interested in his astronomical research and supported in some way the preparation of the Latin versions of his works. In spite of the absence of most of Levi's tables from the Latin version of his *Astronomy* [15], it was nevertheless likely that a version of his tables for eclipses existed in a language other than Hebrew, as Levi composed them at the request of this Christian circle. These tables and their corresponding canons are indeed

---

[9] Cf. Goldstein (1974a), 20; Renan-Neubauer (1893), 630. For the Latin text of chapter 99 see Appendix, 99:1.

[10] See section 4.2, ad [32]-[33].

[11] Mancha (1992a), 289-297.

[12] Goldstein (1979), especially 109-117.

[13] Goldstein (1979), 151-155.

[14] Goldstein (1979), p. 118, 100:5-6, and p. 122, 100:81-87.

[15] The Latin manuscripts only preserve 3 of the 44 tables in the Hebrew text (Goldstein [1974a], 80; according to the marginal note in the Hebrew manuscript where it is preserved, I assume that table 10 is not Levi's), but one of them is the table for the solar correction depending on Levi's solar model II, which is absent from the Hebrew manuscripts. This lack is difficult to explain (and likely due in part to Levi's sudden death, which interrupted the composition of the Latin version): the revision of his lunar theory that Levi was doing after 1339 could account for the absence of tables dependent on lunar equations, but not for that of the tables for the mean motions.

preserved in the manuscripts Scaligerani 46 and 46* of the Universiteitsbibliotheek in Leiden. They were redacted not in Latin but in Provençal, and their existence has never been mentioned in the bibliography on Levi.

2. On the manuscripts

The manuscript Scal. 46 has two parts. The first (ff. 1-53, 220 mm x 160 mm), written in France [16] by various hands in the XVth century, contains astronomical and astrological texts and tables in Latin (see a summary description in Molhuysen [1910], 13-14). The second (ff. 54-73, 205 mm x 150 mm) contains Levi's tables (ff. 54^v-67^r) and canons (ff. 68^r-73^v), written by a single hand, different from those which appear in the first part. At the top of fol. 67^v (that is, in the blank page between the tables and the canons) there is a partially legible sentence in Hebrew script and languague: QNYTY ZH HSPR M ... (vocalized: qaniti zeh ha-sepher mi ...: I bought [or: acquired] this book from ...). The so-called Ms. Scal. 46* (a parchment sheet, approx. 300 mm x 290 mm, originally attached to the other tables; see Plate 2, and the description in section 5.A) contains Levi's tables for mean syzygies (in their original circular form) by the same hand.

Three catalogues of the University Library in Leiden refer to Levi's work. The Catalogus vetus (1716) does not mention the author, describes the work simply as "tables", and incorrectly identifies the language as Spanish [17]. Molhuysen (1910, p. 14) attributes text and tables to Levi, repeats the wrong identification of the language, and briefly describes the circular table for mean syzygies [18]. The abbreviated catalogue published in 1932 mentions Levi's tables in describing Ms. Scal. 46, but omits any reference to Ms. Scal. 46*. [19] Attached to the inside cover of the codex is an autograph note by the Spanish scholar Marcelino Menéndez Pelayo (1856-1912), which mentions the author, describes the work as a calendar, and

---

[16] Byvanck (1931), 93. Arms and motto (Non sine causa) of the De Lupis family appear in ff. 1^r, 36^r, 41^r.

[17] Catalogus Bibliothecae Publicae Universitatis Lugduno-Batavae, Lugduni-Batavorum, 1716: "Almanach ordinatum super villam Montispesulani. Sequitur ad Antistitem Michaelem ex Jafar Indi astrologi Astrologia, incipiens ab jüdiciis astrorum. Aliud continens tabulas, & Hispanico sermone descriptionem".

[18] Molhuysen (1910), 14: "II. f. 54-73'. Hispanice, 165 x 110. Litt. init. rubrae. f. 54-73 Levi ben Gerson (Leonis de Bagnols) Tabulae astronomicae. Altera tabula astronomica, quae 4 circulis constat, maioris quam est ipse codex formae, seorsim conservatur notaturque Scal. 46*. Explicatio tabularum est f. 68-73; inc. Dis leon de bayhnols cet.".

[19] Bibliotheca Academiae Lugduno-Batavae. Catalogus. Deel XIV. Inventaris van de Handschriften. Eerste Afdeeling, Leiden, Universiteits-Bibliotheek, 1932, p. 53 (col. a): "Levi ben Gerson [Leonis de Bagnolo] Tabulae astronomicae".

274

identifies the language as Catalan "with some peculiarities similar to those of the Roussillon dialect" [20]. Zinner (1925, p. 110, Ms. 3187), who inspected the codex in 1924, described the work as *Finsternistafeln mit Erklärung* in the Portuguese language and attributed it to Levi, but he did not mention Ms. Scal. 46*.

There is no reason to date the second part of the codex also from the XVth century, as explicitly Menéndez Pelayo, Molhuysen, and Zinner did, and no paleographical evidence precludes to consider it from the XIVth century. The different sizes and conditions of preservation of the two parts also suggest a different provenance. Taking moreover into account the characteristics and the quality of the copy, it is a likely guess that the copyist worked in Levi's circle and probably during Levi's life.

3. The Provençal Canons and Tables

3.1. As is the case for the Latin versions of the *Prognostication* and the *Astronomy*, there are many significant differences between the Provençal and Hebrew versions of these tables and canons. The canons are a separately redacted version and not a translation of the preserved Hebrew text: the most significant difference between them (the absence from the Provençal version of the fifth section of the Hebrew one, devoted to the computation of eclipses, which, Levi tells us, he added to his first version of his work as a gift for scholars) [21] allows us to establish that the Provençal version represents a redaction of the work earlier than the preserved Hebrew one and, presumably, a response more closely related to the initial Christian request. As for the tables, their headings, variant readings for the entries, and two entire tables (for interpolation in correcting eclipses and for hourly lunar velocity), different in both sets, prove that the Provençal version does not depend on the extant manuscripts of the Hebrew version. The Leiden copy, moreover, preserves Levi's tables 36-39 for mean syzygies in their original circular form (see Plate 2), that has been transformed in the extant Hebrew manuscripts, thus providing an explanation for why Levi called them 'circles'.

---

[20] "Calendario *catalán*, de 6 hojas de introduccion que explican el manejo de las tablas, su utilidad etc. Comienza: *Dis Leon de Banhols* ... Las tablas son once, y además una *roda* como el autor dice. Quizá deba atribuirse este calendario á Leon de Banyols, y (si no me equivoco) la lengua del ms. ofrece particularidades semejantes á las del dialecto hablado en el Rosellón y tierras comarcanas. Siglo XV. [alia manu:] Scripsit Menendez Pelayo n. Nov. 1877".

[21] The fifth section is described in the Hebrew version as follows: "We have further composed a fifth chapter that we give as a gift for scholars in which we inform them in a simple way how to know the calculation of solar and lunar eclipses, and the true position of the moon for any day of the days of the moon by means of some tables that we composed for this purpose" (Ms. *A*, f. 13b; see also Carlebach [1910], 40).

3.2. By comparison of the surviving texts, it can be concluded that from the end of 1335 Levi produced at least four successive redactions of the canons for his syzygy and eclipse tables:

– R_1: the Provençal version, which includes an introduction and four sections devoted to (i) a description of the tables mentioned in the other sections (tables 36, 37, 38, 39, 40a, 41a, 42a, 43a, 44a, 4, and 7), [22] and instructions for finding (ii) the date and time of mean syzygies, (iii) the time of true syzygies from mean syzygies, and (iv) the true positions of the sun, the mid-heaven and the ascendent, at the time of the true syzygies or any other time;

– R_2: the Hebrew version preserved in manuscripts A, B, and M, which adds to R_1 (besides minor modifications in sections 1-4) a fifth section with instructions for computing lunar and solar eclipses and true positions of the moon for any day of the month. This section describes tables 20, 21, 35 (col. VI), 30, 31, 32, 33, 34, 22 (with col. IVa), 27, 23, and 45, and includes a worked example for the solar eclipse of 3 March 1337 (Goldstein [1974a], 77-79; [1979], 151-155);

– R_3: the Hebrew version preserved in manuscripts P and Q as chapter 99 of the *Astronomy*, where tables 13 to 19, 40, 41, 42, 43, and 22 (with col. IV) replace tables 20, 21, 40a, 41a, 42a, 43a, and 22 (IVa); in it, moreover, a method for domification (North [1986], xi) not mentioned in previous versions is described in section 4, and a computation of the solar eclipse of 26 June 1321 [23] replaces the worked example of version R_2;

– R_4: the preserved Latin version of chapter 99 of the *Astronomy*, which is substantially identical to R_3, except that, at the end of section 4, the description of 12 tables for finding the cusps of astrological houses (AT) for the latitude of Orange (that, according to Levi, he composed as an additional gift for the Christians [24]) is absent from redaction R_3.

As for the tables themselves, Table C-1 displays them in the same order they are described in the corresponding canons or they appear in the different versions of the set [T_1 = tables described in the Provençal canons; T_2 = Provençal tables; T_3 = Hebrew canons and tables in Ms. M; T_4 = canons and tables in the Hebrew version of chapter 99 in Mss. P and Q; T_5 = tables as described in chapter 99 of the Latin version]. This table illustrates the successive modifications of the set, and it suggests that the tables of the Provençal version (T_2) constitute an intermediate stage between redactions R_1 and R_2 of the canons: it contains not only the tables

---

[22] Unless otherwise indicated, I refer to Levi's tables by the number with which they appear in Goldstein (1974a).

[23] This eclipse was also computed by Levi in chapter 80 of the *Astronomy*, and recomputed in chapter 100 with his second solar model (Goldstein [1979], 123-125). The computation of it found in chapter 99 is made using the first solar model but is more detailed than that of chapter 80 (see Appendix, 99: 236-274).

[24] See Appendix, 99: 156-163.

mentioned in the Provençal canons (T_1) but also the tables 20, 21, 35, VI, 30, 31, 32, 33, 34_1, 27, 22_1, 23, and 45, used in section 5 of redaction R_2 for the computation of eclipses with two significant differences: instead of table 22.IVa the Provençal version includes a table for hourly lunar velocity identical to al-Battānī's, and table 34 is computed to seconds instead of to minutes as in T_3 and T_4.

TABLE C-1

T_1 36 37 38 39 40_1 41_1 42, 43, 44, 4 7 -

T_2 36 37 38 39 40_1 41, 42, 43, 44_1 4 7 -   20   21 35vi 30 31 32 33 34_1 27 22_1 23 45

T_3 36 37 38 39 40_1 41_1 42_1 43_1 44_1 4 7 -   20   21 35vi 30 31 32 33 34_2 27 22_2 23 45

T_4 36 37 38 39 40_2 41_2 42_2 43_2 44_2 4 7 -   13 to 19 35   30 31 32 33 34_2 27 22_3 23 45

T_5 36 37 38 39 40_2 41_2 42_2 43_2 44_2 4 7 AT 13 to 19 35   30 31 32 33 34_2 27 22_3 23 45

The content of the Provençal canons provides, in my opinion, an answer to the question about the content of the Christian request to Levi: without the tables and canons for eclipses, the work is a calendar which gives dates and times of mean and true syzygies, true positions of the sun, and other related astrological data, and consequently (as Levi tells us in the introduction to the canons) a way to know exactly which dates satisfy the astronomical requirements of Easter: the Sunday immediately after (and not on) the 14th day of the Paschal moon, reckoned from the day of the new moon inclusive, where the Paschal moon is the calendar moon whose 14th day falls on, or is the next following, the vernal equinox, taken as 21 March [25]. Moreover, this interpretation is confirmed by Levi himself in the text of the Provençal canons: sentence [10] (see below, section 4.2) asserts that "the second benefit that follows [from an exact calculation of true syzygies] concerns the moveable feasts of the Christians", which can only be established after determining Easter.

3.3. Levi's work on philosophical and exegetical matters seems to have been interrupted from December 1332 to December 1337, [26] and from the dated

---

[25] North (1983), 76.
[26] Touati (1973), 52.

observations mentioned in the *Astronomy* it can be easily inferred that this time was devoted to astronomical research [27]. Moreover, the period of Levi's greatest observational activity – as reflected in the text – begins in April 1333 and ends with the aforementioned lunar eclipse of October 1335 (37 observations out of a total of 79 mentioned in the text, 24 of them related to lunar and solar theory). From all this, a plausible inference is that the apparent increase of Levi's astronomical activity in the first months of 1333 followed closely the Christian request of the tables for syzygies.

As for the method of composition of the Provençal text, from the close parallelism in content and style between it and the preserved Hebrew version in the sections where they overlap, an obvious possibility is that it was translated aloud into Provençal from a lost Hebrew draft by Levi himself and then written down by a Christian scribe (that is, a procedure identical with that which I suggested for the composition of the Latin version of the *Astronomy*). But if we take into account that the Latin version contains a great number of passages, sometimes even chapters, redacted in a way entirely different from the Hebrew one (preserving, however, the same content), that it includes long passages absent from the Hebrew one [28], and we suppose that Levi was capable of redacting orally in Provençal (which was his mother language), at least on astronomical subjects, the hypothetical lost text becomes superfluous. Otherwise, we would be forced to assume that another Hebrew version of the *Astronomy*, different from the preserved one (and also lost), was used to compose the Latin version.

3.4. From a careful examination of the surviving versions of Levi's astronomical work some general conclusions follow. First, we need to revise the suitability of the concepts 'original text' and 'translation' in referring to them: Levi's *scriptorium* prepared, produced, and delivered successive (Provençal, Hebrew, Latin) redactions of the results of his research, and there is no reason to prefer any of them over the others *on linguistic grounds*; each redaction represents a different stage of Levi's research and constitutes a valuable source of information that has to be considered. Secondly, that the non Hebrew versions are part of a program undertaken in collaboration with Levi himself for making available the results of his astronomical research to the aforementioned Christian circle, and that the

---

[27] According to the colophon of the copy of Levi's *Astronomy* preserved in Ms. *P* (Goldstein [1974a], 74), the work was finished the 21 *Kislev* 5089 (= 24 November 1328). However, if a first version of the text was really ready at this date, it was very different from the extant one, of which all the chapters devoted or related to solar, lunar, and planetary theory (more than 64% of the text) were composed after 1335.

[28] For some examples, see Mancha (1992b), 41-43, to which we must add a list of 42 zodiacal stars including their description, magnitude, and ecliptical coordinates for 1336, given by Levi in the seventh section of the Latin version of chapter 62 (Vat. Lat. 3098, ff. 55^{rb-vb}).

278

relations of Levi with Christians were not occasional and already probably existed *at least* eight or nine years before the election of Clement VI in 1342. [29] Lastly, the discovery of the Provençal version of Levi's tables for eclipses seems also to confirm the use of the Provençal as the intermediate language between Levi and Christians, in general, and between Levi and his collaborator Peter of Alexandria, in particular.

4. *Canons*

As far as I know, Levi's Provençal text has not been mentioned in any work on medieval Occitan literature, perhaps on account of the incorrect identification of the language noted in section 2. [30] Only a small number of scientific texts in Provençal are preserved, and among them astronomy is a rather uncommon subject: excepting a few elementary chapters devoted to it in the *Elucidari de las propietaz de totas res naturals* – a translation of Bartholomeus Anglicus' *Opus de proprietatibus rerum* made for Gaston II (1315-1343), count of Foix –, only some short astrological or calendaric notes survive [31]. Thus, because of its philological significance, I have preferred to present a line-by-line transcription of the text and to respect the spelling and punctuation of the manuscript (of which f. 68^r is

---

[29] The obvious supposition that the request of the tables preceded their redaction by some period of time does not say anything about the date of the beginning of Levi's relationship with this Christian circle. Despite the prediction addressed in 1339 to Benedict XII, elected in 1334, it is undecidable if he was included among the "many great and noble Christians"; taking into account their personalities and interests (see Guillemain [1966], 134-140), a most likely candidate as central figure of the circle seems to be Pierre Roger. Despite Clement VI's request to John of Murs, there is no evidence to support the attribution of any kind of special competence (or knowledge) in astronomical matters (or texts) to him or to any other known person related to him (including Peter of Alexandria): otherwise it would be difficult to explain the persistence in the Latin version of Levi's *Astronomy* of some peculiarities of the Hebrew one (for instance, the assertion that al-Battānī's value for the maximum lunar latitude is 4;30°), obviously derived from Levi's Hebrew sources. No evidence has yet come to light in support of Goldstein's conjectures that Levi's brother, Solomon, (1) served as physician at the papal court, and (2) that it was through him that Christians at that court heard of Levi's achievements in astronomy. All that is known about Solomon is that he was a physician, and that he collaborated in the Latin traslation of Levi's *Prognostication* with Peter of Alexandria. For archival documents concerning Levi and Solomon, see Shatzmiller (1991).

[30] Levi's text is written in a Provençal without distinguishable dialectal features, and its non-astronomical vocabulary can be found in lexicons such as those of Anglade (1921a), and Brunel (1926).

[31] See, for instance, Brunel (1935), no. 63, 119, 154, 208-209, 248 and 265; Anglade (1921b), 248-251. Sesiano (1984) contains an analysis and partial edition of the only known mathematical manuscript in the Occitan language, an arithmetical treatise composed *ca.* 1430.

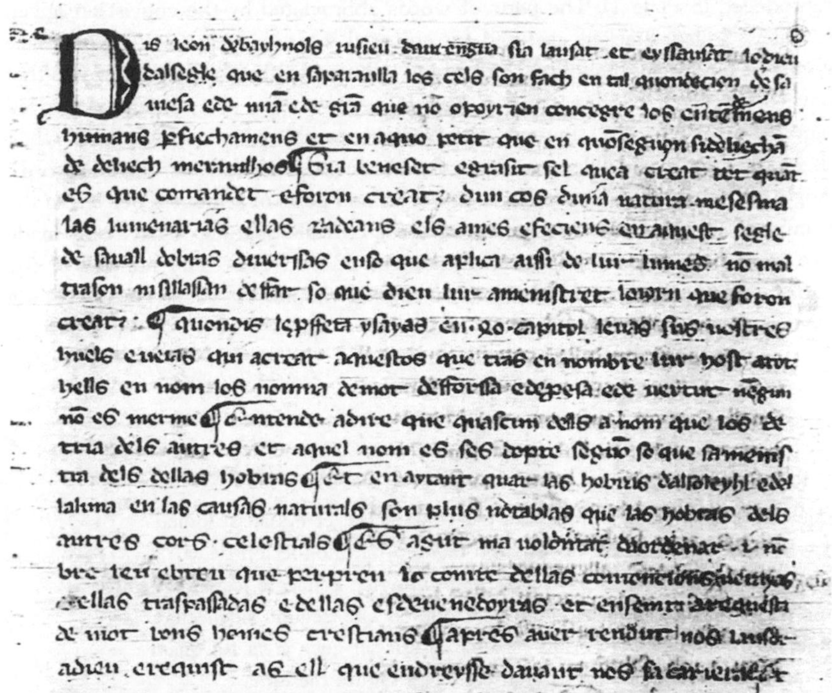

Plate 1. – Leiden, Universiteitsbibliotheek, Ms. Scal. 46, f. 68ʳ: first page of the Provençal version of Levi's canons.

reproduced in Plate 1). The parts of words abbreviated by the copyist are given in italics. Square brackets are used for editorial insertions: they enclose sentence numbers, number of the folios, some letters omitted by the copyist, and words that have disappeared because the last lines of many folios are partially destroyed (apparently by humidity). In most cases I have been able to reconstruct them; when this has not been possible, I have used periods ([...]) to indicate illegible words. Angle brackets enclose words, parts of words, and parts of sentences deleted in the manuscript by the copyist. As indicated in the critical apparatus, I have only made two minor corrections in the text (sentences 99 and 121). The number of the corresponding sentences in the Latin version of chapter 99 of the *Astronomy* are given in the right margin.

4.1. *Transcription*

[*f. 68ʳ*]

[1] **Dis** leon debayhnols iusieu daurengua sia lausat et eyssausat lodieu
dalsegle que en saparaulla los cels son fach en tal quondecion de sa
uiesa ede mi*sericordi*a ede gra*ci*a que no*n* opoyrien concegre los ente*n*demens
humans pe*r*fiechamens et en aquo petit que en quo*n*seguon sideliecha*n*
de deliech merauilhos ¶ [2] Sia beneset egrasit sel quea creat tot qua*n*t
es que comandet eforon creatz dun cos duna natura mesesma
las lumenarias ellas radeans els ames efeciens en aquest segle
de sauall dobras diuerssas enso que aplica aissi de lur lumes / no*n* mal
trason nisillassam deffar so que dieu lur amenistret loiorn que foron
creatz. ¶ [3] quondis lep*r*offeta ysayas en .40. capitol leuas sus uostres
huels eueias qui acreat aquestos que tras en nombre lur host atotz
hells en nom los nonma demot defforssa edep*r*oesa ede uertut ne*n*gun
no*n* es merme ¶ [4] Entende adire que quascun dells a nom que los de
tria dels autres et aquel nom es ses dopte seguo*n* so que samenis
tra dels dellas hobras ¶ [5] Et en aytant quar las hobras dalsoleyhl edel
laluna en las causas naturals son plus notablas que las hobras dels
autres cors celestials ¶ Es agut ma uolontat daordenar .i. no*m*
bre leu ebreu que perpren lo comte dellas conioncions esoposi[ci]ons uerayas
e dellas traspassadas e dellas esdeuenedoyras et ensemt arequesta
de mot bons homes crestians ¶ apres auer rendut nos lausde
adieu erequist as ell que endreysse dauant nos sa cariera ¶ [6] Et
aparnos que en aquest saber saquistan .4. bens [7] luns hes que
en aytant quar tot saber es desirat als humans ¶ et es lo saber
qual es en losobgiet plus noble plus desirat ¶ Et es manifest que
saber aquest nombre hes mot desirat quar notable hes que los
cors celestials son plus nobles que aquestes dessauall iosta aysso

que sen cec daquest saber deprofiech en condicions humanas per
so que uesem en aquestos mouemens denoblesa damenestracion ede
drechura ses deffaylhiment ¶ quondis tolomeu alpermier desson magest [8] esso
en que mais senpot penre deprofiech en aquest quas es loquonte dellas
conioncions es oposicions uerayas dellas lumenarias ¶ quar aysso sa
bem plus complidamens quenungua dels autres nombres dels mo
uemens dels autres quors celestials tant que non si atroba deffalhi
ment en ren en tot aquest lonc tems que son notadas annos es
periencias daquestos nombres que a [... ...] 2000 anns ¶ [9] Ensso en
[f. 68ᵛ]
los autres mouemens satroba deffalhiment enon pauc en los
nombres que nos an amenistrat los permies ¶ [10] El seguon ben
que sen encec hes dellas festas mudablas als crestians quar lo
nombre della luna a alqun seyhnal en aquellas festas ¶ Epero aiusi
eus hes necessari atotas lur festas quar totas son en cert dal
mes lunar ¶ [11] El ters ben que sen encec hes enllas causas natu
rals et especialmens en las causas que perten art natura ayssi
quon hes art demedecina ello lauorar della terra edels ueieietieus
¶ En quar loben que sen encec en aquestas causas hes notori acler
gues hes allaxs ¶ [12] El quart ben hes que cels que entendon e
que samenistron dellas conioncions dellas planetas edellas lumena
rias edellur reguardamens alqunas hobras en aquest cegle de
sauall segon lauertut que dieu lur donet quant las creet non
poirian iuiar nulha causa sinon sabian lotems della conioncion de
las lumenarias o dellas oposicions dellas. enassi quon an decla
rat en lur libres ¶ [13] Eque sia uer que los quors celestials sobre
dichs aian seyhnal en aquestas hobras dcsauall ia pareysserie
que sa cordes allentendement della ley. demoyse edels profetas ¶ [14] dis
el genesin en lacreacion dellas lumenarias edellas redeans que fo
san asseyhnals hes asseyhnorear en iorn hes en nuech ¶ [15] E dis
dieu a iob sabes tu las costumas del cel nias a hordenat son re
giment enterra ¶ [16] Aisso son los bens que apareisson en aquest no
mbre per qual que uie que hom laia ¶ [17] Enpero en lo penre daquest
nombre per lauie que nos auem as ordenat am la iuda de dieu si a
trobara de profiech en especial cell que direm ¶ [18] so es assaber quello es
declarat a cels que sabon en aquesta ciencia que lo aquouen as aqu
els que uolran contar sert aquest nombre que sapian lo luoc <de>
dellauga del soleyl ella quantitat della equacion del soleyl ella quan
titat della equacion della luna ella quantitat della equacion dels
iorns eque sapia[n] sert so que hes mais lo quors della luna que
lo quors dal soleyl allas horas que hes la conioncion holla o
posicion auant miei iorn ho apres ¶ [19] Et en totas aquestas cau

sas trobam alcun deffalhiment en los nombres que nos an
aordenat los permies eper aysso que quar annos es agut ma
nifest dals eclipces lunars essolars que non sen seguon iosta
so que aurien as[... ...] lo conte dels permies ¶ Auem tre
[f. 69ʳ]
baylhat mot entotas aquestas causas et auem fach .i. estur
ment quenos amostrat per esperiencia am demostracion iauma
trical on hes lauga del soleyl et auem lo trobat mot luehn
dal luoc en que degra esser seguon lo nombre dals permies ¶ [20] e
quar auem aiostat anostras esperiencias las esperiencias
dels permies nos es agut reuelat la quantitat dal mouem
ent dellauga ueraya mens ¶ [21] e en laquantitat della equacion del
soleyl trobam als permies diuer<citat> sas hopenions et apres
mot detrebayhl nos hes estat reuelat dals eclipces lunars sa
quantitat ¶ [22] E en la quantitat della equacion dellaluna trobam al
quna diuercitat la qual poirie adure a error entro a 29 me
nutz en las conioncions e en los trasllux dellas lumenarias per que
nos hes agut necessari aquerer lespera don si podon amenistrar
aquestos mouemens que uesem alla luna en las conioncions es
als traslux hes en los autres iorns dalmes don nos hes agut
manifest que la luna non a ipecicle ni excentricum en lacondi
cion que an pausat los permies ¶ [23] hes annos enseyhnat laueritat
daisso alqun esturment que auem per aquesta causa lo qual nos
a mostrat desperiencia am demostracion iaumatriqual que la
luna non es maier annostres huels en los quartz que en las
conioncions si non lo 25.quen. della. ni hes maier a nostres hu
els altems que serie qui lipausaua ipecicle quant hes en .180.
gras dellipecicle que en lo comens dellipecicle ¶ [24] Et auem trobat
atressi lecquacion della luna non seguent lorde dal nombre dellas
taulas quals son as hordenadas per los permies an sen luehna en
alquns luocs non ies pauc [25] per que nos aquouengut aquerer les
pera de que sen segua tot quant uesem della luna dediuercitatz
demouemens nidesaquantitat della ¶ [26] et auem latrobada am lagu
da dedieu en condicion acordant am tot quant si ues della lu
na per esperiencia enaissi quon auem declarat en nostre libre
loqual auem fach per aquesta ciencia ¶ [27] Ellequacion dels iorns
auem trobat en las taulas las quals an as ordenat los permies
en uia de deffalhiment per tres rasons luna per lo luoc dellauga
ellautra per la equacion del soleyl lautra per las acencions dal zodi
ac en lespera drecha per que nos a aquouengut as aordenar a
quest nombre en uia aquordant [... ...]t ses deffalhiment
[f. 69ᵛ]

¶ [28] Essaber so que hes mays lo quors della luna quello quors dal
soleyl allas horas que serie la conioncion ho la oposicion auant
miech iorn ho apres a gran duricia cio uoliam contar per
lauie que an as ordenat los permies essen en cegrie deffalhiment
non pauc et especial ment quar hes mot doras auant miech
iorn ho apres [29] per que nos aquouengut atrebaylhar mot en
aquest nombre perso que auengues lo nombre sert ses nengun
deffalhiment ¶ [30] E pos que aysso hes enayci. manifest es que
so que auem trebaylhat en aquest nombre non hes en uan eper
aysso quar non es depte que trobar aquest nombre cert
en .una. conioncio o en .una. hoposicion hes nobla quausa am tot so que
uenrie am duricia ¶ [31] per que deu esser mot plus noble <sel que>
cell que a aordenat aquest nombre cert en tota conion
cion ho oposicion traspassada o esdeuenedoyra ¶ Et en quaras
plus noble ses tota comparacion que cia as ordenat aqu
est <nombre> nombre atotz tems ses nulha duricia per que
rendem mot de gracias adieu que nos a endreyssat en
tot aysso ¶ Sia son nom eyssausat sobre tota benediction ellauzor
[32] eper so que non ridones merauilhas si nostre conte non sacorda
amlo conte dals permies so es adire lo conte egual dalmes lunar
ni lomouement della diuercitat. fam uos saber que apres gran
trebaylh auem trobat la quantitat dal mes lunar .29. iorns
e .12. oras e .44. menutz et una part de .1138. dun hora ¶ [33] Et
atressi auem trobat lomouement menor umpauc que non
auicm pausat los permies ¶ [34] Et en apres aysso entendem adeue   99:5
sir nostras paraullas en ladeclaracion daquest nombre en 4 can
nons [35] lo permier canno sera adechar lobra dellas taulas que a   :6
uem as hordenadas en rodas en aquest nombre ¶ [36] losseguon can   :7
no hes adeclarar quon sabrem daquellas taulas qual que conioncion
ho oposicion meiana que queram dellas lumenarias ¶ [37] loters can   :8
no sera adeclarar quon sabrem della conioncion hodella oposi
cion meiana la ueraya ¶ [38]lo quart canno sera adeclarar quon   :9
sabrem lo luoc ueray dal soleyl altems della conioncio odella
oposicio ueraya ello gra qual es del zodiac al miei dal cel en
aquel tems ella cendent en aquel tems en aquest orizon en
que nos em daurengua que es luehn dellorien .9. horas e
[f. 70ʳ]
.46. menutz [39] et endreyssarem ambaisso en aquest canno assaber tot
aysso en los autres orizons es assaber lo luoc ueray dal soleyl en
qual que tems que queram
[40] Lopermier canno sapias que larasis daquest nombre hes. apres   :12
1320. anns dels anns delencarnacio en la cieutat daurengua e

lan que comenssa en la ciutat daurengua a .25. demartz nos lo
comenssam en aquestas taulas lopermier iorn demartz [41] et hes
as ordenat en la roda permiera qual es maior et hes dous de :13
foras las conioncions ellas oposi[ci]ons meianas en aquest orizon
entro .8. ans solars en los quals .8. ans a .99. conioncions e .99. oposicions [42] e
las conioncions son escrichas en tencha negra ellas oposicions en :14
tencha uermeylha esson escrich en .4. espasis. [43] [En] lo permier espasi hes :15-6
escrich lomes solar el nombre dels iorns dell en qual esdeuen la conion :17
cion olla oposi[ci]on ¶ En lo ceguon espasi son escrichas las ferias en :18
que esdeuen la conioncion olla oposicion ¶ En lo ters espasi son es :19
crichas las horas eguals. quals son apres aquel miey iorn ¶ En :20
lo quart espasi son escrich los menutz dellas horas ¶ [44] Ededrech qua :21
da una conioncion hes oposicion hes escrich lo luoc della luna en a
quel tems dal mouement della diue[r]citat en .3. espasis en loper :22
mier son escrich los seyhnals et en lo seguon los gras et en
loters los menutz dels gras ¶ [45] El nombre dels anns son escrich
en aquesta roda de .i. entro .8. sobre la lihna permiera daquel ann
[46] Et en la roda laqual uen apres que hes la seguonna .a. .69. :25
lihnas e quada lihna della seruis atot lo tems della permiera
roda lo qual hes .8. anns solars e .i. iorn e .12. horas e .41. me
nut[z] epres de .13. seguons e deueis si a .5. espasis ¶ [47] en lo permi :26
er es escrich lo nombre dellas lihnas de .i. entro .69 ¶ en lo :27
ceguon espasi son escrich los iorns per aiustar amlos. iorns
dalmes ho per sostrayre ells dells ¶ Et en lo ters espasi son es :28
crichas las ferias per aiustar las am las autras ferias oper sos
trayre las dellas ¶ En lo quart espasi son escrichas las horas
¶ Et en lo sinquen son escrich los menutz dellas horas ¶ [48] Et aque :29
sta roda perpren .552. anns e .105. iorns e .11. horas e .24. me
nutz ¶ [49] ededrech quada lihna hes escrich lo luoc della luna dal :30
moue[me]nt della diuercitat en aquel tems
[50] Et en la roda qual li uen apres [que hes] la terssa a .40. lihnas :31
[f. 70ᵛ]
e quada lihna daquella roda perpren tot lotems della seguonna ro
da edeuesis si a .5. espasis en lamaniera della seguonna roda ¶ [51] per :32
que sen cec daysso que aquesta roda perpren lo tems de .22091. :31
ann[s] e .201. iorn[s] [52] edauar quada lihna hes escrich lo luoc della lu :33
na dal mouement della diuercitat en aquel tems
[53] Et en la roda queuen apres qual hes la quarta a .28. lih :34
nas eperpren quada lihna daquesta roda tot lo tems de
la terssa roda que hes lo tems .22091. ann[s]. e .201. iorn[s] edeue :35
sis si a .3. espasis quar non .ya. horas nimenutz per que hes
lotems daquesta quarta roda .618563. anns e .156 iorns [54] ede :34

drech quada lihna hes escrich lo luoc della luna dal mouement   :36
della diuercitat en aquel tems ¶ [55] Et en apres aysso reuoluant   :37
totas aquestas rodas enpero en doas reuolucions yaura de ior[n]s
sobre an[n]s .313. iorns per lobices ¶ [56] E en lomouement della di
uerssitat della luna aiustaras .8. gras per quada reuolession
detotas las rodas complidas./ [57] Ellas autras taulas que asordenat a   :38
uem en aysso ¶ la permiera perpren lequacion que peruen en a
quest nombre en horas et en menutz doras per lo iorn dellan en
que uen la conioncio olla oposicio meiana ¶ [58] la seguonna taulla per   :42
pren alquna equacion en menutz dora per lo iorn dellan en
que ses deuen la conioncion olla oposicio meiana ¶ [59] la terssa tau   :43
la perpren la equacion qual ses deuen en horas et en me[n]utz
doras per lo luoc della luna dal mouement della diuerssitat
¶ [60] la quar taula perpren alquna equacio que sesdeuen en menutz doras   :45
per lo iorn dellan eper lo luoc della luna dal mouement della diuer
citat ensems: ¶ [61] En la sinquena taula decham assaber lo luoc   :47
ueray dal soleyl en qual quetems quom uuelha ¶ [62] En la seyse   :49
na taula son las assencions dal zodiac en lespera drecha: ¶ [63] En
la setena taula son las assencions dal zodiac en lorizon dau   :51
rengua en que hes aut lo pol .44. gras mens .16. seguons
[64] Lo seguon cannon si uoles saber alquna conioncio .o. oposici   :53
on meiana apres esser passat lan de .1320. del anns
dellen carnacion eque prennas lo millesme lopermier iorn de
martz quar en aquest nombre uuol esser pres en aquel
iorn tu atrobaras a hordenat en lapermiera roda entro .8.
an[n]s sollars [65] essi loltems q]ue queres hes apres .1328. anns tu   .54
[f. 71ʳ]
sostrayras dals anns que tenes .1320. [66] el remanent <el rema   :55
nent> si es mens que lo tems della seguonna roda tu odeuesi
ras per .8. anns e .i. iorn e .12. oras e .41. menut dora. e .13.
seguons en maniera que reste della deuesion mens de .8. anns
e .i. iorn edemiei e .41. menut e .13. seguons ello nonbre qu
al uenra della deuesion hes lo nombre dellas lihnas qual dei
sendras en la seguonna roda ¶ [67] entende adire que si aquel nom   :56
bre hes .10. tu penras la lihna desena della seguonna roda. [68] e
serua aquella lihna en que fenis lo nombre. [69] el remanent   :57
dals anns edals mes mens de .8. anns e .i. iorn e .12. horas
e .41. menut e demiei tu querras en lapermiera roda apres
lo tems della [permiera] conioncion della si queres conioncio / o apres lo
tems della permiera oposicion della si queres oposicion / etroba
ras la lihna [que] queres dalmes edellan: serua aquesta lihna [70] et   :58
en aquestas doas lihnas seruadas trobaras la conioncion olla

oposi[ci]on meiana qual queres aiustant so que es en aquestas
doas lihnas seruadas quada quausa am son cemblant so es assa
ber iorns am iorns efferias am ferias et horas amhoras
emenut amenut essi ya dellas horas .24. tu enfaras .i. iorn   :59
e aiustaras lo am los iorns es am las ferias ¶ [*71*] Essi lotems de que   :62
queres la conioncio olla oposicion hes auant lotems della rais
sobre dicha tu sabras per̃mieramens quans an̄s equans mes
hes auant lo tems della permiera conioncion della permiera
roda si queres conioncion ho auant lapermiera oposicion della
si queres oposicion ¶ [*72*] Essi aquel tems hes mens quel tems de
la seguon̄na roda deuesis lo per .8. anns e .i. iorn e .12. oras
e .41. menut e .13. seguons esso que ti yssira della deuesion pren̄
[*73*] essi ti sobra ren aiusta mais .i. sobre la quantitat queti uen
ra della deuesion essi non̄ yssobra non̄ tiqual ren aiustar [*74*] et aytan̄t
quon sera lo nombre del quocient deyssent en las lihnas della se
guon̄na roda serua la lihna en que fenira lo nombre [*75*] esso que so   :65
bra della deuesion si ren yssobra sostrai .o. de .8. anns e .i. iorn
e .12. oras e .41. menut e .13. seguons [*76*] esso queti sobrara serua
e quer o en lapermiera roda apres lotems della permiera conioncion̄ :66
si queres conioncion odella per̃miera oposicion si queres oposicio
ella lihna en qua fenira lonombre s[erua [*77*] e sostra]i tot quant a
[*f. 71ᵛ*]
en la lihna seruada della seguon̄na roda edesso que hes en la lihna
seruada della permiera roda esso que sobrara hes lotems della conion :67
cion odella oposicion meiana que queres ¶ [*78*] Edaisso hes clar ses :68
trebayl quon husaras si lotems de que queres la conioncion o
la oposicion es luehn del tems della rais sobre dicha mais q*ue*
per lotems della roda seguon̄na quar adoncs querras en lamanie
ra sobre dicha aquella conioncion o aquella oposicion en .3. daque
stas rodas. o en .4. seguon que fenira lo nombre en quascuna del
las ¶ [*79*] E trobaras en quascuna dellas la lihna qual seruaras en lam
aniera sobredicha ¶ [*80*] essi lotems hes apres la rais tu aiustaras tot :69
quant hes en las lihnas seruadas euen̄ra ti la conioncio olla opo
sicion meiana qual queres ¶ [*81*] Essi hes auant lo tems della rays sos
trairas desso que trobaras en la lihna seruada della permiera ro
da tot quant .a. en las autras lihnas seruadas: Esso que sobra
ra hes la conioncion olla oposicion meiana qual queres ¶ [*82*] Et
en aquellas mesesmas lihnas seruadas trobaras lo luoc della :70
luna dal mouement della diuercitat en aquella maniera mesma
que yas trobat lotems della conioncion odella oposicion ¶ [*83*] En
tende adire que aiustes so que troba <ra> ras daquest nombre
en totas las lihnas seruadas si hes apres lotems della rais

o que sostrayas desso quetrobaras en la lihna seruada della per
miera roda so que trobaras en totas las autras lihnas serua
das daquest nombre. si lotems hes auant la rais ¶ [*84*] Essapias   :71
que en aquesta condicio qual dicha auem no*n* tiqual a quomtar
los a*n*ns que son escrichs en las rodas mas tan solament los ior*n*s
els mes ellas horas els menutz eueiras on ta duira lo no*m*bre
dal tems dellan ¶ [*85*] Enpe*r*o situ contauas de la reuolucion della per   :72
miera roda .8. an*n*s tan solament ellaissa[nt] lafraccion: Ella re
uolucion della seguon*n*a Roda .552. an*n*s giquent las efraccio*n*s
Ella reuolucion della terssa Roda .22091. an*n*s laissant los au
tres an*n*s ellas fraccions adoncs ti qualrie contar los an*n*s
ellas fraccions que trobaries en las lihnas seruadas: Et
adoncs sabras qual an*n* seruira lono*m*bre ni en qual iorn della*n*   :73
[*86*] Lo ters canno [po]s tu sabes lotems della co*n*ioncion o della o   :76
posicion meia[na e loluoc] della luna dal mouement della diue
rcitat en aqu[est tems leug]ierament podes saber la conioncio
[*f. 72ʳ*]
holla oposicion ueraya: [*87*] que tu queras en las taulas dellequacion   :77
qual sestalua p*er* lo iorn dellan aquel iorn. epenras las horas els
menutz que trobaras en aquel iorn en aquestas doas taulas et
escrieuras las sotz las horas della conioncion o dela oposicion
meiana ¶ [*88*] Equerras mais en la taula dellequacion que ses tal   :78
ua p*er* lo mouement della diuercitat lo luoc della luna dal moue
ment della diuercitat en aquel tems que queres. epenras
so que hes en aquel <tems> luoc dellas horas edels menutz e
escrieuras las horas apres las horas dessus dichas ellos menutz
am los menutz dessus dich en la maniera sobre dicha ¶ [*89*] E   :79
mais en lataula dellequacion qual sestalua p*er* lo iorn dellan
ep*er* lomouement della diuercitat en cems lequacion as aquel
iorn et as aquel luoc dalmouement della diuercitat ence
ms ¶ E escrieuras los menut ellora quals trobaras am lalre
<autres> dessus dich en la maniera sobredicha ¶ [*90*] E pueis aius
ta tot quant as escrich dels iorns edellas ferias edellas
horas edels menutz essi ya .24. oras fay en .i. iorn et aiu[s]ta
lo am los autres ecreys la feria atressi daquel iorn essi ya
de menut .60. fay en huna hora es a iusta la am las autras
horas dessus dichas essi ya demenutz que no*n* mo*n*te lo complim
ent duna hora conta los / ediras as aitantos ior[n]s daitalmes as aitantas
horas as aitantos menutz ¶ [*91*] Edetot aysso sostray .i. iorn esso que   :80
ti sobrara hes lo tems della conioncion o della oposicion ue
raya ¶ [*92*] Essapias que los iorns que son en la seguo*n*nataula.   :85
dels iorns dellan sen peihno*n* de .i. iorn a .119. ans ellas .3.

partz de .i. an ¶ [93] Entende adire que quant seran passat .119. anns   :86
ellas .3. partz de .i. an apres la Rais tornara so que era de
quacion al seguon iorn demartz tornara alpermier iorn demartz.
Et enaissi en totz los autres iorns dellan ¶ [94] Ello quontrari day   :87
so serie en so que serie auans lo tems della Rais ¶ [95] Els iorns   :88
que son en lapermiera taula et en la quarta sen peihnon a .43.
anns de .i. iorn en lamaniera sobre dicha
[96] Lo quart canno si uoles saber lo luoc ueray del soleyl al tems   :93
della conioncion o della o posicion ueraya [tu] querras en
la taula sinquena ¶ Edaisso sabras at[ressi lo luoc] ueray della lu   :94
na al tems della conioncion edella [oposicion ueray]a ¶ [97] E ssapias   :95
[f. 72ᵛ]
que aquella taula hes aordenada al luoc dal soleyl ueray aquada
miey iorn dellan de .1321. dellencarnacion en lorizon daurengua
[98] e qui uol saber per aquella taula lo luoc ueray del soleyl en qual   :96
que tems que lo quera quouen li hostar dels anns dellenquar
nassion .1320. esso que sobrara tan solament sobre quasers   :97
o .4. ho mens de .4. serua. [99] essi lo seruat hes .i. lo nom   :98
bre que trobaras en la taula hes lo luoc dal soleyl aquada
miei iorn daquel an ¶ [100] Essi lo seruat hes .2. aquel nombre   :99
que trobaras en lataula hes .6. oras apres quada miey iorn da
quel ann ¶ [101] E si lo seruat hes .3. a quest nombre sera en lataula   :100
.12. oras apres quada miey iorn daquel an ¶ [102] Essi loseruat es   :101
.4. sera aquel nombre .6. oras dauant quada miey iorn daq[u]el
an ¶ [103] E ssapias que per quadann complit apres .1320. tiqual   :102
que aiustes am so que trobaras en aquesta taulla .30. seguons eper
quada Reuolucion de .4. anns aiustaras .2. menutz. et en
aquesta maniera ti regiras entro .43. anns apres larazis
¶ [104] Enpero quant seran passatz .43. anns apres larays tu dey   :103
sendras .i. iorn en aquesta taula [105] entende adire que so que   :104
ytrobaras al seguon iorn demartz sera lopermier iorn demartz
essostrayras desso quetrobaras aqui .37. menutz e .54. seguons   :105
[106] e husaras en. los anns daquella reuolucion es en ayssi quon as   :106
husat en los anns della permiera reuolucion apres que seras
deysendut .i. iorn. es auras sostrach desso que trobaras aqui
.37. menutz e .54. seguons ¶ [107] Eycemple que si son passatz .4.   :107
ans daquesta seguonna Reuolucion tu deysendras en aquesta tau
la .i. iorn ennon sostrayras dal nombre que ssi trobara. mas .35.
menutz .e. .54. seguons quals loti conuenrie acreycer .2. me
nutz per lareuolucion que hes passada de .4. anns es aytal faras
a quada Reuolucion de .43. anns quals uenrien apres lo tems
della Rays so hes assaber .1320. ¶ [108] Entende adire que per quada

Reuolucion de .43. an*n*s deyssendras i. iorn en aquesta taula e
sostray desso que sitrobara *per* quada reuolucion de .43. an*n*s <que
seran denfra la Reuolucion dels .43. an*n*s.> .37. menutz e .54. se
guons een la [reuolu]cion dels .4. an*n*s que seran denfra la reuolu
cion dels .43. an[ns ... ...] en lamaniera sobredicha ¶ [*109*] Ello con   :109
trari detot [... ... ...]em faras en los an*n*s daua*n*t la rays:
[*f. 73ʳ*]

¶ Entende adire que en loluoc que aiustauas sostrayras es al luoc   :110
que sostrasias aiustaras es en los luocs que deyssendies als
iorns dellan pinaras ¶ [*110*] E pueys que tu sabes lo luoc ueray del   :111
soleyl alponch della conioncion ho della oposicion ueraya quer
en la seysena taula las assencions dal. gra en que lo soleyl. hes
en lorisont drech essi la conioncion ho la oposicion hes al   :112
ponch demiey iorn logra en que lo soleyl hes. hes lo cap de
la mayson desena ¶ [*111*] quer en la setena taula lo. gra. dal. zodi   :115
ac que sas assencions son eguals allas assencions dal. gra
en que le soleyl hes en lorison drech hes aquel. gra. hes la   :116
sendent en aquel tems en aquest orison en que son as
ordenadas aquestas taulas ¶ [*112*] E ssi la conioncio ho laopos[i]cion   :113
ueraya hes auant miey iorn ho apres multiplica las oras
els menutz que hes auant o apres *per* .15. gras esso que ti
yssira della multiplicacion aiusta am las assencions dalgra en
que le soleyl hes en lorison drech si hes apres miey iorn /
ho*n*nosostray si es auant miey iorn / [*113*] el remant son las as   :114
cencions dal gra dal cap dellamayson desena dellorizon drech
ellas assenciono dell assendent en aquest orizon en que son as
ordenadas aquestas taulas ¶ [*114*] etrobaras *per* aquestas assencions
la maniera sobre dicha lo gra. dal. zodiac que es cap della dese
na mayso*n* ello gra < dal zodiac > que es assendent en aquest ori
son ¶ E daysso poyras saber las .12. maysons en lamaniera
que an escrich los p*er*miers ¶ [*115*] E sso am que si poiram saber   :118
sert plus leugieramens hes am lestralabi ¶ [*116*] Essi la lon.   :151
guesa dal tieu orison es diuerssa alla longuesa daquest orison
daurengua essabes ladiuercitat leugiera mens poyras saber
daquest nombre lotems della conioncion odella oposicion ueraya
en lotieu orison ¶ [*117*] Essapias que aytant quant hes ladiuerssitat
della longuesa aytant sera ladiuercitat dal tems della co*n*ioncio*n*
odella oposicion al tieu orison ¶ [*118*] Essi lo tieu orison hes orien   :151
tal daquest orison daurengua aiustaras aquella diuercitat
am lotems que ti yssira daquestas taulas della conioncion ode
la oposicion ueraya e auras lo tems [sert] en lo <tems> tieu ho
rison ¶ [*119*] Essies accidental dellor[ison daureng]ua tu las en   :151

sostrayras ¶ [*120*] Eyssemple sillo [tieu horison es plu]s oriental
[*f. 73ʳ*]
que aquel daurengua .iᵃ. hora aiusta aquella hora am lotems de
la concioncio odella oposicion ueraya qual ta yssira daquest nom
bre ¶ hossi es plus accidental .iᵃ· hora sostray la en ¶ [*121*] Essi lotieu   :153
orison hes diuers daquest orison daurengua en larguesa qual
ti assaber permiera ment las assencions dal zodiac en lotieu ho
rison dellas taulas que yan as ordenat los permies essabras
pueys daysso lassendent en lo tieu orison all tems della concioncion
odella oposicion ueraya en laman[i]era sobre dicha ¶ [*122*] Es atressis   :155
tenseh[n]aran leugieramens aquestas taulas la longuesa del tieu
orison si acertas dun eclipsse della luna am lestralabi o am
loautre esturment am que si puescan saber las horas / que aysso
tenssehnara lotems dal miei delleclipci lo qual hes tem[s] della opo
sicion ueraya ¶ [*123*] Eueiras quant .a. dediuercitat entre aquel
tems ello tems queti yssira daquesta taula es aquo hes la
diuerssitat della longuesa qual hes entre lotieu orison hes
aquel daurengua

---

5. esoposions *add. mg.* 14. redeans: *a.m. add. s.l.* planetas. 18. dellauga: *a.m. add. s.l.* altitudinis;
equacion dels iorns: *a.m. add. mg.* equacionem ueram. 19. lauga: *a.m. add. s.l.* altitudo. 30.
concioncio *add. s.l.* 36. taulas *add. s.l.* 42. oposicions e las *add. mg.* 60. equacio *add. s.l.* 85. de
add. s.l. 89. ellora *add. s.l.* 99. taula: *ms.* tauaula. 121. lassendent en lo tieu: *ms.* lassendent el
tieu.

4.2. Translation

[f. 68ʳ] [1] Leon of Bagnols, a Jew from Orange, said: let the God of the world
be praised and exalted; by his word heavens were made in such a condition of
wisdom, and mercy, and grace, that human understandings cannot perfectly know
them, but over what little they know, they rejoice with a marvelous delight. [2]
Let Him be blessed and honored who created everything that exists, who ordered,
and of a body of one and the same substance the luminaries and the radiant
[bodies] were made, efficient souls in this lower world of diverse actions by means
of their lights, and they neither weary nor get tired of doing what God decreed
for them the day they were created. [3] As the prophet Isaiah says in chapter
40[:26]: "Lift up your eyes and see who created these, who brings forth their host
in number, He named all of them with a name, with His great strength, and might,
and power not one of them is missing". [4] He means that each of them has a name
that distinguishes it from the others, and beyond doubt that name is according to
what follows from their actions. [5] And inasmuch as the actions of the sun and

the moon on natural things are more noteworthy than the actions of the other celestial bodies, it has been my wish to arrange an easy and brief [method of] computation which allows us to calculate the true conjunctions and oppositions, both past and future, [and I have] compiled [these tables] at the request of many noble Christians, after having praised God and having petitioned Him as the one who straightens His way before us. [6] And it seems to us that with this knowledge four benefits can be obtained.

[7] The first, since all knowledge is desired by human beings, and the knowledge whose object is noblest is the most desired, it is manifest that to know this computation is much desired because it is evident that the celestial bodies are nobler than those of the lower world, together with the benefit that follows for the human condition from what we see in those motions of nobility, ordering, and rightness without error, as Ptolemy says in the first [book] of his Almagest. [8] And what can be most profitable in this is the calculation of the true conjunctions and oppositions of the luminaries, because we know now more completely than ever about the other calculations of the motions of the other celestial bodies, in such a way that we do not find any error whatsoever in this long time during which experiences of these computations are known to us [which is about] 2000 years. [9] But for [f. 68ᵛ] the other motions we find an error, which is not small, in the computations we have received from the Ancients.

[10] The second benefit that follows concerns the movable feasts of the Christians, because the computation of the moon is a sign for these feasts, and it is necessary for them for [determining] all their feasts insofar as all of them depend on the lunar month.

[11] The third benefit which follows concerns the natural things, and especially the things that belong to the art of nature, as the art of medicine and the working of lands and of plants, and moreover the benefit that follows with regard to these things is evident for clerics and laymen.

[12] The fourth benefit is for those who know about the conjunctions of the planets and the luminaries, and their aspects, and derive from them some actions in this lower world according to the power that God gave them when He created them, because they could not judge anything unless they knew the time of the conjunctions of the luminaries, and of their oppositions – as they have explained in their books. [13] And that it may be true that the aforementioned celestial bodies are signs for these operations in the lower world would seem to be in agreement with the understanding of the Law of Moses and the Prophets. [14] The book of Genesis says concerning the creation of the luminaries and the radiant [bodies], that they are to be signs and to govern day and night [Gen. 1:16]. [15] And God says to Job [38:33]: "Do you know the habits of the heavens and have you established their government on the earth?"

[16] These are the benefits which follow from this computation using any procedure whatsoever. [17] However, making this computation by the procedure

we have established with the help of God you will find as an additional benefit what we will say; [18] that is, it has been declared to those who know this science that it is necessary for those who want to do this computation with certainty to know the position of the solar apogee, and the amount of the [maximum] solar equation, and the amount of the lunar equation, and the equation of time, and to know with certainty the excess of the course of the moon over the course of the sun in the hours before or after noon at which the conjunction or opposition occurs. [19] And in all these things we find some error in the computations established by the Ancients, and for this reason – because it has been evident for us from lunar and solar eclipses, which do not occur in agreement with the computations made according to the Ancients – we have [f. 69^r] worked hard on these things and we have made an instrument which has shown us by experience and geometrical demonstration where the apogee of the sun is [located], and we have found it very far from the position where it ought to be according to the computation of the Ancients. [20] And by adding our experiences to the experiences of the Ancients, the true amount of the motion of the apogee has been revealed to us. [21] We found different opinions among the Ancients about the [maximum] equation of the sun, and after hard work its amount has been revealed to us from [the observation of] lunar eclipses. [22] In the amount of the equation of the moon we find some discrepancy, which could produce an error of up to 29 minutes in [the computation of] the conjunctions and the oppositions of·the luminaries; therefore it has been necessary for us to investigate the [arrangement of] sphere[s] from which we can derive the motions we see in the moon at conjunctions, oppositions, and in the other days of the month, whence it has been evident for us that the moon has no epicycle and [also] no eccentric in the way assumed by the Ancients.

[23] An instrument that we have made for this matter has taught us the truth of it, and it has shown us by experience and geometrical demonstration that the moon is greater to our eyes at the quadratures than at the conjunctions only by 1/25 of it[s diameter], nor is it greater to our eyes – if an epicycle is supposed – at the moment when it is at 180° on the epicycle than at the beginning of the epicycle. [24] We have also found that the equation of the moon does not agree with the amounts in the tables composed by the Ancients, and the discrepancy at some places is not small; [25] therefore, we considered it appropriate to investigate the [arrangement of] sphere[s] from which follows all that we see about the diversity of motions of the moon and its amounts. [26] We have found it with the assistance of God in a way that agrees with all that we see of the Moon by experience, as we have explained in our book that we have devoted to this science. [27] We have found errors in the equation of time in the tables composed by the Ancients due to three reasons: one [is] due to the position of the apogee of the sun, another [is] due to the equation of the sun, and another [is] due to the ascensions of the Zodiac at *sphera recta*, and from this we have considered it

appropriate to do this computation in a way that agrees [...] without error. [f. 69ᵛ] [28] And to know how much the course of the moon exceeds the course of the sun at conjunction or opposition in the hours before or after noon would be very difficult to compute according to the method that the Ancients established, and an error that is not small would follow especially when there are many hours before or after noon.

[29] For this reason it was necessary for us to work very much on the [method of] calculation in order that you may compute exactly without any error. [30] And it being so, it is evident that the work we have made on this computation is not in vain, since there is no doubt that to find an exact [method of] computation for only one conjunction or opposition is a great achievement even if it is reached with effort; [31] therefore, it is necessarily greater to find this exact computation for any conjunction or opposition in the past or future, and it is even greater, beyond any comparison, to find that [method of] computation for any time without any effort, whereby we give many thanks to God who has guided us through this [task]; let His name be exalted above all blessing and praise. [32] And to avoid your wondering that our numbers do not agree with the numbers of the Ancients in that which concerns to the mean lunar [i.e., synodic] month and the motion in anomaly, let us inform you that after much work we have found that the length of the lunar month is 29 days, 12 hours, 44 minutes, and 1/1138 of an hour. [33] And so we have found this motion to be a little lesser than the Ancients had assumed.

[34] Thus we intend to divide our words for explaining this computation into four canons. [35] The first canon is devoted to explaining how to use the tables we have arranged in circles for this computation. [36] In the second canon we will declare how to find from the tables any mean conjunction or opposition of the luminaries we wish. [37] In the third canon we will explain how to find from the mean conjunction or opposition the true ones. [38] The fourth canon will be devoted to explain how to find the true position of the sun at the time of the true conjunction or opposition, and the degree of the zodiac that is in mid-heaven at that time, and the ascendant at that time at this horizon of Orange where we are, whose distance from the Orient is 9 hours [f. 70ʳ] and 46 minutes. [39] And at the end of the canon we will describe how to know all that for other horizons, and how to know the true position of the sun at any time we wish.

[40] The first canon. You should know that the epoch for this calculation is at the end of 1320 years of the years of the Incarnation in the town of Orange, and the year that begins in the town of Orange on the 25ᵗʰ of March begins in these tables for us on the first day of March. [41] And in the first circle, which is the greatest and the outermost, the mean conjunctions and oppositions for this horizon are arranged up to 8 solar years; in these 8 years there are 99 conjunctions and 99 oppositions. [42] And the conjunctions are written in black ink and the oppositions in red ink, and they are written in four columns. [43] In the first column is written

the solar month [and] the number of the days within it in which the conjunction or the opposition occurs; in the second column are written the weekdays in which the conjunction or the opposition occurs; in the third column are written the equal hours that have elapsed after that noon; in the fourth column are written the minutes of hours. [44] And opposite to each conjunction and opposition is written in three columns the position of the moon according to the motion in anomaly for that time: in the first are written the signs, in the second the degrees, and in the third the minutes of degrees. [45] The number of the years are written in this circle from 1 to 8 above the first row of each year.

[46] And in the circle that comes next, which is the second one, there are 69 rows, and each row of it corresponds to the whole time of the first circle, which is 8 solar years, 1 day, 12 hours, 41 minutes and nearly 13 seconds, and it is divided into five columns. [47] In the first is written the number of rows from 1 to 69; in the second column are written the days to be added to the days of the month, or to be subtracted from them; in the third column are written the weekdays to be added to the other weekdays, or to be subtracted from them; in the fourth column are written the hours; and in the fifth are written the minutes of hours. [48] And this circle covers 552 years, 105 days, 11 hours, and 24 minutes. [49] And opposite to each line is written the position of the moon according to the motion in anomaly for that time.

[50] And in the circle that comes next, which is the third, there are 40 rows, [f. 70ᵛ] and each row of this circle covers all the time of the second circle, and it is divided into five columns in the same way as in the second circle. [51] From this it follows that this circle covers a time of 22091 years and 201 days. [52] And at the end of each row is written the position of the moon according to the motion in anomaly for that time.

[53] And in the circle that comes next, which is the fourth, there are 28 rows, and each row of this circle covers all the time of the third circle, which is 22091 years and 201 days, and it is divided in three columns because in it there are not hours or minutes; therefore the time of this fourth circle is 618563 years and 156 days. [54] And opposite to each row is written the position of the moon according to the motion in anomaly for that time. [55] And after this there begins again the revolution of all these circles, but after two revolutions there will be an excess of 313 days over the years because of the leap year. [56] And for the motion in anomaly of the moon you should add 8 degrees after each revolution of all these circles.

[57] In the other tables that we have composed in this, the first contains the equation in hours and minutes of hours which, in this computation, is due to the day of the year when the mean conjunction or opposition occurs. [58] The second table contains a certain equation in minutes of hours due to the day of the year when the mean conjunction or opposition occurs. [59] The third table contains the equation in hours and minutes of hours which is due to the position of the moon

according to the motion in anomaly. [60] The fourth table contains a certain equation in minutes of hours due to the day of the year together with the position of the moon according to the motion in anomaly. [61] In the fifth table we present the true position of the sun for any time you wish. [62] In the sixth table are the ascensions of the Zodiac at *sphera recta*. [63] In the seventh table are the ascensions of the Zodiac for the horizon of Orange, in which the altitude of the pole is 44 degrees less 16 seconds.

[64] The second canon. If you seek a mean conjunction or opposition that takes place after the year 1320 of the years of the Incarnation – and you take the beginning [of the year] on the first day of March, because it must be so considered in this computation –, for up to 8 solar years you will find it arranged in the first circle. [65] And if the time [for which] you seek [a conjunction or opposition] is after 1328 years, you [f. 71ʳ] must subtract 1320 from those years you have; [66] if the remainder is less than the period of the second circle, divide it by 8 years, 1 day, 12 hours, 41 minutes and 13 seconds, in such a way that there remains from the division less than 8 years, one day and a half, 41 minutes, and 13 seconds, and the number which will result from the division is the number of rows you will descend in the second circle; [67] that is to say, if the number is 10, you will take the tenth row of the second circle. [68] And keep the row where the computation ends. [69] Seek with the remainder of the years and the months – which is less than 8 years, 1 day, 12 hours, and 41 minutes and a half – in the first circle after the time of the first conjunction [written] in it, if you seek a conjunction, or after the time of the first opposition [written] in it, if you seek an opposition, and you will find the row of the month and the year you seek. Keep this row. [70] And with these two kept rows you will find the mean conjunction or opposition you seek by adding what there is in these two kept rows, each quantity with that of the same kind; that is, days to days, weekdays to weekdays, hours to hours, and minutes to minutes; and if you obtain 24 hours, consider them as one day, adding it to the days and the weekdays.

[71] And if the time for which you seek the conjunction or the opposition is before the aforementioned epoch, you must know first how many years and how many months it is before the time of the first conjunction of the first circle, if you seek a conjunction, or before the first opposition of it, if you seek an opposition. [72] And if this time is less than the time of the second circle, divide it by 8 years, 1 day, 12 hours, 41 minutes, and 13 seconds, and take the result from the division. [73] If something remains, you must add 1 to the result from the division, and if nothing remains, you must add nothing. [74] And descend in the second circle as many rows as the number of the quotient, and keep the row where the computation ends. [75] And subtract the remainder from the division, if something remains, from 8 years, 1 day, 12 hours, 41 minutes, and 13 seconds; [76] keep the remainder [from the subtraction] and seek it in the first circle after the time of the first conjunction, if you seek a conjunction, or of the first opposition, if you seek

an opposition, and keep the row where the computation ends. [77] And subtract all there is [f. 71ᵛ] in the row kept from the second circle from what is in the row kept from the first circle, and the remainder is the time of the mean conjunction or opposition you seek.

[78] From this, without effort, it is clear how to proceed if the time for which you seek the conjunction or the opposition is far from the time of the aforementioned epoch by an amount greater than the time of the second circle, because then you will seek that conjunction or opposition in the aforementioned way in three of these circles, or in four, depending on which circle the computation ends. [79] And in each of them you will find the row that you will keep in the aforementioned way. [80] And if the time is after the epoch, add together all there is in the kept rows and you will obtain the mean conjunction or opposition you seek. [81] And if it is before the time of the epoch, subtract from what you have found in the kept row from the first circle [the sum of] all that is in the other kept rows, and the remainder is the mean conjunction or opposition you seek.

[82] And in these same rows that you kept you will find the position of the moon according to the motion in anomaly in the same way you have found the time of the conjunction or the opposition; [83] this means, you must add together what you have found from this computation in all the kept rows, if it is after the time of the epoch, or subtract from what you found in the row kept from the first circle what you have found from this computation in all the other kept rows, if the time is before the epoch.

[84] Know that for the aforementioned case you need not compute the years written in the circles, but only the days, the months, the hours, and the minutes, and you will see that you obtain the number of the time of the year. [85] But if you computed from the revolution of the first circle only 8 years leaving aside the fractions, and from the revolution of the second circle 552 years leaving aside the fractions, and from the revolution of the third circle 22091 years leaving aside the other years and the fractions, then you need to compute the years and the fractions that you have found in the kept rows, and then you will know the year and the day of the year resulting from the computation.

[86] The third canon. Since you know the time of the mean conjunction or opposition and the position of the moon according to the motion in anomaly at that time, you can easily find the true conjunction [f. 72ʳ] or opposition. [87] Seek that day in the tables of the equation due to the day of the year and take the hours and minutes that in these two tables correspond to that day, and write them under the hours of the mean conjunction or opposition. [88] Moreover, seek in the table of the equation due to the motion in anomaly the position of the moon according to the motion in anomaly at the time you seek, and take the hours and minutes that you find in this place, and write the hours under the aforementioned hours and the minutes under the aforementioned minutes in the way previously explained. [89] Moreover, in the table of the equation due to the day of the year

and the motion in anomaly together, [seek] the equation corresponding to that day and to that motion in anomaly together, and write the hour and the minutes you find with the others aforementioned in the aforementioned way. [90] And then add the days, the weekdays, the hours, and the minutes that you have written; but if there are 24 hours, replace them with one day and add it to the others, and increase also the weekday of that day; and if there are 60 minutes, replace them with one hour and add it to the other aforementioned hours; and if there are minutes which do not reach one hour, count them; and say: so many days of such a month and so many hours and so many minutes. [91] And from all that subtract one day, and the remainder is the time of the true conjunction or opposition. [92] And know that the days of the second table of the days of the year will be late by one day after 119 years and three [fourth] parts of a year; [93] that is to say, when 119 years and three [fourth] parts of a year after the epoch have passed, the equation which [in the table] corresponds to the first day of March will correspond to the second day of March, and similarly for the other days of the year. [94] And the opposite [correction] must be made for the time before the date of the epoch. [95] The days of the first and fourth tables will be late by one day after 43 years in the aforementioned way.

[96] The fourth canon. If you wish to know the true position of the sun at the time of the true conjunction or opposition, seek in the fifth table, and from it you will also know the true position of the moon at the time of the true conjunction or opposition. [97] And know [f. 72ᵛ] that this table is arranged for the position of the true sun at each noon of the year 1321 of the Incarnation for the horizon of Orange. [98] And whoever wants to know by means of this table the true position of the sun at any time he wishes, needs to subtract 1320 from the years of the Incarnation, and, after subtracting 4 repeatedly from the result, to keep only the remainder, 4 or less than 4. [99] And if what has been kept is 1, the number you find in the table is the position of the sun at each noon of that year. [100] And if what has been kept is 2, the number you find in the table is [the position of the sun] 6 hours after each noon of that year. [101] If what has been kept is 3, the number in the table will correspond to [the position of the sun] 12 hours after each noon of that year. [102] And if what has been kept is 4, the number is [the position of the sun] 6 hours before each noon of that year.

[103] And know that for each year that has gone by since 1320 you must add 30 seconds to what you find in this table, and for each revolution of 4 years you must add 2 minutes, and similarly until you reach 43 years after the epoch. [104] However, when 43 years have passed after the epoch, descend one day in this table; [105] this means, what you find in it for the second day of March will correspond to the first day of March, and subtract from it 37 minutes and 54 seconds. [106] And proceed with the years of that revolution in the same way you did with the years of the first revolution after you have descended one day and subtracted from what you found there 37 minutes and 54 seconds. [107] For example: if 4 years

of this second revolution have passed, descend one day in this table and subtract from what you find in it only 35 minutes and 54 seconds, because you must add 2 minutes for the completed revolution of 4 years, and similarly for each revolution of 43 years which will come after the time of the epoch, that is, 1320; [108] that is to say, for each revolution of 43 years descend one day in this table, and subtract from what you find 37 minutes and 54 seconds for each revolution of 43 years, and similarly in each revolution of 4 years until the revolution of 43 years has been completed. [109] And do the opposite of all that [...] for the years before the epoch; [f. 73ʳ] this means, where you have added you will subtract, and where you have subtracted you will add, and where you have descended with the days of the year you will ascend.

[110] And after you know the true position of the sun at the exact time of the true conjunction or opposition, seek in the sixth table the right ascension of the degree in which the sun is [located], and if the conjunction or opposition is exactly at noon, the degree in which the sun is [located] is the beginning of the tenth house. [111] Seek in the seventh table the degree of the zodiac whose ascensions are equal to the right ascensions of the degree in which is [located] the sun, and that degree is the ascendant for that time for the horizon at which these tables are composed. [112] And if the true conjunction or opposition takes place before or after noon, multiply the hours and the minutes before or after [noon] by 15 degrees, and add the product to the right ascension of the degree in which is [located] the sun, if it takes place after noon, or subtract it, if it takes place before noon; [113] the result is the right ascension of the degree of the beginning of the tenth house and the ascension of the ascendent for this horizon for which these tables are composed. [114] And similarly from these ascensions you will find the degree of the zodiac which is the beginning of the tenth house and the degree of the zodiac which is the ascendent at this horizon, and from it you can find the 12 houses in the way explained by the Ancients. [115] And the easiest way to find them exactly is by using an astrolabe.

[116] And if the longitude of your horizon is different from the longitude of this horizon of Orange, and you know the difference, you will easily be able to find from this number the time of the true conjunction or opposition at your horizon. [117] And know that the difference in time of the conjunction or opposition at your horizon will be as much as the difference in longitude. [118] And if your horizon is to the east of this horizon of Orange, add that difference to the time for the true conjunction or opposition which will result from these tables, and you will obtain the exact time at your horizon. [119] And if it is to the west of the horizon of Orange, subtract them. [120] For example: if your horizon is [f. 73ᵛ] one hour to the east of that of Orange, add this hour to the time of the true conjunction or opposition which will result from this computation; or, if it is one hour to the west, subtract it. [121] And if your horizon is different in latitude from the horizon of Orange, you must first find the ascensions of the zodiac for your

horizon from the tables composed by the Ancients, and from it you will then find the ascendant for your horizon at the time of the true conjunction or opposition in the aforementioned way. [122] And these tables will also allow you to determine easily the longitude of your horizon if you observe a lunar eclipse with an astrolabe or with another instrument with which the hours can be known, and they will also allow you to determine the time of the eclipse-middle, which is the time of true opposition. [123] And note the amount of the difference between this time and the time that results from these tables; and it is the difference in longitude between your horizon and that of Orange.

4.3 Commentary

Ad [1]. 'A Jew of Orange' is missing in the Hebrew version (cf. Carlebach [1910], 35).

Ad [1]-[4]. These sentences are similar to sentences 5-14 of chapter 2 of the *Astronomy* (Goldstein [1985], 24-25), where Isaiah 40:26 is also quoted [32].

Ad [7]. This sentence is also developed in chapter 2 of the *Astronomy* (Goldstein [1985], 24, sentences 1-4). [33] The passage ascribed here to Ptolemy occurs in *Almagest* I.1 (Toomer [1984], 36-37: "With regard to virtuous conduct in practical actions and character, this science, above all things, could make men see clearly;

---

[32] The passage was rendered into Latin as follows: "Prophete etiam nos adducunt ad inquirendum profunde in ista scientia, quia ex ea multum in Dei cognitionem dirigimur ex hoc quia apparebit ex ista notitia quod stelle et omnia celestia corpora sunt Dei uerbo creata, sicut ostendetur in sequentibus huius libri. Item quia habemus ex istorum notitia aliquam cognitionem mirabilis sapientie Dei ac eius potentie maxime, quia tam nobilia corpora in ista dispositione mirabili sapientia sua creauit et influit ab eis diuersa ex quibus ista inferiora perficiuntur, et eorum suppletur defectus quamuis diuerse essentie, scilicet quinte, existant, nec in materia nec in qualitatibus cum istis communicet, et insuper sine labore et tedio continuant ea ad quem sunt ordinata per dispositionem diuinam. Propter que dicit Ysaias propheta: 'Leuate in celo oculos uestros et uidete quis creauit hec, qui eduxit in numero militiam eorum et omnes ex nomine uocat pro multitudine fortitudinis et roboris uirtutisque eius nec unum reliquum fuit'. Intentio autem prophete est quod leuemus oculos nostros ad uidendum et considerandum celestia corpora, ex quibus quis est eorum creator aliquam cognitionem habebimus, quia ipsorum consideratio nos dirigat ad habendum notitiam quod sunt creata et per consequens nos dirigat in notitiam Creatoris. Et ex hijs que apparent in eis nos dirigunt in notitiam mirabilis sapientie eius que in ipsorum creatione apparet et mirabilis eius potentie, quia numero qui erat necessarius inferioribus istis produxit eorum militiam sicut quod nec unum deficit. Et secundum hoc eis imposuit nomina ex quibus ad inuicem distinguntur et diuersam influentiam Dei uirtutis necessariam istis inferioribus habent" (Vat. Lat. 3098, f. 3[ra-b]).

[33] The corresponding Latin version reads: "Et ad hoc mouemur tum quia predicta scientia in se est digna et nobilis, tum quia nos dirigit in scientias alias. Quod sit multum nobilis in se ipsa

from the constancy, order, symmetry and calm which are associated with the divine, it makes its followers lovers of this divine beauty, accustoming them and reforming their natures, as it were, to a similar spiritual state").

Ad [8]. Levi asserts repeatedly in the *Astronomy* (*e.g.*, chapter 17: Goldstein [1985], 105-106, or chapter 43: see below, comments to sentence [22]) that Ptolemy's models for some planets do not agree with their observed sizes. However, with regard to their motion in longitude and latitude, in the Hebrew version of chapter 18 Levi points out that it is not possible to use observations to refute Ptolemy's models without first determining the mean position of the Sun, adding that even after having done it, discrepancies found by observation could be attributed to errors in the position of the apogees, or in their motion in longitude (superior planets) or anomaly (inferior planets), or in the determination of the positions of the fixed stars with which they are observed (Goldstein [1985], 108-109). In chapter 46 we are told that "this also led us at first to accept the views of the ancients concerning the positions of the fixed stars, the apogees for each planet, as well as the positions in longitude and anomaly, as we already mentioned, for we did not determine the position of the mean Sun perfectly until the year 1335 according to the Christian reckoning" (Goldstein [1988], 387:36). [34] Moreover, Levi did not determine with accuracy the center of vision – a crucial point insofar as most of Levi's observations were made with the Jacob staff – until the end of 1335. [35] Thus, although the early dated planetary observations reported in the

---

manifestum est ex conditione subiecti, quia secundum gradum nobilitatis subiecti est gradus nobilitatis scientie. Notum est autem quod subiectum istius scientie est corpus celeste quod inter cetera naturalia corpora optinet principatum et cuius forma est nobilior qualibet naturali in tantum ut corpus istud nobile ad alia naturalia corpora comparatum ipsa sine ulla proportione excedat et forma eius comparationem non habeat ad formas alias naturales; ymo non habent unitatem equiuocatione carentem. Et istud est declaratum in prima philosophia et scientia naturali perfecte. Quod nos uero dirigat in scientias alias satis est euidens. Nam ad methaphysicam [et] scientiam naturalem nos mirabili directione adducit, fructus enim et finis totius methaphysicalis notitie est aprehensio illarum formarum sublimium que corpora ista mouent et aspectus earum ad omnium causam primam. Et ob hoc methaphysica sapientia diuina uocatur. Ipsa etiam directione aliquali nos dirigit ad philosophiam moralem et hoc luce est clarius in principio Almagesti" (Vat. Lat. 3098, f. 3ra).

[34] Latin text: "Et ex predictis iam docuimus inueniri loca stellarum fixarum quantum sufficit, quod potest fieri postquam est habita scientia medij motus solis modo quo superius docuimus, quia sine hoc non possumus hanc inquisitionem complere, ut dictum est supra. Et quia dictam scientiam nundum habueramus superius, coacti fuimus primo recipere ab antiquis illud quod ab eis recepimus de motibus supradictis, quia perfecte locum medij motus solis non habuimus usque ad annum incarnationis Christi 1335, ut in futuro patebit cum de sole loquemur" (Vat. Lat. 3098, f. 38rb).

[35] "Et sciendum quod multum laborauimus circa experientias predictas nec dispositionem consentientem inuenire potuimus, quia computabamus loca stellarum fixarum secundum compu-

Astronomy occurred in 1325, the sentence in the Provençal text ("we do not find any error whatsoever in this long time") suggests that the *terminus post quem* for Levi's research on planetary models is also the end of 1335, and it can be interpreted as a simple way of avoiding a careful explanation of a problem still not examined (which is beside the point in a work devoted only to the computation of syzygies). In the preseved copies of the Hebrew version of Levi's *Astronomy*, the planetary theory (chapters 103 to 136 according to the table of contents) is incomplete; the text of 11 of these 34 chapters is simply lacking (104, 105, 107, 108, 111, 112, 115, 116, 119, 120, and 122) whereas other 3 are apparently unfinished (118, 121, and 122). Furthermore, chapters 109, 113, and 117, that contain dated observations of Saturn, Jupiter, and Mars, respectively, from which Levi inferred that it was necessary to correct the positions of the apogees of these planets, and also the motion in longitude of Jupiter and the eccentricity of Mars (Goldstein [1988], 390, 394, and 396), were beyond doubt redacted after January 1339. Consequently, it is very likely, in my opinion, that Levi's planetary models were never finished. Levi's troubles with planetary theory [36] were due to an initial condition for his lunar model (the correction at 0° of anomaly is always 0°), which he apparently based on observational arguments [37], and later extended to the

---

tum Albeteni, quousque peruenimus ad eclipsim lunarem que fuit anno Christi 1335, die 3 octubris, et tunc inuenimus Aldebaram in minori quam esset consequens ad computum Albeteni circa 0;22°. Et tunc iudicauimus quod motus stellarum fixarum a tempore Albeteni usque ad nostrum erat minor in quantitate predicta quam esset consequens ad Albeteni, cum hoc quod centrum uisus non erat nobis in principio scitum perfecte, quousque multum minutiando peruenimus ad eius perfectam notitiam predicte experientie tempore. Et istud fuit complementum cum eo quod innouauimus in sole et luna anno Christi 1335 in mense decembris. Sit Creator omnium benedictus, qui ad hanc perfectionem nos duxit sua pietate et gratia. Amen." (Vat. Lat. 3098, f. 81^{rb-va}.)

[36] "Et habebamus magnum appetitum habendi scientiam horum antequam haberemus omnes experientias necessarias ad complendum istam inquisitionem perfecte, quia scientia quam inueneramus in generali de dispositione cuiuslibet planete timebamus deficere si defecissemus in uita antequam esset in quolibet planetarum in speciali completa. Et laborauimus cum argumentis ingenij modo quo declarabitur infra in tantum quod aliquando inueniebamus dispositionem concordem cum nostris experientijs ac etiam antiquorum et cum multis alijs quas inueniebamus post illas. Et quando gaudeabamus ex istis experimentis credendo ipsa punc-taliter esse inuenta, extrema gaudij tristitia occupabat, quia inueniebamus experientiam unam que ualde elongabatur ab experientijs dispositionis inuente in illo planeta de quo agebatur. Et quia tunc non ponebamus diametrorum latitudinem in principio motus diuersitatis in 36° capitulo multum a ueritate elongabamur ista de causa, et plus circa principium motus diuersitatis in Marte" (chapter 46, Vat. Lat. 3098, ff. 37vb-38ra; cf. Goldstein [1988], 386).

[37] "Item apparet quod dispositio nostra sit uera ex eo quod de latitudine diametrorum motus diuersitatis uidemus, que nullo modo cum dispositione Ptolomei concordat, ut patet ex duabus experientijs superius recitatis; quarum una [on June 24, 1333] a Ptolomeo discordat de 1;42°,

302

planetary models. According to chapter 100 (Goldstein [1979], 118-120, and 123), observations made in 1337 and 1339 (the latest one: 11 December 1339) suggested that it was probably necessary to accept (a) some correction for the Moon at 0° of anomaly, and (b) that at 180° of anomaly the Moon is nearer to the earth than at 0° of anomaly – a variation in distance which cannot be produced by the model [38]. Levi then redacted chapters 101 and 102 (the latter incomplete), devoted to the preliminary work on a new lunar model [39].

Ad [10]-[11]. The Christians are mentioned only in the second part of the corresponding sentence in the Hebrew version, where "scholars" replaces "clerics" (Carlebach [1910], 37).

alia [on January 17, 1334] circa 1;1°. Item ex sua dispositione propter latitudinem diametrorum in principio motus diuersitatis sequeretur equatio; quam equationem non inuenimus in una experientia quam habuimus stante luna in locis medijs inter augem et eius oppositum, quia ibi equationem non inuenimus nisi illa que ad illum locum pertinebat pro motu diuersitatis" (chapter 76, Vat. Lat. 3098, f. 64ᵛ). The first two observations are discussed in chapter 71 (latest observation mentioned: February 1334), but I have found no evidence which allows us to date the last one. Levi's expression "latitudo diametrorum" (inclination of diameters) must be considered in the context of a physical interpretation of Ptolemy's models, in which the epicyclic mean apogee is determined by the line joining the centers of the epicycle and the equant, producing thus an oscillatory motion of the diameter of the epicycle (see comments to sentence [22]). On Levi's lunar model see Goldstein (1972), and (1974a).

[38] See, for instance, the following passage of the Latin version of chapter 100: "Et ex experientia quinte eclipsis lunaris supradicte [= 26 Jan. 1339] uidetur quod tempus eius a medio usque ad finem fuit maius in aliquo quam sequeretur ex ista doctrina; *quod si est uerum necesse esset ponere propinquiorem lunam nobis in 180° motus diuersitatis quam in eius principio.* Et istud nos mouet quod circa huius inuestigationem intendamus proposse, quod et nunc facimus. Et ad dirigendum in uiam ueram dispositionem lune inueniendi faciliter cum omnia proposita inquirenda de luna inquisiuerimus sciuerimusque perfecte, talem dispositionem lune proponimus ordinare cuius equationes consentant radicibus Ptolomei; quo facto facile erit deuenire in dispositione ista fienda in quantitates latitudinum diametrorum quas per experientias inueniemus... et hoc statim fiet in sequenti capitulo" (Vat. Lat. 3098, f. 90ᵛ). Although in both versions of chapter 100 the problem seems still undecided, other chapters provide evidence enough to think that Levi finally decided to modify his lunar model. Thus, at the end of chapter 76, after having stated that observations confirm that there is no inclination of the diameters of the spheres of the motion in anomaly at the beginning of this motion, a marginal note adds: "Nobis finaliter fuit notum ex eclipsibus solaribus et lunaribus quod necesse est *ponere aliam latitudinem diametrorum in principio motus diuersitatis,* que potest poni per modum spere latitudinis, ut dictum est supra" (Vat. Lat. 3098, f. 64ᵛ). See also Appendix, 99:3.

[39] Levi's theory of planetary distances developed in chapters 130-135 (Goldstein [1986]) requires only parameters contained in the preliminary (and incomplete) chapters 103 (Venus), 106 (Mercury), 110 (Saturn), 114 (Jupiter), and 118 (Mars), and these parameters are not related to the problem of the correction at 0° of anomaly. Moreover, the distances given in these chapters depend on the solar model I and on the lunar model described in chapter 71; see Mancha (1996), section 3.4. Consequently, chapters 130-135 were composed before, and not after, the aforementioned observations of 1339.

Ad [12]-[13]. Astrology is never mentioned in the Hebrew and Latin versions of the *Astronomy*. The influence of the celestial spheres on the sublunary world, stated here in sentences [2]-[5], repeated by Levi in chapter 2 of the *Astronomy* (cf. note 33 above, and Goldstein [1985], 24-25), and developed in chapters 7 and 8 of the *Wars of the Lord*, V, 2 (Goldstein [1976], 222) are a commonplace in medieval philosophy, with which even Maimonides (*Guide*, II.10), a well known opponent of astrology, agreed. The only known works of Levi on astrology are two predictions, one dedicated to Pope Benedict XII (apparently lost, written in 1339), and the other to Clement VI (Goldstein and Pingree [1990], esp. 4 and 32); in the introduction to the extant text Levi points out that, nevertheless, these influences "can be changed in two ways: first by man in respect to those things which are subject to free will, secondly by the grace of God" (*ibid.*, 11 and 30). It is thus a plausible hypothesis that on astrological matters Levi's motivation was rather that of the Christian circle than his own.

Ad [19]-[20]. The instrument to which Levi refers is the Jacob Staff, called by him the "revealer of profundities" (Goldstein [1985], 71), and "instrumentum reuelator secretorum and baculus Jacob" in the Latin version (Mancha [1992], 36); Levi devoted chapters 6 to 11 of the *Astronomy* to its construction and different uses (Goldstein [1985], 51-81). In the Prologue to the *Tractatus instrumenti astronomie* Levi tells that the discrepancy between observation and computation according to Ptolemy's models for the magnitude of the solar and lunar eclipses of 26 June and 9 July 1321, respectively, motivated his invention of an instrument, easy to construct, which allowed accurate observations [40]. In a marginal note to the Provençal tables the solar apogee is located at Cnc 2;39° in 1321 (Cnc 2;38° according to chapter 99:195 of the Latin version of the *Astronomy*: see Appendix). In chapter 57 it is located "near the end of the third degree of Cancer" in 1334, [41]

---

[40] "Ego uero considerans dictam controuersiam tam magnam inter antiquos philosophos necnon inter modernos, et inueniens diuersitatem non modicam ab eo quod erat consequens ad sententiam Ptolomei in digitis eclipsatis in luna tempore eclipsis lunaris que fuit anno incarnationis Christi 1321, 9 die Julij, et in sole tempore eclipsis solaris precedentis inmediate dictam lunarem, quarum quantitas fuit longe maior quam esse debuisset secundum Ptolomei doctrinam, motus fui ad inquirendum instrumentum ueridicum quod nos duceret in ueritatem omnium predictorum. Et Deus sui gratia oculos meos aperuit ad inueniendum unum instrumentum leuis facture, quod ducit faciliter in ueram notitiam omnium predictorum et aliorum multorum magis desideratorum scibilium circa celestia corpora" (Brescia, Bibl. Civica Queriniana, A. IV. 11, f. 1ʳ; Klagenfurt, Bischöflische Bibl., XXX.b.7, f. 23ᵛ; Wien, Nationalbibliothek, 5277, f. 41ᵛ).

[41] "Et scias quod de hoc multiplicauimus experientias ualde sepe et semper inuenimus augem circa finem 3 gradi Cancri, dato quod ex aliqua experientia aliquando inuenerimus eam in modico uariari in plus uel in minus. Et hiis omnibus consideratis iudicauimus eam esse in fine 3 gradi Cancri in istius experientie tempore, cum etiam quia hoc experientijs Ptolomei et Albeteni consentit, ut declarabitur in futuro" (Vat. Lat. 3098, f. 46ʳᵃ).

304

and the motion of the solar apogee in this time interval agrees very well with Levi's parameter, 1° in approximately 43 Egyptian years (see commentary to Table XIV in section 5). [42]

Ad [21]. We already know that an eccentricity of 2;14 was derived in chapter 56 from observations made in 1334 using the Jacob Staff (Mancha [1992a], 289-297); consequently, I interpret Levi's reference here to lunar eclipses to mean that those he observed confirmed the aforementioned value. Levi claims in chapter 59 that the most sensitive tests for the solar eccentricity are eclipses that occur when the Sun is near the place of its maximum equation [43]. Of the five lunar eclipses observed by Levi before March 1337 and reported in chapter 80 of the *Astronomy*, the only one that closely satisfies such a condition took place on 3 October 1335, for which the true solar longitude is given as Libra 17;3,5° (where the solar correction is only 0;7° smaller than its maximum, 2;8°: see Goldstein [1979], 143-144). This eclipse played a significant role in the *Astronomy*, since Levi also used it to derive his value for the mean synodic month (see comments to sentences [32]-[33]), the mean solar distance (2140 t.r.) which is almost twice the Ptolemaic value (see Mancha [1993], 109 and 117-118), and the rate for precession (1° in 67 Egyptian years, 123 days, and 16;40[h]) slightly different from al-Battānī's (Goldstein [1975], 36).

Ad [22]. The Hebrew version does not mention the error of up to 29 minutes which can be produced in the computation of the syzygies (Carlebach [1910], 38). Besides the observational arguments mentioned in this sentence and the following one (see below), in chapter 43 of the *Astronomy* Levi rejected the existence of the epicyclic models described in Ptolemy's *Planetary Hypotheses* because (1) the spots in the surface of the Moon appear always to us in the same disposition (an argument widely used in the Latin West from the XIIIth to the XVIth centuries that appeared, as far as I know, for the first time in Bacon's *Opus tertium*, ca. 1267; cf.

---

[42] Levi assigned such a position to the solar apogee (accurately, Cnc 2;56,28°) by means of an iterative procedure similar to Ptolemy's method in *Almagest*, X.7, from solar observations made on 24 April and 4 August 1334 (see Mancha [1998], section 3.6).

[43] "Et expedit scire quod sicut inquisitio sciendi temporis quantitatem quo tota reuolutio solis completur fieri non potest perfecte prius quam de motu lune tractetur, sic est de motu augis solis; quod in hoc quod est hic declaratum de loco augis est aliquod dubium, ut dictum est supra, quousque hoc sit perfecte ex eclipsibus solaribus et lunaribus declaratum. Et tunc declarabitur etiam illud quod de hoc diximus multum consentire experientijs antiquorum. Et inde etiam habebimus adiutorium ad inueniendum quantitatem equationis solis finalis, quia quando habebimus experientias eclipsium stante uero loco solis prope locum equationis maioris tempus quo ueniet eclipsis nos docebit quantitatem equationis maioris. Verbi gratia, si erit eclipsis stante sole in Piscibus uel Ariete, erit notum quod quanto erit maior solis equatio tanto tardior erit eclipsis, et econtrario sole in Vergine stante uel Libra. Et ideo est notum quod quando habebimus duas eclipses hoc modo nos uere ducent in quantitatem equationis solis maioris, et sic fecimus ut recitabimus in futuro" (Vat. Lat. 3098, f. 47[vb]).

Duhem [1909], 133; Gabbey [1991], 115); (2) the apparent planetary diameters do not vary according to what would be expected from Ptolemy's models; (3) the alignment of the epicyclic mean apogee with the equant point requires to move the epicyclic sphere by turns in opposite directions producing so a point where the epicycle is at rest, something excluded by natural philosophy; and (4) the Ptolemaic small circles which account, in the inferior planets, for the inclination (*egklisis*) of the apogee-perigee diameter of the epicycle, and the slant (*loxōsis*) of a second diameter laying between the mean longitudes of the epicycle and perpendicular to the first, are physically impossible [44]. Argument 2, for the case of the Moon, was also used in the fourteenth century by Ibn al-Shāṭir, who produced a model in which this effect was eliminated (cf. Roberts, 1957), and some inconsistencies in

---

[44] "Item per experientiam et demonstrationem geometricalem declarabitur magis quod est impossibile poni epiciclos modo quo posuit Ptolomeus rationibus multis. Prima ratio est quia in luna continue unam maculam nos uidemus semper in statu eodem et in eodem loco ipsius, que est uera et non fantasia uel apparentia solum, ut declarabitur in futuro. Et hoc est etiam declaratum in philosophia geometricali, ut dicit Philosophus in libro *Celi et mundi*. Et si in luna esset epiciclus modo quo posuit Ptolomeus, esset necessarium quod dicta macula non uideretur in ea semper modo eodem, set esset aliquando in parte nostri aspectus, aliquando in parte nobis occulta, aliquando partim uideretur a nobis et partim non [...] Secunda ratio est quia si epiciclis in erraticis planetis ponantur taliter quod ex eis sequatur diuersitas uisa in motibus eorumdem ut posuit Ptolomeus, erit necessarium quod diameter eorum in aliquo uideatur maior multo ipso stante in depresiori loco epicicli quantitate uisa in eo ipso stante in altiori loco ipsius, in tantum quod sequitur quod diametri Veneris et Martis in uno maiores dictorum locorum quam in alio plus quam in sextuplo uiderentur, et diameter lune quasi in duplo. Et hoc per experientiam non uidemus in aliquo predictorum, ymo quantitates diametro-rum eorum diuersificari non de uno loco ad alium nisi modicum [...] Tertia demonstratio geometricalis impossibilitatem epiciclorum ostendens est ista: quia in latitudine diametrorum epiciclorum, que est necessaria poni secundum dispositionem Ptolomei ut ex ea sequetur illud quod nos uidemus de motibus planetarum errantium, impossibilitas inuenitur; quia latitudinem istorum diametrorum semper cssc ad centrum deferentis sine positione alterius motus neces-sario est omnino, quia diameter terminans maiorem et minorem distantiam epicicli semper in directo centri deferentis, ut patet cuilibet intuenti, et hoc est superius clare ostensum. Set Ptolomeus fuit constrictus, ex eo quod de experientia inuenit de motibus planetarum erran-tium, quod in omnibus poneret dictam diametrum dirigi ad alium punctum quam ad centrum predictum. Et non declarauit modum quo posset poni in diametro talis motus, nec mirum, quia impossibile est ipsum ullius nature ymaginationi subesse. Nam et aliqui naturales cognouerunt inconuenientia motus istius quem posuit in predictis diametris, et per rationes naturales quod est impossibile probauerunt; quia motus diametrorum istorum per totum girum non circuit, sed partim circuit et retrocedit. Et motus qui per totum girum non circuit, set partim circuit et retrocedit, describit punctum in actu in quo necesse est quietem, ut in octauo *Physicorum* Aristotiles declarauit, quam corpori celesti inesse impossibile declaratur in scientia naturali [...] Quarta ratio quia in latitudinibus Veneris et Mercurij – que necessarie sunt poni modo quo eas posuit Ptolomeus ponendo epiciclos, taliter quod ex eis sequatur illud quod de eorum latitudinibus nos uidemus – inuenitur illud quod nostre ymaginationi aliquo modo subesse non potest ..." (Vat. Lat. 3098, ff. 33[ra-b], and 34[rb-va]).

306

Ptolemy's theory of planetary distances and sizes were also pointed out by al-'Urḍī in the thirteenth century (Saliba, 1979; Goldstein & Swerdlow, 1970). Arguments 3 and 4 were used at least since al-Jūzjānī's *Kayffiyat tarkīb al-aflāk* and Ibn al-Haytham's *al-Shukūk ʿalā Baṭlamyūs* (eleventh century; see Sabra, 1979; Saliba, 1980), although the most likely sources of Levi are Maimonides' *Guide of the Perplexed* (II.24; Munk [1856-1866], II, 190) and the Hebrew version of Averroes' *Epitome of the Almagest* (Lay [1991], III, 63-65, 128-131; I, 36-38, 49-54). As for Ptolemaic eccentrics, and leaving aside arguments derived from planetary apparent diameters, in chapter 43 Levi repeated Maimonides' objection (*Guide*, II.24; Munk [1856-1866], II, 187) that the centers of many planetary spheres must participate in the motions of the inferior ones, since they lay between the spheres of the Moon and Mars, and avoided it by assuming a fluid of variable thickness between planetary orbs (Goldstein, 1986; Mancha [1996], section 3.4).

Ad [23]. From Ptolemy's lunar model it follows that the diameter of the Moon should be close to twofold greater at quadrature at 180° of anomaly than at opposition at 0° of anomaly. In several chapters of the *Astronomy* Levi repeatedly claims that, according to his observations, the diameter of the Moon at quadrature (s_q) is only slightly greater than its diameter at syzygies (s_s), and that there is no perceptible variation observing its diameter at 0° and 180° of anomaly [45]. However, nowhere in the text is a numerical value given for the "very small" difference between the apparent diameters of the Moon at syzygy and quadrature, here stated as

---

[45] "Iam diximus aliqua uerba de luna ex quibus sine dubio demonstratur quod spere eius non possunt esse secundum dispositionem secundum quam eas posuit Ptolomeus, quia secundum eius doctrinam diameter lune in aliquibus temporibus in quartis deberet uideri maiori quasi in duplo quam in aliquibus temporibus in oppositionibus, scilicet quando esset in oppositionibus in maiori altitudine epicicli et in quartis in depressione maiori eiusdem; quod non est uerum. Nam accepimus multas experientias in temporibus istis cum nostro instrumento, in quo error esse non potest, nec inuenimus diuersitatem predictam in uno tempore quam in alio nisi modicum esse maiorem" (chapter 17, Vat. Lat. 3098, f. 13rb; Goldstein [1985], 105); "Et de hoc habuimus experientiam ueram in luna cum nostro instrumento superius nominato ex diuersitate radiorum transeuntium per fenestram, et non inuenimus diferentiam in quantitate diametri radij lune de uno loco ad alium nisi modicam ualde" (chapter 43, Vat. Lat. 3098, f. 33rb); "Quarta probatio sumatur ex quantitate diametri lune quam in oppositionibus et quartis et in diuersis locis motus diuersitatis uidemus, quam quantitatem diuersam notabiliter non uidemus; in ueritatem quantitatis predicte possumus deuenire cum instrumento baculi superius nominati. Et quando quesiuerimus cum probationibus antedictis inueniemus necessario quod ex eis concluditur quod luna non est in epiciclo firmata nec in spera excentrica modo quo posuit Ptolomeus" (chapter 72, Vat. Lat. 3098, f. 63^v); "Set ut inuenimus diameter non est maior in quartis quam in oppositionibus nisi modicum, nec in 180° motus diuersitatis quam in principio maior in aliquo notabili inuenitur; quare sequitur quod luna epiciclum non habeat nec speram excentricam modo quo posuit Ptolomeus" (chapter 75, Vat. Lat. 3098, f. 64va).

$$s_s = s_q - (s_q / 25).$$ (1)

However, it can be recomputed from other values given in chapter 92 (Vat. Lat. 3098, f. 78ra), where Levi asserts that at the time of the lunar eclipse of 3 October 1335 the angular radius of the Moon was 0;13,55,30°; that is, $s_s = 0;27,51°$. In the same chapter, the radius of the Moon is 0;15,9,32 terrestrial radii, and the distance between the center of the earth and the convexity of the highest sphere of the Moon is 62;48,42,36 t.r. Consequently, the distance between the centers of the earth and the Moon at syzygy (d_s) is

$$d_s = 62;48,42,36 - 0;15,9,32 = 62;33,33,4 \text{ t.r.}$$ (2)

Since, according to Levi's lunar model III, the ratio of the relative distance between the centers of the Moon and the earth at syzygy and that relative distance at quadrature is 65;16 / 62;44, the distance between the centers of the earth and the Moon at quadrature (d_q) is 60;7,51,23 t.r. Since the angular diameter of the Moon is measured from the surface of the earth, we subtract 1 t.r. from d_s and d_q, and the result is

$$s_q / 0;27,51 = 59;7,51,23 / 61;33,33,4.$$ (3)

Thus, $s_q = 0;28,59,37°$ ($\approx$ 0;29°), and this satisfies equation (1) very well (for an edition of the Latin version of chapters 90-92 of the *Astronomy*, see Mancha [1993]).

Ad [24]-[26]. For a comparison of Ptolemy's and Levi's lunar equations, cf. Goldstein [1974a], 55-66, and Goldstein [1974b], 280-281. In sentence 26 Levi refers to chapter 71 of the *Astronomy*. For some examples of comparison of lunar positions computed according to Ptolemy's and Levi's models with observations made by Levi himself, see Goldstein [1974a], 72-73 and 85-86.

Ad [32]-[33]. In chapter 64 of the *Astronomy* the length of the mean synodic month is given as 29;31,50,7,54,25,3,32^{d}, or 29^{d} 12;44^{h} and nearly 1/1138 of an hour. Levi also tells us that his value is a little lesser than Hipparchus' and Ptolemy's value, given as 29;31,50,8,9,20^{d}, which is very near to the value with which agreed "our ancient scholars", that is, 29;31,50,8,20^{d} or 29^{d}, 12^{h}, and 793/1080 of an hour [46] (an allusion to Abraham Bar Ḥiyya, Abraham Ibn Ezra, and

---

[46] "Et dico quod Ptolomeus declarauit experientijs antiquorum et suis, et Abarcas ante eum declarauit hoc idem, scilicet, quod tempus medij mensis lunaris est 29;31,50,8,9,20^{d}. Et nos inuenimus istum computum ita ueritati propinquum quod in toto de cursu temporis a Ptolomeo usque ad presens non inuenitur defectus nisi 0;12°, in quibus inuenimus distantiam

Maimonides; see Millás Vallicrosa [1959], 55; Millás Vallicrosa [1947], 98-99, and Neugebauer [1967], 114, respectively). Levi derived his value for the mean synodic month dividing the time elapsed between the lunar eclipse of 2/3 Choiak, Hadrian 19, observed by Ptolemy (*Almagest*, IV.6; Toomer [1984], 198, 203: 139 AD, Oct. 20/21), and the lunar eclipse of 3 October 1335, observed by himself, $438647;24,57,30^d$, by 14854 months (accurately computed $29;31,50,7,54,25, 3,35,41...$). [47] Ptolemy (*Almagest*, IV.3; Toomer [1984], 175-176) says that Hipparchus, dividing $126007^d + 1^h$ by 4267 synodic months, found the mean length

lune a sole in tempore nostro maiorem quam esse deberet secundum computum Ptolomei. Et iste computus quasi consentit computui cui consenserunt sapientes nostri antiqui, qui ponebant tempus mensis lunaris 29 dies, 12 horas, 793 puncta, atribuendo 1080 puncta hore cuilibet, que sunt $29;31,50,8,20^d$; qui computus excedit computum Ptolomei in $0;0,0,0,10,40^d$ [...] Set nos in hoc considerantes subtiliter per experientias antiquorum et nostras, ut declarabitur in futuro, ista inquisitione completa inuenimus tempus medij mensis lunaris $29;31,50,7,54,25,3,32^d$ [...] Et secundum computum nostrum esset mensis lunaris 29 dies, $12;44^h$ et circa unum punctum atribuendo hore 1138 puncta" (Vat. Lat. 3098, f. 57rb). On the Babylonian origin of the division of the hour into 1080 parts, see Neugebauer (1967), 117.

[47] "Et dico quod nos in scriptis nomine sapientis magistri Abrae Abenazerij quod uilla Montis Pesulanj est occidentalis Yerusalem $2;16^h$. Et est notum quod inter Montem Pesulanum et Auraycam non est in longitudine notabilis diuersitas, quia dicte uille sunt multum propinque, et in latitudine est maior earum diuersitas, quia Allexandria est occidentalis Yerusalem circa $0;20^h$; sequitur quod Aurayca sit occidentalis Allexandrie circa $1;56^h$. Quo posito dicimus quod Ptolomeus recitauit in Almagesti 6 capitulo quarti libri in secunda eclipsium quas uidit in Allexandria quod tunc erat uerus locus solis in Libra $25;10°$. Et hinc est notum quod tunc erat medius locus solis in Libra $26;42°$, quia secundum computum Ptolomei solis equatio erat in illo loco $1;32°$. Et ibi dicit Ptolomeus quod medius locus lune in Ariete tunc erat $29;30°$; quare est notum quod distantia medij loci lune a medio loco solis erat $182;48°$. Et hoc fuit post Ptolomei radicem, ut recitat ipse ibidem, 881 annis egiptiacis, 91 diebus, 10 horis equalibus, 30 minutis. Et tempus inter Ptolomei radicem et tropicum estiualem superius recitatum fuit 2081 anni egiptiaci, 262 dies, 16 hore equales, 49 minuta. Et quia eclipsis ultimo recitata, que fuit anno Christi 1335, die 2 mensis octubris, 15 horis et circa 9 minutis post equalem meridiem in oriçonte Aurayce, fuit post predictum tropicum 1 anno egiptiaco, 111 diebus, $0;16^h$, est notum quod ista oppositio fuit post Ptolomei radicem 2083 annis egiptiacis, 8 diebus, $17;5^h$. Et tunc fuit distantia medij loci lune a medio loco solis $181;14,18°$; a quo tempore subtracto tempore elapso inter dictam radicem et eclipsim quam recitat Ptolomeus, remanet tempus medium inter eclipsim Ptolomei et nostram 1201 anni, 282 dies, 6 hore, 35 minuta. Et est notum modico studio quod istud tempus reuolutiones mediorum mensium lunarium completas includeret, si distantia lune a sole fuisset tanta in ista eclipsi quanta fuit in prima; set secunda deficiebat a prima respectu distantie in $1;43,42°$, que sunt motus distantie in 3 horis et circa 24 minutis. Et ideo est notum quod si $3;24^h$ adderentur medio tempore inter dictas duas eclipses recluderentur reuolutiones perfecte mensium mediorum lunarium quarum tempus esset 1201 anni egiptiaci, 282 dies, 9 horas, 59 minuta; quod est $438647;24,57,30^d$. Et est notum modico studio quod istud tempus includit 14854 menses lunares. Et si diuidamus istud tempus per numerum lunarium mensium, sequitur quod tempus medij mensis lunaris sit $29;31,50,7,54,25,3,32^d$. Et hoc est minus eo quod posuit Ptolomeus $0;0,0,0,14,54,56,28^d$, quod minus induceret magnum errorem in computu coniunctionum et oppositionum" (chapter 82, Vat. Lat. 3098, f. 70ra).

of the synodic months "approximately 29;31,50,8,20 days". The division, in fact, does not yield this result (Aaboe, 1955), but 29;31,50,8,9,20^d, and this is the value ascribed by Levi to Ptolemy. It has been suggested that Copernicus (*De revolutionibus*, IV, 4) was the first to check Ptolemy's account and to give the "correct" result (Neugebauer [1975], 310) following in this point Regiomontanus's interpretation (Swerdlow & Neugebauer [1984], 199). However, a more plausible explanation (which also avoids Levi's apparent inconsistency in attributing to Ptolemy the "corrected" value and to his Hebrew predecessors the Ptolemaic one) is the following. Levi, like Copernicus, merely took the figure ...8,9,20 from the Hebrew and Latin, respectively, translations of the *Almagest*, both deriving from al-Ḥajjāj's Arabic translation, which has ...8,9,20 (Ms. Leiden, Or. 680, f. 50b:6). It is known that Gerard of Cremona's translation – see, for instance, the printed edition, Venetiis, Lichtenstein, 1515, f. 36^r – was made using al-Ḥajjāj's version for Books I-IX, and Ishāq-Thābit's version for Books X-XIII. Ishāq-Thābit's version has Ptolemy's figure ...8,20 (Tunis, Bibliothèque Nationale, 07116, f. 53b:19-20; there is at this place in the manuscript a marginal note which reads "in the translation of al-Ḥajjāj nine fourths and twenty fifths", thus confirming both readings). Jacob Anatoli's Hebrew version of the *Almagest* also gives ...8,9,20 (Turin, Biblioteca Nazionale Universitaria, A.II.10 [Peyron catal., no. 30], f. 40^v:27), and this agrees with the apparent dependence of the Hebrew version on the Latin one by Cremona (Zonta [1993], 332). Georg of Trebizunt's Latin translation, apparently made from the Greek, has also ...8,9,20 (Venetiis, Junta, 1528, f. 33^r). Al-Bīrunī's *al-Qānūn al-Masʿūdī* (al-Bīrunī [1954-1956], book 7, ch. 2; vol. 2, p. 730) gives ...8,9,20,13, a value very close to the accurate result (126007;2,30 / 4267 = 29;31,50,8,9,20, 12,22...). A similar value is found in Ibn Yunus' *al-Zīj al-Ḥākimī*, where the length of Muḥarram is given as 29;31,50,8,9,24^d (Delambre [1819], 96; Ms. Leiden, Cod. Or. 143, p. 20; however, on p. 12 the mean synodic month is 29;31,50,8,9,20^d, and the value at both places for 12 months is 354;22,1,37,52,48^d, which derives from ...8,9,20); this same value, ...8,9,24, is used by Abū Shāker in his *Chronology* (Neugebauer [1987], 280), and it is attributed to Ptolemy by al-Biṭrūjī (Goldstein [1971], 145).

Ad [38]-[39]. Orange is omitted in the corresponding passage of the Hebrew manuscripts (Carlebach [1910], 40). The Hebrew version modifies sentence 39 as indicated in note 17 of the Introduction. The tables alluded here are those numbered IX-XXI in section 5.

Ad [40]. Levi's year begins on 1 March; therefore, the epoch of the tables is noon, the last day of February, 1320 (= 28 February 1321 in our style). In the medieval civil calendar it was common to use the *stylus communis* which placed new year's day on 1 January (*stylus circumcisionis*), but the so-called *stylus annunciationis* or *stylus incarnationis*, which placed new year's day on 25 March (the feast of Annunciation and the old Roman vernal equinox), was also widely used. After the Gregorian reform of the calendar (1582), 1 January was adopted as new year's day by Roman Catholic countries (cf. Pedersen [1983], 62). The sentence which informs us that in Orange the new year's day was 25 March is absent from the Hebrew version (Carlebach [1910], 41).

Ad [41]. Levi's choice of 99 lunations as the time covered by Circle 1 is obviously reminiscent of the *octaëteris*, a period well known to Greek astronomers (Neugebauer [1975], 568 and 584-585), used in some Eastern canons of the Alexandrian church (Dionysius, Hyppolitus; cf. Pedersen [1983], 31-39), which stated that 8 solar years = 99 lunar months.

Ad [46]-[53]. Using Levi's value for the mean synodic month, 29;31,50,7,54,25,3,32^d, the period of each one of the four circles, accurately computed, is 8 years, 1 day, 12;41,13,6,56,... hours; 552 years, 105 days, 11;24,4,58,... hours; 22091 years, 201 days, 6;3,19,... hours, and 618563 years, 156 days, 7;32,57,... hours, respectively. Although the period of the third circle is given in the text only to days, Levi used the fractions of days to compute the whole period of the fourth circle.

Ad [56]. The first line of a Circle 5 (not constructed by Levi) should be identical to line 28 of Circle 4 (years 323, days 156, weekday 7, motion in anomaly 8°: cf. Goldstein [1974a] 228). Thus, the second line should read 'years 646, days 313 [= (156 × 2) + 1 leap day], weekday 7, motion in anomaly 16°'.

Ad [57]-[63]. The description of these tables (numbered II-VIII in section 5) is slightly more developed in the Hebrew version (Ms. *A*, f. 14b). The latitude of Orange is given in [63] as 43;59,44° N (modern: 44;8°).

Ad [65]-[85]. Let us consider the following example of computation using Levi's tables for mean syzygies. We wish to know the time of the first mean opposition on April 1888 and the corresponding lunar motion in anomaly. From 28 February 1320 until the beginning of April 1888 in the Julian calendar there elapsed 567 Julian years + 31 days, that is, 567;5,5 Julian years. Dividing this number by 552;17,19 (the period of the second Circle) we obtain 1;1,36, and we keep the first row of Circle 3 (Goldstein [1974a], 227): years 0, days 105, weekday 4, hour 11;24, motion in anomaly 10^s 13;24°. The remainder is 14;47,46 years, from which we subtract the period of the first Circle, 8;0,15 years, and the result is 6;47,31 years. This implies keeping the first row of Circle 2 (Goldstein, *ibid.*, 226): days 1, weekday 4, hour 12;41, motion in anomaly 1^s 5;51°, and also row no. 21 of year 7 of Circle 1 (Goldstein, *ibid.*, 224): 28 December, weekday 2, hour 14;32, motion in anomaly 6^s 3;58°. Thus, adding these data, we find

days	weekday	hour	motion in anomaly
105	4	11;24	10^s 13;24°
1	4	12;41	1^s 5;51°
	2	14;32	6^s 3;58°
107	3	14;37	5^s 23;13°

which, when added to 28 December, leads to 14;37 hours after mean noon of Tuesday, 14 April 1888 (Julian), and to 173;13° for the lunar mean anomaly at this time.

Ad [69]. "A half" is obviously a copyist's error instead of the correct reading: "13 seconds" (as in sentences [46], [66], [72], and [75]).

Ad [85]. It is necessary to correct the Provençal text and to read 22080 instead of 22091, as in the Hebrew manuscripts, because the period of the third Circle, 22091^y, 201^d, $6;13^h = (552^y \times 40) + [(105^d + 11;24^h) \times 40]$, and $552 \times 40 = 22080$. Without the correction, moreover, the Provençal assertion "leaving aside the other years" does not make sense. Levi's instructions are more detailed in the Hebrew version, where this sentence continues: "But however you computed your computation, you ought to be careful with the years in the third and fourth circles, that is, if the years in these two circles are 3 or 4 in a cycle of 4 years, and there is nothing of these 3 or 4 years in a cycle of 4 years in the first circle, then subtract 1 day on account of the leap day from the days of the year and of the week" (Ms. A, f. 15b).

Ad [87]-[91]. For finding the true opposition from the data for the mean opposition obtained above we enter the four tables numbered by Goldstein ([1974a], 230, 232, 234, and 239-240) 40a, 41a, 42a, and 43a, with the day of the month and the motion in anomaly. According to sentence [95] we must enter table 40a with 1 April, instead of 14 April, and the shift is to account for the number of 43-year cycles that have elapsed from the epoch (see below, comments to sentence [95]); so, we obtain $17;42^h$. With 10 April, instead of 14 April, to account for the number of cycles of 119 3/4 years that have elapsed, we enter table 41a and obtain $0;2^h$. Interpolating in table 42a, 173;13° of anomaly yields $10;50^h$, and entering with 1 April and 173;13° of anomaly in table 43a we obtain by double interpolation $0;3^h$. Thus, $17;42^h + 0;2^h + 10;50^h + 0;3^h = 28;47^h$. According to sentence [91] we must subtract one day from this amount, and the result is $4;47^h$ which, when added to the time of the mean opposition, yields the time of the true opposition: $19;24^h$, 14 April 1888, Tuesday.

Ad [92]-[94]. According to chapter 65 of the Astronomy [48], the daily mean solar motion is 0;59,8,20,8,44,6,3,14°, that Levi derived from the mean solar position at

---

[48] "... remanet medius locus solis tempore experientie Timocrati in Piscibus 11;16,3°. Et iam declaratum est quod medius locus solis tempore experientie nostre erat in Geminis 29;53,32°. Sequitur quod sol peragrauit super reuolutiones completas in toto tempore medio experientiarum istarum 108;37,29°. Et est notum modico studio quod cursus solis in isto tempore medio fuerit 1627 reuolutiones complete, 108;37,29°. Et quando istud totum diuidatur per numerum dierum inter istas duas experientias elapsarum, qui sunt 594358;22,2,30 dies, sequitur quod medius motus solis in una die sit 0;59,8,20,8,44,6,3,14°" (Vat. Lat. 3098, ff. 57vb-58ra).

the time of the occultation of Spica by the Moon reported by Timocharis (*Almagest*, VII.3; Toomer [1984], 335-336: – 293 Mar. 9/10), computed as Pisces 11;16,3°, and the mean solar position at the summer solstice of 1334 (Monday, 13 June, 14;53^h after mean noon), computed as Gemini 29;53,32°. The solar motion in this interval is given as 1627 complete revolutions and 108;37,29°, and the time elapsed between both dates as 594358;22,2,30^d (that, computing accurately, yield 0;59,8,20,8,44,6,3,12,29,...$^{o/d}$). Levi's value for the daily mean solar motion leads to a tropical year of 365;14,29,55,56,6,30,...^{d} (= 365^d 5;47,58,22,26,36,...^{h}). The difference between this value and the Julian year (365;15^d) will produce 1 day after

$$1 \ / \ (365;15 - 365;14,29,55,56,6,30,...) = 119^y \ 266^d \ 11;50,55,...^h \qquad (4)$$

In the same chapter [49], Levi gives 119^y, 266^d, 10;19^h – derived from dividing the daily mean solar motion by the excess over 360° of the solar motion in 365;15^d, namely 0;0,29,38,10,27,51,10,58,30° (the accurate result is also 119^y, 266^d, 11;50,55,...^{h}) –, rounded to 119;45 (= 119^y, 273^d, 22;30^h) in the Provençal text. Jacob ben David Bonjorn (Yomtob Poel), who composed a set of astronomical tables for latitude 42;30° (Perpignan) with epoch 1361, strongly depending on Levi's tables and parameters, rounded Levi's parameter to 120, and that implies a tropical year of 365^d 5;48^h (Chabás [1992], 198-199; on Bonjorn's tables see also Chabás, 1991). [50]

---

[49] "Et residuum sui motus in 365 diebus et quarta parte diei est 0;0,20,38,10,27,51,10,58,30°. Et [si] diuidatur cursus solis in una die per istud residuum, inuenietur in quot annis istud residuum faciet unam diem. Et inuenietur quod hoc faciet in annis 119, 266 diebus, 10 horis et circa 19 minuta hore, attribuendo cuilibet dies 365 et quarta parte diei" (Vat. Lat. 3098, f. 58ra).

[50] Bonjorn's value for the mean synodic month is, in my opinion, identical to Levi's. Taking Levi's value with four sexagesimal places, Bonjorn's cycle c results from 29;31,50,7,54^d × 383;30 = 3,8,44;58,55,29,39^d, which expressed in Egyptian years yields 31 Ey, 9^d, 23;34,11,51,35,59^h. It is not true that Levi computed his mean syzygy tables with a mean synodic month of 29;31,50^d (Chabás [1991], 292); as it is proved below (section 5, comments to Table I) Levi used 29;31,50,7,54,25,3,32^d. Moreover, I would suggest some corrections to Chabás's interpretation of two significant passages of Bonjorn's canons, which assert that the entries in his table for the solar corrections and the lunar argument of latitude must be shifted 1 day each 66 years (Chabás [1992], 194-195, sentences 92-93), and that the true solar positions given in his table for syzygies must be also shifted 1 day each 120 years (Chabás [1992], 198-199, sentence 109). These rules for using the tables are not related to precession (as asserted in Chabás [1991], 292, and Chabás [1992], 79-81 and 93-95), since it could not affect tropical positions which are used in Bonjorn's tables. The second one, as we have seen, is almost identical to Levi's rule in sentences [92]-[94]. The first one deserves additional comments. It corresponds to Levi's rules for using his tables for computing the time of the true syzygies from the time of the mean ones, and his table for true solar positions, after a number N of 43-year cycles have elapsed (see below, comments to Levi's sentences [95] and

Ad [95]. The entries in tables II and V of the Provençal version (= Tables 40a and 43a in Goldstein [1974a], 230 and 239-240) depend on the solar corrections. If we consider that in 43 Julian years the motion of the solar apogee, according to Levi, is 0;59,10°, it is obvious that in Levi's table for the solar equation of center, c,

$$c(\alpha - 0;59,10°) \approx c(\alpha - 1°). \tag{5}$$

However, the aforementioned tables are arranged for the days of the Julian calendar, and the solar corrections needed to compute their entries were probably derived, as Goldstein (*ibid.*, 140) suggested, from the true solar longitudes given in table VI of the Provençal version (= table 44a in Goldstein [1974a], 244-245). In this case, the correction Levi says is to be made after 43 years — for instance, the entries that in the table correspond to 1 March are to be used for 2 March — is only an approximate procedure to account for the motion of the solar apogee in that time.

Ad [97]-[109]. Levi explains in these sentences the corrections to be applied to the true solar positions given in the table for each day of 1321 (Table VI) in order to extend the usefulness of the table to previous or later years. The first correction (sentence 103) accounts for the shift between Julian years of 365;15^d and calendar years of 365 or 366 days, and it must be applied for finding the solar position within the 43-year solar cycle. After 43 years, the solar apogee moves approximately 1°, and Levi's second correction (sentences 104-109) intends to account for it. According to Levi, the mean solar motion in 365;15^d is 360;0,29,38,10,27,51,10,58,30° (Goldstein [1974a], 106), and these 0;0,29,38° are rounded by Levi to 0;0,30° per year, and 0;2° each 4 years (accurately computed, 0;1,58,32°); obviously, to constant increases of 0;0,30° per year in the mean longitudes do not correspond identical increases in the solar corrections, but the difference is quite small and can be disregarded. Let λ be the true solar position at noon of a calendar day d_1. At noon of day d_2 (= d_1 + 365^d) the true position of the Sun will be λ - 0;15°, since the mean solar motion in 365^d is approximately 359;45°. Therefore, the position obtained from the table after applying Levi's rule (to add 0;0,30°) will be reached by the Sun approximately 6^h after noon. At noon of d_3 (= d_2 + 365^d), the true position will be approximately λ - 0;30°, and the

[104] ff.), which account for the motion of the solar apogee. However, Bonjorn's use of a 66-year cycle shows that he did not adopt Levi's motion of the solar apogee (which is independent of precession) but al-Battānī's (which is identical with his parameter for precession). Consequently, Chabás ([1991], 293) corrections to the time of the first mean opposition of April 1888 are superfluous (see the correct data in the comments to Levi's sentences [65]-[85]).

position obtained will correspond to approximately 12^h after noon. However, if d_4 = d_3 + 366^d, at noon of that day the true solar position will be λ + 0;15°, and the Sun will be at the degree obtained 6^h before noon. As d_5 = d_4 + 365^d, λ will be reached again at noon. As Levi's table gives true solar positions for 1321, and 1323 was a leap year, d_2, d_3, d_4, and d_5 represent days of years whose remainder is, respectively, 2, 3, 4, and 1, in Levi's sentences [98]-[102]. Levi's procedure provides good results as the following example proves. Using Levi's table 11 for mean solar positions (Goldstein [1974a], 170) and Cnc 2;11° for the solar apogee, the computed true solar longitudes on 1 March 1301 and 1315 are, respectively, 348;49,25° and 348;26,33° (or 348;26,22° taking into account the motion of the solar apogee in 14 years). Using Levi's rule, we obtain for the last date 348;49,25 + (0;0,30 × 14) = 348;56,25. And, as 1315 – 1300 = 15 ($\equiv$ 3 mod. 4), according to sentence [101] this position corresponds to 12 hours after noon of that day, which is correct. The approximation, however, is worse according as the number of 43-year cycles elapsed after the epoch increases, as we will see below.

Levi's second correction (to descend one day in the table for each 43 years that have elapsed, and to subtract 0;37,54° from the position which corresponds to that day) can be explained as follows. After 43 Julian years, computing accurately, the increase in the mean longitude is 0;21,14,21,...°, not the 0;21,30° that follows from 0;0,30° × 43. However, after 43 Egyptian years, 232 days, and 6;46 hours, the solar apogee has moved 1°, and the solar anomalies decrease by the same amount. Therefore, in 43 Julian years the motion of the solar apogee is

$$15705;45^d \text{ / } 15927;16,55^d = 0;59,9,55,39,51,52° \approx 0;59,10°. \qquad (6)$$

If we take the entry in the table corresponding to the day which follows the day we seek (sentence [104]), this corresponds to an increase of 1 day in the mean solar longitude (that is, 0;59,8,20,8,44,6,3,14° $\approx$ 0;59,8°). However, if we consider 0;59,10° $\approx$ 0;59,8°, since the increase in the mean solar longitude in 43 Julian years is 0;21,14°, after having descended one day in the table we must subtract from that position the difference

$$0;59,8° - 0;21,14° = 0;37,54° \qquad (7)$$

to account at the same time for the changes due to the increase of the mean longitude and the motion of the solar apogee.

Levi's sentences [98] and [107]-[108] suggest that the procedure can be applied for any number N of 43-year cycles, and the supposed validity of the associated tables for mean syzygies over a very long period (more than 600,000 years) supports this interpretation. However, the positions found by means of Levi's approximative procedure differ in a significant way from the accurately computed positions for values relatively small of N, as can be proved.

Let λ_0 and λ_n be the true solar positions on days 0 and n of year 0 of the first cycle of 43 years, and μ_0 and $c(\alpha)$ the mean longitude and the solar correction corresponding to day 0. Thus,

$$\lambda_0 = \mu_0 + c(\alpha) \tag{8}$$

$$\lambda_n = \mu_0 + n \cdot 0;59,8 + c(\alpha + n \cdot 0;59,8). \tag{9}$$

Let $\lambda_{1(N)}$ be the true solar positions on day 1 after N cycles of 43 years. Thus, assuming that $0;59,8°$ is the motion of the solar apogee in 43 Julian years,

$$\lambda_{1(N)} = \mu_0 + N \cdot 0;21,14 + 0;59,8 + c(\alpha + N \cdot 0;21,14 + 0;59,8 - N \cdot 0;59,8). \tag{10}$$

Therefore, on the day n after N cycles of 43 years,

$$\lambda_{n(N)} = \mu_0 + N \cdot 0;21,14 + n \cdot 0;59,8 + c(\alpha + N \cdot 0;21,14 +$$
$$+ n \cdot 0;59,8 - N \cdot 0;59,8) =$$
$$= \mu_0 + N \cdot 0;21,14 + n \cdot 0;59,8 + c(\alpha + n \cdot 0;59,8 - N \cdot 0;37,54) =$$
$$= \mu_0 + (n + N) \cdot 0;59,8 - N \cdot 0;37,54 + c(\alpha + [n + N] \cdot 0;59,8 -$$
$$N \cdot 1;37,2). \tag{11}$$

But according to Levi's procedure,

$$\lambda_{n(N)} = \lambda_{n+N(0)} - N \cdot 0;37,54 -$$
$$= \mu_0 + (n + N) \cdot 0;59,8 + c(\alpha + [n + N] \cdot 0;59,8) - N \cdot 0;37,54 =$$
$$= \mu_0 + (n + N) \cdot 0;59,8 - N \cdot 0;37,54 + c(\alpha + [n + N] \cdot 0;59,8). \tag{12}$$

Equations (11) and (12) are identical except for the last term, where the arguments of anomaly differ by $N \cdot 1;37,2°$. Thus, the maximum absolute difference between accurate computation and Levi's procedure for year 1 after N cycles is given by $c(N \cdot 1;37,2°)$, and for small values of N the most noticeable discrepancies will occur near perigee, where they already reach $1;1,4°$ (more than 1 day) for $N = 17$. The maximum absolute errors occur when the last terms in both equations produce the maximum correction but of opposite sign, although because of the symmetry of the eccentric, it is also clear that there exists at least a day within the year for which the difference produced by $c(N \cdot 1;37,2°)$ almost vanishes. In short, if

$$e_t = |c(\alpha + [n + N] \cdot 0;59,8) - c(\alpha + [n + N] \cdot 0;59,8 - N \cdot 1;37,2)|, \tag{13}$$

316

since $0° \leq \alpha \leq 360°$, and in Levi's first solar model $c_{max} = 2;8°$,

$$0° \leq e_t \leq 4;16°. \tag{14}$$

Inside a given year y within a cycle N, the deviation e_y between accurate computation and Levi's procedure is

$$0° \leq e_y \leq c(N \cdot 1;37,2°). \tag{15}$$

Ad [110]-[111]. The ecliptic longitude of the point at which the tenth house begins, λ_{10}, is called in the middle ages *medium caelum*, and defined as the culminating point of the ecliptic, that is, where it intersects the meridian. The ascendent (A) is the rising point of the ecliptic and the beginning of the first house, λ_1. If ρ is the oblique ascension, and α_{10} is the right ascension of λ_{10}, $\rho(A) = \alpha_{10}$.

Ad [114]-[115]. The "method of the Ancients" is probably an allusion to the so-called *standard method* (North [1986], 3-5), often ascribed to 'Alchabitius' in the late Middle Ages. In chapter 99 of the Hebrew and Latin versions of the *Astronomy* this sentence is replaced by a long passage (see Appendix, 99:119-150) where Levi explains another method for domification which leads to identical results that the one called by North ([1986], 21-27) the *hour-lines (fixed boundary) method*, explained by Abraham Ibn Ezra in his *Sefer ha-moledot (Liber nativitatum)*, whose two versions were composed in 1148 and 1154. The same procedure was also used by Levi in his *Prognostication for the conjunction of 1345* (Goldstein and Pingree [1990], 41-45) for calculating the cusps of the astrological houses. As we have seen, some sentences, absent from the Hebrew version, are added by Levi in the Latin version of chapter 99 at the end of the fourth *dictio* (99:156-163), in which are described 12 tables for astrological houses for the horizon of Orange, which "were not included in the tables that we made at the request of some noble Christians", and that Levi computed (or intended to compute) as an additional gift to that was promised.

Ad [122]. The identification made here between eclipse-middle and true opposition suggests that columns I, II, IV, and V of Provençal tables XII and XIII were computed assuming that the lunar orbit and the ecliptic are parallel (see the commentary on these tables in section 5.B). The assertion is corrected in the Hebrew version of the canons: "And these tables allow you to determine the longitude of the horizon in which you are [located] when you fix precisely the times of a lunar eclipse with the observational instrument, for then you will determine the time of eclipse-middle which is the time of opposition approximately" (Ms. M, f. 6b: 15-17). In the *Astronomy*, Levi devoted chapter 79 (Vat. Lat. 65vb-66ra) to a demonstration that eclipse-middle and true opposition coincide only when the Moon has no latitude; if it is not so, and the Moon is situated before the nodes, the time between the beginning of the eclipse and the eclipse-middle will be shorter than the time from eclipse-middle to its end, and viceversa if the Moon is placed after the node. At the end of the chapter Levi adds that it has not been pointed out by any of the "Ancients" [51]. Although there are medieval texts where the distinction is made [52], many others ignored it (for example al-Khwārizmī, cf. Neugebauer [1962], 66-69; Ibn al-Muthannā, cf. Goldstein [1967], 111). Ptolemy discussed the matter in the *Almagest* (VI.7), computed a maximum difference of ·1/16th of an equinoctial hour, and, concluding that "scrupulous accuracy about such a small amount is a sign of vain conceit rather than love of truth", dismissed it (Toomer [1984], 296-298; Neugebauer [1975], 83). The difference was also ignored by al-Battānī, Abraham Bar Ḥiyya, and Ibn Ezra, to whom (with Ptolemy) Levi probably referred with the expression "the Ancients" (see Nallino [1899-1907], I, 98:9; Millás Vallicrosa [1959], 76:2-3; Millás Vallicrosa [1947], 165:15-16). In the canons for his tables with epoch 1361, Bonjorn in turn dismissed Levi's remarks in chapter 79 on the significance of the distinction [53].

---

[51] "Expedit quod hic demonstremus quod non est medium eclipsis tempore uere oppositionis nisi quando luna latitudinem non haberet. Set quando luna latitudinem habet et est ante caput et cauda draconis, tempus a principio eclipsis usque ad medium est brcuius quam tempus quod est a medio usque ad finem. Et ex ista inquisitione sequitur magna utilitas, quia sine hoc non possumus scire tempus uere oppositionis ex eo quod scimus tempus medij eclipsis. Et si econtrario, econtrario [...] Et hoc est quod declarare uolebamus in isto capitulo, ut ex eo nos possimus iuuare ad tempus uere oppositionis sciendum per scientiam quam de tempore medij eclipsis habemus, uel principij eius uel finis. Et hic est unum mirabile declaratum, quod nunquam uidit aliquis antiquorum quorum uerba ad notitiam nostram peruenerunt, qui semper medium eclipsis esse in uere oppositionis tempore posuerunt. Et hoc est ualde utile ad inueniendum ueritatem eorum que inuestigare intendimus" (Vat. Lat. 3098, ff. 65vb-66ra).

[52] For instance, William of Saint-Cloud in the introduction to his *Almanach planetarum* (1292) carefully distinguished between true opposition and eclipse-middle, and estimated the maximum difference as about 0;10h: "Quando enim luna accedit ad nodum, siue ad caput siue ad caudam, tunc hora uere oppositionis precedit horam medie eclipsis; quando uero a nodo recedit econtrario est, scilicet quod hora medie eclipsis precedit horam uere oppositionis. [...] Et quanto luna est remotior a nodo in medio eclipsis tanto est maior diuersitas in tempore inter horam medie eclipsis et horam uere oppositionis; ista tamen diuersitas siue antecedendo siue consequendo non excedit 10 minuta hore uel 12" (Paris, B.N. n.a. lat. 1242, ff. 41vb).

[53] "... quia non est diversitas sensibilis inter medium oppositionis et medium eclipsis, sicut declarat Ptolomeus capitulo VII° libri sexti Almagesti. Et licet iam gloriatus fuerit in se de hoc ille sapiens rabi Levi, eo quia subtiliavit istam diversitatem et curavit de ea et dixit in fine verborum suorum super hoc capitulo 79 sui libri de astrologia, quem nominavit Bella Domini, talia verba: 'et hic est declaratum factum mirabilem valde de quo non curarunt antiqui quarum verba pervenerunt ad nos'; hoc ille, sed tu iam vides Ptolomeum declarasse istam diversitatem in capitulo predicto et dedit tibi causam quare non curavit se ad hoc subtiliare, scilicet propter parvitatem quantitatis eius, et dixit ibi: 'et non extimet aliquis nos ignorasse et cetera'. Item ibi: 'et causa que proibet nos curare de istis arcubus in partibus libri nostri est quia sunt parvi et diversitas eorum est insensibilis et cetera'; dixit etiam ibi quod 'subtiliare in simili istius quantitatis est superfluitas et intricatio et non diligere veritatem', hoc ille" (Chabás [1991], 207-209: 155-156).

5. Description of the Tables

In the following pages the tables of the Provençal version are identified with their corresponding number in Goldstein's edition, abbreviated sometimes as G; the titles and the headings of the columns, numbered also according to Goldstein's edition, are given in Provençal followed by translation. The variant readings are marked with an asterisk if the Provençal reading is better than the preserved Hebrew one accepted by Goldstein, and with two asterisks if it is a trivial copyist's error. Most of the variants not marked are alternative readings (varying only by one unit in the last digit) that are as acceptable as those of the Hebrew manuscripts. If the Provençal variant reading is identical to that which appears in some of the Hebrew manuscripts (but different from that accepted by Goldstein), the corresponding Hebrew manuscript is indicated between brackets. The analysis of the variant readings shows that the extant copies of the tables can be distributed into two classes: (a) which consists of the Provençal copy and M, and (b) which consists of P and Q. From Goldstein (1974a), 78-79, it is evident that A is an intermediate manuscript (it contains tables 20-21 as in M, but tables 40-44 as in P and Q).

A) Ms. Scal. 46*

TABLE I [= Tables 36-39; see Plate 2]. No title.

Circle 1 (= Table 36, 1-8): columns I and II are lacking as in the Hebrew manuscripts, excepting A. As in the description of this table in the Provençal canons and chapter 99 of the *Astronomy*, the entries in column III are written in red ink (oppositions) and black ink (conjunctions). Headings of columns IV and V: *feria* [weekday] / *oras, menutz* [hours, minutes]. Circle 2 (= Table 37): headings: column II: *mes* [month; read: days]; column III: *feria*; column IV: *oras, menutz*. No headings for the columns in Circles 3 and 4 (= Tables 38 and 39). Circles 2, 3, and 4 are superimposed plates of decreasing radii where the different columns of entries are concentrically written. Plates of Circles 2, 3, and 4 can rotate upon plate of Circle 1 around their common center in such a way that any entry of Circles 2, 3, and 4 can be aligned with any other entry of Circle 1, in order to facilitate the addition or subtraction of these entries when the computation of mean syzygies requires the use of the four Circles (see the text of the Canons, sentences 64 to 81). In Circles 1, 2, and 3 the columns for the lunar motion in anomaly (VI) are written after column V, but they remain hidden under the plates of Circles 2, 3, and 4, respectively.

In the microfilm I have used some entries are partially illegible: those of lines 9-25 in column V of Circle 1, year 4; lines 2-12 in column V of Circle 1, year 5; lines 13-19 in column V of Circle 1, year 8; lines 1-4 in column V of Circle 3; and lines 2, 4-6, 8, and 17 in column V of Circle 4. Variant readings for the rest of the table: circle 1, year 1, column V, line 17: 7;44 (= all the Hebrew Mss.); year 2, column V, line 8: 19;15**; year 3, column V, line 20: 2;50**; year 6, column III,

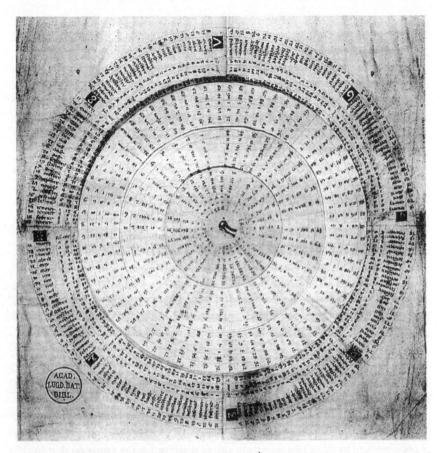

Plate 2. – Leiden, Universiteitsbibliotheek, Ms. Scal. 46*: Levi's tables for mean syzygies, called in the canons first, second, third, and fourth Circles.

line 6: 17 May**; year 7, column V, line 8: 16;46 (= M); year 8, column V, line 24: 1;49 (= M). Circle 2, column IV, line 57: 3;50**. Circle 4, column II, line 24: 276 (= all the Hebrew Mss.); column III, line 16: 198**; column IV, line 23: 5**.

The hour in Circle 1 (column V) was computed using Levi's value for the length of the mean synodic month to the fourth sexagesimal place, $29;31,50,7,54^d$: the entry in line no. 24 of year 8 is $0;49^h = 0;52^h + [29;31,50,7,54^d \times 98] = 0;52^h + 23;57^h$. Column IV of Circle 2 was computed using $12;41,13^h$ [$\times 69 = 11;23,57^h \approx 11;24^h$: text, line 69]. With $11;23,57^h$, line 40 of Circle 3 ought to be $23;58^h$ instead of $0;00^h$ (text).

The motion in anomaly (column VI in Circles 1 and 3, and column V in Circles 2 and 4) was also computed using $29;31,50,7,54^d$. In Circle 1, $29;31,50,7,54^d \times 98.5 = 18;18,22,...°$, [$\approx 18;18°$: text, line 25, year 8]. After 99 lunations, the parameter to four sexagesimal places yields $35;50,52,24,...°$, rounded by Levi to $35;51°$ (line 1, Circle 2). Column V of Circle 2 seems to have been computed with some approximation: line 69 reads $10^s 13;24°$, whereas $35;51° \times 69 = 10^s 13;39°$; using $31;50,52,24°$, after 69 times 99 lunations the motion in anomaly is $10^s 13;30,15°$. Column VI in Circle 3, $9^s 26°$, was computed from line 69 of Circle 2; using $10^s 13;30°$ the accurate result is 10^s; from this, line 28 of Circle 4 ought to be $0^s 4;59°$. Using the long parameters ($13;3,53,55,55,33,30^{o/d}$ and $29;31,50,7,54,25,3,32^d$), for line 28 of Circle 4, recomputation yields $0^s 4;29,52°$.

B) Ms. Scal. 46, ff. 54^v-67^r.

Tables II and III, ff. 54^v-55^r [= Tables 40a and 41a]. Fol. 54^v: *lapermera taula* [the first table]; fol. 55^r: *seguonna taula* [second table].

Common title: [54^v] *En aquestas doas paiemas. son las doas taulas dellas equacions per lo iorn* [55^r] *dellan en que uen la conioncio ho la oposicion e quascuna paiema es taula per sci* [In these two pages are the two tables for the equation due to the day / of the year on which occurs the conjunction or the opposition. And each page is a table in itself].

Fol. 54^v: heading of column I: *iorns de mes* [days of the month]; headings$_1$ of columns II-XIII: *martz; abril; may; iun; iull; aost; setembre; octembre; nouembre; desembre; ienoier; febrier* [March; April; May; June; July; August; September; October; November; December; January; February]; heading$_2$ of columns II-XIII: *oras, menutz* [hours, minutes].

Fol. 55^r: heading of column I: *iorns de mes*; headings$_1$ of columns II-XIII: *martz; abril; may; iun; iull; aost; cetembre; octembre; nouembre; desembre; ienoier; febrier*; heading$_2$ of columns II-XIII: *menutz*.

Variant readings for the entries in table II: March, day 26: 17;44, day 27: 17;44, day 30: 17;45. April, day 11: 17;32, day 12: 17;30. November, day 7: 11;22*, day 28: 12;35. December, day 7: 13;7, day 15: 13;36. January, day 25: 15;58. February, day 10: 16;45.

Variant readings for the entries in table III: March, day 23: 6. October, day 13: 2, day 14: 2, day 20: 1, day 21: 1. January, day 2: 17, day 31: omits**.

TABLE IV, ff. 55ᵛ-56ʳ [= Table 42a]. Fol. 55ᵛ: *terssa taula* [third table]; fol. 56ʳ: *terssa taula* [third table].

Common title: [55ᵛ] *En aquestas doas pagemas es lataula della equassion per lo luoc* [56ʳ] *della luna dal mouement della diuercitat al tems della conioncion odella oposicion* [In these two pages is the table for the equation due to the position / of the Moon according to the motion in anomaly at the time of the conjunction or the opposition].

Fol. 55ᵛ: heading of column I: *gras* [degrees]; headings₁ of columns II-VII: *sehnal 0; sehnal 1;... sehnal 5* [sign 0; sign 1;... sign 5]; heading₂ of columns II-VII: *oras, menutz*.

Fol. 56ʳ: heading of column I: *gras*; headings₁ of columns II-VII: *sehnal 6; sehnal 7;... sehnal 11*; heading₂ of columns II-VII: *oras, menutz*.

Variant readings: 0ˢ 20°: 13;9*. 1ˢ 10°: 16;7, 12°: 16;22, 19°: 17;12. 3ˢ 13°: 19;7, 14°: 19;5, 17°: 18;57. 5ˢ 14°: 12;18, 27°: 10;12. 6ˢ 9°: 8;14, 14°: 7;26, 15°: 7;16, 16°: 7;7, 27°: 5;26, 28°: 5;17, 29°: 5;9, 30°: 5;0. 7ˢ 4°: 4;26, 5°: 4;18, 15°: 3;0. 8ˢ 16°: 0;22, 17°: 0;20, 18°: 0;17, 19°: 0;18**. 9ˢ 18°: 0;22. 11ˢ 9°: 6;8*, 13°: 6;47.

TABLE V, ff. 56ᵛ-58ʳ [= Table 43a]. Ff. 56ᵛ, 57ʳ, 57ᵛ, 58ʳ: *quarta taula* [fourth table].

Common title: [56ᵛ] *En aquestas doas pagemas es en las doas dapres es la taula della* [57ʳ] *equacion per lo iorn dellan eper lo mouement delladiuercitat encems* [57ᵛ] *compliment della taula dauant dicha en longuesa* [58ʳ] *compliment della taula mesma en longuesa e en larguesa* [In these two pages and in the two following ones is the table for the / equation due to the day of the year and to the motion in anomaly together / Completion of the aforementioned table in length / Completion of the same table in length and width].

Fol. 56ᵛ: column I: *Martz, abril, May, iun, iull, aost*; heading of column II: *iorn de mes*; headings₁ of columns III-XIV: *sehnal 0, gras 0; sehnal 0, gras 15; sehnal 1, gras 0;... sehnal 5, gras 15*; heading₂ of columns III-XIV: *oras, menutz*.

Fol. 57ʳ: column I: *Martz, abril, May, iun, iull, aost*; heading of column II: *iorn de mes*; headings₁ of columns III-XIV: *sehnal 6, gras 0;... sehnal 11, gras 15*; heading₂ of columns III-XIV: *oras, menutz*.

Fol. 57ᵛ: column I: *setembre, octembre, nouembre, desembre, ienoier, febrier*; heading of column II: *iorn de mes*; headings₁ of columns III-XIV: *sehnal 0, gras 0; sehnal 0, gras 15; sehnal 1, gras 0;... sehnal 5, gras 15*; heading₂ of columns III-XIV: *oras, menutz*.

Fol. 58ʳ: column I: *setembre, octembre, nouembre, desembre, ienoier, febrier*; heading of column II: *iorn de mes*; headings₁ of columns III-XIV: *sehnal 6, gras 0;... sehnal 11, gras 15*; heading₂ of columns III-XIV: *oras, menutz*.

Variant readings: 0^s $0°$, August 10: 0;25. 0^s $15°$, March 11: 1;7, September 4: 1;21**, September 10: 1;21**, December 22: 0;46*. 1^s $0°$, June 28: 0;38, September 10: 0;23. 1^s $15°$, September 4: 1;27**, September 10: 1;27**, February 28: 0;57. 2^s $0°$, January 10: 0;45, January 16: 0;46. 3^s $15°$, June 28: 0;30. 4^s $0°$, May 4: 0;17, September 10: 1;0. 4^s $15°$, September 4: 1;5. 5^s $0°$, July 4: 0;46. 5^s $15°$, September 4: 1;18. 6^s $15°$, March 29: 0;12. 7^s $0°$, November 10: 1;16. 7^s $15°$, March 1: 1;25**, March 5: 1;24**, June 4: 0;48. 8^s $0°$, March 5: 0;30, June 4: 0;48*. 8^s $15°$, March 1: 1;35**, March 5: 1;35**. 9^s $0°$, June 4: 0;46. 9^s $15°$, March 1: 1;47*, March 5: 1;47*. 10^s $15°$, May 10: 0;57, September 10: 0;33. 11^s $0°$, September 4: 0;27. 11^s $15°$, June 10: 0;48, August 16: 0;27, August 22: 0;26, January 28: 1;2*, February 4: 1;4*, February 10: 1;6*.

TABLE VI, ff. 58^v-59^r [= Table 44a]. Fol. 58^v: *taula cinquena* [fifth table]; f. 59^r: *taula cinquena* [fifth table].

Title: [58^v] *taula del luoc ueray dal soleyl en quascun iorn dellan* [59^r] *compliment della taula dauan dicha en larguesa* [Table for the true position of the Sun for each day of the year / Completion of the aforementioned table in width].

Fol. 58^v: heading of column I: *iors de mes*; headings$_1$ of columns II-VII: *Martz/pices*; *abrill/aries*; *May/taurus*; *iun/geminis*; *iull/quancer*; *aost/leo*; heading$_2$ of columns II-VII: *gras, menutz, segons*.

Fol. 59^r: heading of column I: *iors de mes*; headings$_1$ of columns II-VII: *se[te]mbre/uirguo*; *octembre/libra*; *nouembre/scorpio*; *desembre/sagitarius*; *ienoyer/ cabricornus*; *febrier/aquarius*; heading$_2$ of columns II-VII: *gras, menutz, segons*.

In columns II-VII of ff. 58^v and 59^r, the name of the sign which follows to that which appears in the heading is written above the first entry corresponding to it; variants: *uirgo, cabricorn.*

Variant readings: March 4: 21;58,11*, March 6: 23;57,9*, March 8: 25;56,1*, March 18: 5;48,9*. April 9: 27;17,52*, April 24: 11;47,56. May 14: 0;58,30 (= M), May 15: 1;55,53 (= M), May 31: 17;10,51 (= M). June 6: 22;53,11*, June 7: 23;50,13*, June 14: 0;29,19 (= M), June 16: 2;23,26, June 17: 3;20,24 (= M). June 28: 13;47,44*. July 29: 13;21,12**. August 18: 2;37,49*, August 19: 3;35,52*. September 6: 21;6,50*. October 1: 15;46,18*, October 3: 17;45,47, October 17: 1;45,25. November 28: 14;21,4. January 16: 4;25,57, January 22: 10;31,49*, January 28: 16;37,2*, January 30: 18;38,32*. February 8: 27;43,37, February 12: 1;45,2*, February 17: 6;46,9*.

TABLE VII, ff. 59^v-60^r [= Table 4]. Fol. 59^v: *taula seysena* [sixth table]; f. 60^r: *taula seysena* [sixth table].

Title: [59^v] *taula dellas assencions dall zodiac en lespera drecha* [60^r] *compliment della taula dauant dicha en larguesa* [Table for the zodiacal ascensions at *sphaera recta* / Completion of the aforementioned table in width].

Fol. 59^v: heading of column I: *gras de sehnal*; headings$_1$ of columns II-VII: *cabricorn*; *aquarius*; *picis*; *aries*; *taurus*; *geminis*; heading$_2$ of columns II-VII: *gras*, *menutz*.

Fol. 60^r: heading of column I: *gras de sehnal*; headings$_1$ of columns II-VII: *quancer*; *leo*; *uirgo*; *libra*; *scorpio*; *sagitarius*; heading$_2$ of columns II-VII: *gras*, *menutz*.

Variant readings: Leo 27$^o_;$: 239;16**.

TABLE VIII, ff. 60^v-61^r [= Table 7]. Fol. 60^v: *taula setena* [seventh table]; f. 61^r: *taula setena* [seventh table].

Title: [60^v] *taula dellas assencions dal zodiac en lautesa de .44. gras* [61^r] *compliment della dauant dicha en larguesa* [Table for the zodiacal ascensions at latitude 44° / Completion of the aforementioned table in length].

Fol. 60^v: heading of column I: *gras desehnal*; headings$_1$ of columns II-VII: *aries*; *taurus*; *geminis*; *quancer*; *leo*; *uirgo*; heading$_2$ of columns II-VII: *gras*, *menutz*.

Fol. 61^r: heading of column I: *gras desehnal*; headings$_1$ of columns II-VII: *libra*; *scorpio*; *sagitarius*; *cabricorn*; *aquarius; picis*; heading$_2$ of columns II-VII: *gras*, *menutz*.

Variant readings: Virgo 6°: 148;28**. Pisces 24°: 356;44**.

TABLE IX, f. 61^v [= Table 20]: *taula .i.* [Table 1].

Title: *taula dal mouement egual dal cap dal draguon lo qual ua per lo zodiac dorient en ocscident* [Table for the equal motion of the ascending node, which moves through the zodiac from east to west].

Heading of columns I, III, and V: *simples anns* [single years]; heading of columns II, IV, and VI: *sehnals, gras, menutz, segons*.

Variant readings: 4 years: 4s 23;29,19**. 58 years: 3s 18;51,53 (= *M* and *A*). 59 years: 5s 7;14,43**.

TABLE X, f. 62^r [= Table 21, columns I and II]: *taula .2.* [Table 2].
TABLE XI, f. 62^r [= Table 21, columns III-X]: *taula .3.* [Table 3].

Common title: *taula dal mouement egual dal cap dal draguon en ans cuhlitz et en meses et en iorns et en horas et en menutz dora* [Table for the equal motion of the ascending node in collected years, and months, and days, and hours, and minutes of an hour].

Heading of column I: *lihna dans aiustatz de .76. ans cada lihna / ans* [Column of the collected years, each line of 76 years / years]; heading of column II: *sehnal, gras, menutz, segons*; heading of column III: *linha dels mes* [line of the months]; lines of column II: *Martz, abrill, May, iun, iull, aost, cetembre, octembre, nouembre, desembre, ienoier*; heading of column IV: *gras, menutz, segons*; heading of column V: *iorns*; heading of column VI: *gras, menutz, segons*; heading of column VII: *oras*; heading of column VIII: *menutz, segons*; heading of column IX: *menutz doras*; heading of column X: *segons*.

324

Variant readings: column II: as in Ms. *A*, 21,36 appears at the end of the column; columns III and IV: the last entry (February 28 19;19,40) is omitted (= *M*).

TABLE XII, f. 62^v [= Table 30]: *taula .5.* [Table 5].

Title: *taula delleclipce della luna enla maior autesa dal soleyl* [Table for a lunar eclipse at the maximum solar distance].

Headings$_1$: column I: *linha de la largesa della luna al ponch della oposicion* [Column of the lunar latitude at the time of the opposition]; column II: *nombre dels detz eclipsatz* [Number of the eclipsed digits]; column III: *lo nombre dal cos de la luna sobre lo cos dal soleyl en lotems que a entre lo miech de leclipce ella uera oposicio* [Amount of the course of the Moon over the course of the Sun at the time between the middle eclipse and the true opposition]; column IV: *nombre dal cos de la luna sobre lo cos dal soleyl deus que sa comenssa a eclipsar tro que es al miey de leclipce* [Amount of the course of the Moon over the course of the Sun from the beginning of the eclipse to the middle of the eclipse]; column V: *nombre dal cos della luna sobre lo cos dal soleyl deus que es tota eclipsada entro lo miech tems delleclipce* [Amount of the course of the Moon over the course of the Sun from the moment the Moon is totally eclipsed to eclipse-middle]; headings$_2$: columns I, and III-V: *menutz, segons*; column II: *detz*.

Variant readings: column I: lunar latitude corresponding to 15 digits: 0;21,45; column V: entry corresponding to 15 digits: 0;18,45.

TABLE XIII, f. 62^v [= Table 31]: *taula .6.* [Table 6].

Title: *taula delleclipce della luna en la menor autesa dal soleyl* [Table for a lunar eclipse at the minimum solar distance].

Headings$_1$: column I: *linha de la larguesa della luna al ponch della oposicion* [Column of the lunar latitude at the time of the opposition]; column II: *nombre dels detz eclipsatz* [Number of the eclipsed digits]; column III: *lo nombre dal cos della luna sobre lo cos dal soleyl en lotems que a entre lo miech de leclipce ella uera oposicio* [Amount of the course of the Moon over the course of the Sun at the time between the middle eclipse and the true opposition]; column IV: *nombre dal cos de la luna sobre lo cos dal soleyl deus que sa comensa a eclipsar tro que es al miey delleclipce* [Amount of the course of the Moon over the course of the Sun from the beginning of the eclipse to the middle of the eclipse]; column V: *nombre dal cos della luna sobre lo cos dal soleyl deus que es tota eclipsada entro lo miech tems del leclipce* [Amount of the course of the Moon over the course of the Sun from the moment the Moon is totally eclipsed to eclipse-middle]; headings$_2$: columns I, and III-V: *menutz, segons*; column II: *detz*.

Variant readings: column IV: entry corresponding to 10 digits: 0;55,8**; entry corresponding to 18 digits: 0;53,49**.

Goldstein ([1974], 123-128) reconstructed these tables assuming (i) that Levi computed them with an angle $i = 4;30°$ between the ecliptic and the lunar orbit, as column III suggests, and (ii) that, in consequence, z and z', the radii of the shadow when the Sun is at apogee and perigee, are 0;42,44° and 0;41,47°, respectively [54]. Nevertheless, in chapter 94 of the *Astronomy* [55], $z = 0;42,35°$, and $z' = 0;41,37°$, that Levi derived by interpolation from 0;42° and 0;42,25°, the radii of the shadow at the lunar eclipses of 3 October 1335 and 23 October 1333, and from the corresponding relative distances using solar model I. Therefore, as sentence [122] of the Provençal version of the canons suggests, Levi computed Tables XII and XIII assuming, as for Tables XV and XVI, that the ecliptic and the lunar orbit are parallel. Thus, the entries in column I (lunar latitude, β) and column II (digits of the eclipse, d) were computed from equation (see Fig. 1)

$$SM = \beta = 0;56,31 - d \cdot 0;2,19 \qquad (16)$$

where $0;56,31 = z + r = 0;42,35 + 0;13,56$, and $0;2,19 \approx 2r/12 = 0;27,51/12$. Column IV (course of the Moon from the beginning to the middle of the eclipse $= c$) and column V (course of the Moon from the beginning of totality to eclipse-middle $= c'$) were computed from equations (see Figs 1 and 2, respectively)

$$c^2 = BM^2 = BS^2 - SM^2 = (z + r)^2 - \beta^2 \qquad (17)$$

$$c'^2 = BM^2 = BS^2 - SM^2 = (z - r)^2 - \beta^2 \qquad (18)$$

---

[54] With these values for z and z' it is not possible, however, to reconstruct in a satisfactory way column I of the tables if we calculate an entry β from equation $\beta = SM \cos i$ (Goldstein [1974a], 123-125) after having previously computed SM from equation (6) (*ibid.*, 124). So, for instance, for $d = 14$, $SM = 0;24,11$ and $\beta_{14} = 0;24,7$ (text: 0;24,5), and for $d = 20$, $SM = 0;10,15$, and $\beta_{20} = 0;10,13$ (text: 0;10,11). Moreover, we find different values for SM using $\beta = SM \cos i$ and equation (6); for instance, for $d = 10$, according to equation (6), $SM = 0;33,29$ (with $r = 0;13,56$); but with $\beta = SM \cos i$, and β_{10} in the table, $SM = 0;33,27$. The values $z = 0;42,44°$ and $z' = 0;41,47°$, do not allow either to recompute the entries d in column II with equation (6). Chabás ([1991], 310-311; [1992], 137-138), in his reconstruction of Bonjorn's tables for lunar eclipses, also assumes that 0;42,44° and 0;41,47° are Levi's values for z and z'.

[55] "Et ideo est notum quod in quantitate in qua linea SU est sinus anguli 0;42°, semidiameter umbre quando centrum solis est in puncto M [= at perigee] est sinus anguli 0;41,37°, et tanta est semidiameter umbre in isto loco [...] Sequitur quod proportio semidiametri umbre in hoc loco ad lineam TX, que est equalis linee GH, que est sinus 0;42°, sit quasi talis qualis est proportio 0;42,35° and 0;42°" (Vat. Lat. 3098, f. 78[vb]).

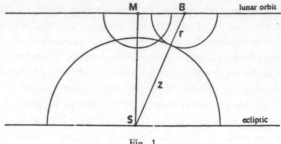

Fig. 1

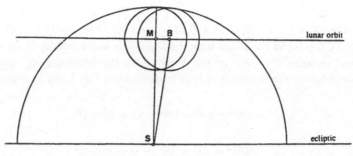

Fig. 2

for table XII and from the same equations with z' instead of z for table XIII. Column III (course of the Moon between eclipse-middle and true opposition, i.e., distance IE in Fig. 3), [56] was probably added later to the tables, and computed from equation

$$IE = DI \; tan \; i \tag{19}$$

taking the values for DI from column I, and using $tan \; i = 0;4,37$ instead of $0;4,43$, a choice explained in chapter 79 of the *Astronomy* as follows [57]. In table XX for

---

[56] Vat. Lat. 3098, f. 65ᵛ.

[57] "Et est notum, ut diximus, quod medium eclipsis est stante luna in puncto I. Sequitur quod uelocitas motus lune [*sic*] inter medium eclipsis et ueram oppositionem [*ms.* dispositionem] sit arcus IE. Et deueniemus in quantitatem arcus IE ex declarato quod angulus IDE est equalis

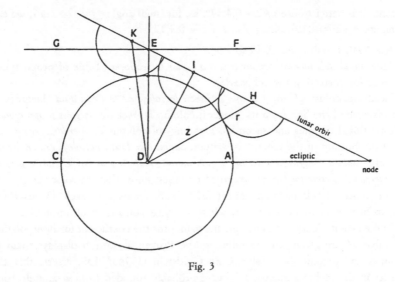

Fig. 3

lunar latitude (= col. VI of table 35 in Goldstein's edition), for arguments of latitude $0° < \omega < 12°$, the line-by-line differences range between $\Delta(\beta) \approx 0;4,42°$ (for $0° < \omega < 5°$, i.e., near the node) and $\Delta(\beta) \approx 0;4,37°$ (for $10° < \omega < 12°$, i.e., near the limit of visibility of the eclipse), and, in these cases, we can consider $\Delta(\beta) \approx \sin i$. To simplify the computation of the table which lists values of IE, we must adopt a value for $\sin i$: if we use $0;4,37°$, the values in the table will differ from the accurately computed ones for positions of the Moon near the node, whereas if we use $0;4,42°$, the differences will occur for values of ω near $12°$. Levi's conclusion is that, as the differences in the first case will be smaller than in the

quasi angulo IEF, quia latitudo quam habet arcus uelocitatis lune tempore eclipsis, qui est arcus HEK, semper est in temporibus eclipsium quasi in proportione eadem; quia latitudo unius gradus ante punctum E in temporibus eclipsium non est minor 0;4,37° nec est maior 0;4,42°, quia latitudo unius gradus circa caput uel caudam est 0;4,42°, et latitudo unius gradus distantis per 12 gradus a capite uel cauda est 0;4,37°. Et declarabitur in futuro quod luna non potest distare a capite uel cauda in temporibus eclipsium ultra 12 graduum nisi parum. Et quia ista differentia est pauca, non dampnat si ponatur ista latitudo in temporibus eclipsium in minori quantitate predicta, que est 13ᵃ pars respectu unius gradus. Sequitur quod linea IE sit scita, et ex hoc erit scitum tempus medium inter medium eclipsis et ueram oppositionem. Et hoc possumus in tabulis eclipsium lunarium ordinare perfectius; intendo dicere quod si arcus HK in ista figura est iuxta caput uel caudam usque in 5°, ponemus proportionem arcus IE ad arcum ED talem qualis est proportio 0;4,42° ad 1°, et si arcus HK in ista figura distet circa 12° a capite uel cauda, ponemus dictam proportionem talis qualis est proportio 0;4,37° ad 1°" (Vat. Lat. 3098, f. 65ᵛᵇ-65[sic]ʳᵃ; the text refers to Fig. 3).

second, it is better to use *sin i* = 0;4,37°; as, for small angles, *sin i* ≈ *tan i*, we can compute equation (19) taking also *tan i* = 0;4,37°. [58]

TABLE XIV, f. 63ʳ: *taula .9*. [Table 9].

Title: *taula dels menutz proporcionals a equar los eclipces* [Table of proportional minutes for correcting the eclipses].

Headings: column I: *los gras que es luhen lo soleyl de la sieua auia* [Degrees of the distance of the Sun from its apogee]; column II: *menutz essegons proporcionals* [Proportional minutes and seconds]. Heading$_2$ of column II: *menutz, segons.*

At the bottom of the table it is written: *Sapias que lauga dalsoleyl era en larasis daquesta taula a .2. gras de quansser e .39. menutz eua .1. gras en .43. ans e pres de 45. menutz dan contatat lan .60. menutz* [You must know that the solar apogee was at the epoch of this table in Cancer 2;39°, and it moves 1° each 43 years and approximately 45 minutes of year, where the year consists of 60 minutes].

In the extant Hebrew manuscripts the values for the coefficient for interpolation in Table 34 are given only to minutes; the Provençal version displays them to minutes and seconds (see Table C-2). Goldstein (1974a), 132, asserts that the entries in the Hebrew manuscripts were probably rounded from accurately computed values, and indeed his recomputed values (given to seconds) for arguments of 60°, 90°, and 120°, agree exactly with the corresponding entries in the Provençal table. The entries in column III of Table C-2 have been computed according to the formula

$$c = (62;14 - d) / 4;28 \tag{20}$$

where *c* is the entry in the table, 62;14 the solar distance at apogee, *d* the solar distance at some anomaly, and 4;28 twice the eccentricity in Levi's solar model I. They agree with the textual values to within half a second, except for those corresponding to arguments of 96° and 150°, which ought to be 0;32,34 and 0;55,50, instead of 0;32,32 and 0;55,43, respectively.

The marginal note at the bottom of the table is absent in the Hebrew manuscripts. The value for the solar apogee stated in it (Cnc 2;39° in 1321) differs slightly from the value given in the Latin text of chapter 99, about Cnc 2;38° (see Appendix, 99:195: "et iam sciuimus per experientias longas et certas quod locus eius erat tempore dicte radicis in Cancro 2° et circa 0;38°"), and Goldstein (1974a), 146, considers that the best recomputed value for the solar apogee, to the nearest minute, for table 44a is Cnc 2;37°. There is also some discrepancy in the values

---

[58] Levi's concern about accuracy is here somewhat surprising: it leds him to prefer 0;4,37 to 0;4,42, but not to recompute column I (which indeed does not list values for the lunar latitude).

	TABLE C-2	
argument	v/text	v/comp.
6 354	0; 0, 9	0; 0, 9.51
12 348	0; 0,38	0; 0,37.94
18 342	0; 1,25	0; 1,25.01
24 336	0; 2,30	0; 2,30.26
30 330	0; 3,53	0; 3,53.04
36 324	0; 5,32	0; 5,32.54
42 318	0; 7,28	0; 7,27.75
48 312	0; 9,38	0; 9,37.52
54 306	0;12, 1	0;12, 0.54
60 300	0;14,35	0;14,35.35
66 294	0;17,21	0;17,20.34
72 288	0;20,14	0;20,13.83
78 282	0;23,14	0;23,13.97
84 276	0;26,19	0;26,18.85
90 270	0;29,26	0;29,26.51
96 264	0;32,32	0;32,34. 9
102 258	0;35,42	0;35,41.95
108 252	0;38,46	0;38,45.59
114 246	0;41,44	0;41,43.75
120 240	0;44,34	0;44,34.41
126 234	0;47,16	0;47,15.61
132 228	0;49,45	0;49,45.47
138 222	0;52, 2	0;52, 2.24
144 216	0;54, 4	0;54, 4. 3
150 210	0;55,43	0;55,50.19
156 204	0;57,19	0;57,18.65
162 198	0;58,28	0;58,28.59
168 192	0;59,19	0;59,19.16
174 186	0;59,50	0;59,49.76
180 180	0;60, 0	0;60, 0

(v/text: entries in the Provençal table; v/comp.: recomputed values)

for the motion of the solar apogee: it is given in the *Astronomy* as 1° in 43 Egyptian years, 232 days, and 6;30 hours (= 43;38,11 years: chapter 19; Goldstein (1984), 113, 19:40; Vat. Lat. 3098, f. 14rb: "motus augis solis est gradus unius in 43 annis, 232 diebus, 6 horis et circa 30 minuta hore, loquendo de annis egiptiacis"), 43 Egyptian years, 232 days, and 6;46 hours (chapter 57; Goldstein [1974b], 94; Vat. Lat. 3098, f. 46rb: "concluditur quod per 1° mouetur in 43 annis egiptiacis, 232 diebus, 6 horis et circa 46 minuta hore"), 43 2/3 Egyptian years (= 43;40 years, or 43 years and 243 days: chapter 60; cf. Goldstein [1974a], 146; Vat. Lat. 3098, f. 47vb: "que spera augis mouetur motu concentrico terre in 43 annis et duabus tertijs annj circa 1°"), and 43 Egyptian years, 221 days, 22 hours (= 43;36,29 years, see Appendix, 99:196).

TABLE XV, f. 63r [= Table 32]: *taula .7.* [Table 7].

Title: *taula delleclipce dal soleyl en sa maior autesa* [Table for the solar eclipse at its maximum distance].

Headings₁: column I: *lihna della larguesa <della> uista alla luna al ponch de la conioncio uista* [Column of the apparent lunar latitude at the moment of the apparent conjunction]; column II: *detz eclipsatz dessoleyl* [Solar eclipsed digits]; column III: *nombre dal cos uist ala luna sobre lo cos dal soleyl deus lo comens de leclipce entro lomaier eclipce* [Amount of the apparent course of the Moon over the course of the Sun between the beginning of the eclipse and the maximum phase]; headings₂: columns I and III: *menutz, segons*; column II: *detz*.

No variant readings.

TABLE XVI, f. 63r [= Table 33]: *taula .8.* [Table 8].

Title: *taula delleclipce dal soleyl en samenor autesa* [Table for the solar eclipse at its minimum distance].

Headings₁: column I: *lihna della larguesa uista a la luna al ponch della conioncio uista* [Column of the apparent lunar latitude at the moment of the apparent conjunction]; column II: *detz eclipsatz del soleyl* [Solar eclipsed digits]; column III: *nombre dal cos uist alla luna sobre lo cos dal soleyl deus lo comens de leclipce en tro lo maier eclipce* [Amount of the apparent course of the Moon over the course of the Sun between the beginning of the eclipse and the maximum phase]; headings₂ columns I and III: *menutz, segons*; column II: *detz, menutz*.

Variant readings: column II, eclipsed digits for latitude 0;0,0°: 11;37.

TABLE XVII, f. 63v [= Table 23]: *taula 12* [Table 12].

Title: *taula della diuercitat dellesgart della luna en lespera del soleyl altems della conioncion o della oposicion* [Table for the lunar parallax in the sphere of the Sun at the time of the conjunction or opposition] .

Headings₁: column I: *ayso son los gras que si luenha la luna dal cenit* [Here are [written] the degrees of the lunar zenith distance]; column II: *ayso es la cantitat della diuercitat dal regart della luna en lespera del soleyl altems della conioncion* [Here is [written] the lunar parallax in the sphere of the Sun at the time of the conjunction]; headings$_2$: column I: *gras*; column II: *menutz, segons*.

Variant readings: column II, parallax for 60° of zenith distance: 0;45,11**.

TABLE XVIII, f. 63^v: *taula 11* [Table 11].

Title: *taula dal cos della luna en .i^a. ora segon los gras en que es dal mouement della diuercitat edal cos dal soleyl en unaora segon los gras que es luhen dellauga e dellas oras del miei iorn en lautesa de .44. gras segon que es luhen lo soleyl dal cap de canser* [Table for the hourly lunar velocity according to the degrees of the motion in anomaly, and for the hourly solar velocity according to the degrees of its distance from the apogee, and for the hours of half-daylight for latitude 44° according to the solar distance from Cancer 0°].

Headings of the columns: columns I-II: *gras*; column III: *cos della luna en .i^a. ora* [Hourly lunar velocity]; column IV: *cos de soleyl en una ora* [Hourly solar velocity]; column V: *oras emenutz dal miei iorn en lautesa de .44. gras* [Hours and minutes of half-daylight for latitude 44°]. Heading$_2$ of columns III-IV: *menutz, segons*; heading$_2$ of column V: *oras, menutz*.

The ordering of columns III and IV is inverted in the Hebrew manuscripts (see Goldstein [1974a], 182), where the column for lunar velocity (Table 22, IV/IVa) is placed after that for solar velocity (Table 22, III). However, the ordering of the Provençal version is repeated in the description of this table in two different passages of the Latin version of the *Astronomy*: chapter 78 (Vat. Lat. 3098, f. 65rb: "Et residuum tabule in spatia 3 diuiditur. In primo scribetur cursus lune in una hora secundum suam distantiam a principio motus diuersitatis in directo cuiuslibet numeri graduum, in secundo cursus solis in una hora secundum distantiam suam ab auge; in tertio scribentur hore medie diei in oriçonte Aurayce secundum suam distantiam a capite Cancri") and chapter 99 (Appendix, 99:171-172), and this proves that it is not due to the copyist. There are two different versions of the column for lunar velocity in the Hebrew manuscripts: IVa, preserved in Mss *M* and *B*, and IV, preserved in Mss *P* and *Q*. Column IVa was computed by Levi using his lunar model III for syzygy (Goldstein [1974a], 113); column IV was computed taking into account the second lunar inequality, but probably using al-Battānī's tables for lunar anomaly instead of his own table derived from lunar model III (Goldstein [1992], 9-10), and they represent two different stages in Levi's research on the matter. The column for lunar velocity in the Provençal version is different from columns IVa and IV in the Hebrew manuscripts, and it reproduces almost exactly al-Battānī's table for lunar velocity (cf. Nallino [1899-1907], II, 88) – the only discrepancies are the entries 0;30,41, 0;31,39, 0;32,42, and 0;33,18 for arguments of 36°, 66°, 90°, and 102°, instead of al-Battānī's values 0;30,43, 0;31,38, 0;32,41, and 0;33,17, respectively.

No variant readings for columns IV and V.

TABLE XIX, ff. 64^r-65^r [= Table 27]: *taula 10* [Table 10].

Title: [64^r] *taula della diuerssitat dal reguart della luna en lespera del soleyl en longuesa e en larguesa estant la luna en lo cap de cada seyhnal e en quantas horas que sien dauant miei iorn o apres* [64^v] *quompliment dellataula dauant dicha en larguesa* [65^r] *quompliment della taula mesesma dauauant dicha en larguesa* [Table for the lunar parallax in the sphere of the Sun in longitude and latitude when the Moon is at the beginning of each [zodiacal] sign and for the hours before or after noon./ Completion of the aforementioned table in width./ Completion of the same aforementioned table in width].

Headings$_1$: f. 64^r, columns I-IV: *Quancer, Leo, Virgo, Libra*; f. 64^v, columns V-VIII: *Scorpius, Sagitarius, Cabricorn, Aquarius*; f. 65^r, columns IX-XII: *Pisses, Aries, Taurus, Geminis*; headings$_2$: columns I-XII: *longuesa, largesa*; headings$_3$ of columns I-XII: *oras, menutz/ menutz, segons/ menutz, segons*.

To the left of 0^h *miey iorn, miech cel,* and *miech iorn* are written in columns I, V, and IX. In columns I-XII *miech cel* is written above the entries 0;0,0 for longitude.

Variant readings: Cancer, 1^h b.n., latitude: 0;19,17. Virgo, 2^h b.n., longitude: 0;27,36 (= *P*), 1^h b.n., lat.: 0;23,30 (= *M*), 6;45^h a.n., long.: 0;21,47. Libra, 4^h b.n., lat.: 0;20,20, 1;40^h a.n., lat.: 0;40,26, 6^h a.n., long.: 0;21,37 (= *M*). Scorpio, 1^h a.n., lat.: 0;45,49 (= *M*), 3^h a.n., long.: 0;10,24, 4^h a.n., long.: 0;16,21, 5^h a.n., long.: 0;20,55, 5;14^h a.n., long.: 0;21,41. Pisces, 5;14^h b.n., long.: 0;21,41 (= *M*), 5^h b.n., long.: 0;20,55 (= *M*), 4^h b.n., long.: 0;16,21 (= *M*), 3^h b.n., long.: 0;10,24 (= *M*), 1^h b.n., lat.: 0;45,49 (= *M*). Aries, 6^h b.n., long.: 0;21,37 (= *M*), 1;40^h b.n., lat.: 0;40,26, 4^h a.n., lat.: 0;20,20. Taurus, 6;45^h b.n., long.: 0;21,47 (= *M*), 4^h b.n., long.: 0;19,38**, 1^h a.n., lat.: 0;23,30 (= *M*), 2^h a.n., long.: 0;27,36. Gemini, 7;23^h b.n., lat.: 0;45,50**, 6^h b.n., lat.: 0;41,35.

The sequence 0;10,24, 0;16,21, 0;20,55, and 0;21,41 for Scorpio from 3^h a.n. to 5;14^h a.n. (repeated in the text for Pisces from 5;14^h b.n. to 3^h b.n.) proves that Ms. *M* represents an intermediate stage between the Provençal version and Mss *P* and *Q*. The last two have the sequence 0;9,56, 0;15,26, 0;19,38, and 0;21,37 for Scorpio and Pisces, whereas Ms. *M* gives the second sequence for Scorpio and the first for Pisces. The recomputation of the entries for Scorpio, 3^h and 5;14^h a.n., by Goldstein (1974a), 122 (0;10,39 and 0;21,50), shows a slightly better agreement with the Provençal version.

TABLE XX, f. 65^v [= Table 35, column VI]: *quarta taula 4* [Table 4].

Title: *taula della larguesa della luna asseptentrio o al meredional* [Table for the lunar latitude, to the north or to the south].

Because of its double symmetry, the table is arranged in the Provençal text in five columns. The heading of the two first (for the degrees of the argument of latitude from 1 to 30, and from 20 to 0, respectively) is *lihna dels nombres dels gras.* Headings$_1$ of columns III-V indicate the signs and their latitude (*sehnals 6 Mere-*

dional / 0 septemtrion, etc. above the column; *sehnals 5 septentrion / 11 meridional*, etc. below). Headings₂ of columns III-V: *gras, menutz, segons*.

Variant readings: 0ˢ 18° 1;23,16 (= *Q*). 2ˢ 7°: 4;8,38**.

TABLE XXI, ff. 66ʳ-67ʳ [= Table 45].

Title: [66ʳ] *taula dal cors ueray della luna aquascun iorn apres la conioncion olla oposi[ci]on meiana entro .14. iorns* [Table for the true motion of the moon for each day after mean conjunction or opposition, until 14 days].

The entries of the table are distributed into six sub-tables (two in each folio), of which the first one has the aforementioned title and the others *dal compliment della taula* (for the completion of the table). Each sub-table has seven columns, the first for the days (*iorns*) and the other six for 10-degree intervals (*sehnal 0, gras 0 / sehnal 0, gras 10 / sehnal 0, gras 20 ...*). Heading₂ of columns II-VII: *gras, menutz, segons*.

Variant readings: 0ˢ 20°, day 2: 24;31,9**. 1ˢ 0°, day 1: 12;20,22 (= *M* and *Q*), day 14: 189;2,50. 2ˢ 0°, day 14: 192;52,17. 3ˢ 0°, day 12: 167;6,55 (= *Q*). 3ˢ 10°, day 8: 113;5,33, day 11: 154;7,16. 4ˢ 0°, day 8: 114;51,21. 4ˢ 10°, day 12: 166;52,31. 4ˢ 20°, day 7: 102;1,45 (= *M*), day 12: 166;3,56. 5ˢ 0°, day 13: 177;4,43 (but 179;4,43 is a misprint in G). 5ˢ 20°, day 2: 30;45,4. 6ˢ 0°, day 1: 15;27,6, day 7: 99;33,21**. 6ˢ 10°, day 2: 30;29,2, day 9: 122;53,3 (= *A*), day 10: 134;45,4, day 12: 158;22,2. 7ˢ 0°, day 12: 154;45,2. 7ˢ 20°, day 4: 55;44,58 (= *M*), day 9: 115;56,23, day 13: 163;45,20. 10ˢ 10°, day 14: 176;9,20. 11ˢ 0°, day 9: 18;18,54**.

APPENDIX

A critical edition of chapter 99 of the Latin version of Levi's Astronomy

1. *Introduction*

Four manuscripts preserve copies of the Latin version of Levi's *Astronomy*: Vat. Lat. 3098 (*A*, the oldest one, middle of the XIVᵗʰ c.); Vat. Lat. 3380 (*B*, XVᵗʰ c.); Lyon, Bibl. Munic. 326 (*L*, XVIᵗʰ c.); and Milano, Ambros. D. 327 (*M*, XVIᵗʰ c.). The *stemma* of the manuscripts is as follows: *B* and *L* have been copied from *A*, and *L* was later corrected in chapters 4 to 11 using the independent *editio* (based on the text represented by *A*) of these chapters entitled *Tractatus instrumenti astronomie* as preserved, for instance, in Ms. Paris, Bibl. Nat. Lat. 7293, ff. 1ʳ-17ᵛ. The fourth and poorest text, *M*, was copied from *L*, and I have omitted its variant readings in the critical apparatus. The only characteristic worthy of mention of *M* is that it preserves an interlinear alternative translation for some passages of the table of contents and chapters 1 to 3 apparently based on the Hebrew text and made in the XVIth century. In *A*, at the end of the *dictio quarta*, half of the column b in f. 82[v] is blank, as well as the *recto* of the following folio, numbered 84; the *dictio quinta* begins in f. 84[va] and ends, after six lines of text, at the top of f. 85[v], and chapter 100 begins in the *recto* of the following folio, numbered 90. These missing five folios still existed when *B* was copied from *A*, as is attested by marginal notes in *B* at the end of the fourth section (*Erant hic folia alba per huiusmodi tabulis faciendis*, f. 203[r]) and at the end of the chapter (*spatium 4[or] foliorum*, f. 206[v]). A blank space for tables is also preserved in *L* (ff. 202[v] and 203[r-v], at the end of the fourth section, and ff. 206[v]-210[v], at the end of the chapter).

In editing the text I have retained the spelling of *A*, with a few exceptions indicated in the critical apparatus. Editorial insertions are enclosed in square brackets, as well as sentence numbers added for ease of reference and comparison with the Provençal text. Paragraphing, punctuation, and capitalization are entirely due to the editor. The description of the chapter given below in sentence [0] is taken from the table of contents of manuscript *A*, f. 2rb. The text of the chapter is found in *A*: ff. 81va-85[v], *B*: ff. 197vb-206[v], and *L*: ff. 199[r]-206[r].

2. *Comments*

Sentences [120]-[145] contain Levi's procedure for computing the cusps of the astrological houses which replaces the "method of the Ancients" in Provençal sentences [114]-[115]. If λ_1, λ_2, ... denote the ecliptic longitudes of the points at which begin the first, second ... houses, and α_1, α_2, ... the corresponding right ascensions, Levi computes λ_2, λ_3, λ_{11}, and λ_{12} in the following way. Let be x the degree of the zodiac which corresponds in Table 4 (Goldstein [1974a], 160) to $\alpha_1 + 1/3(\alpha_1 - \alpha_4)$. If $h_d(x)$ is the half-daylight corresponding to x in Table 5 (Goldstein, *ibid.*, 161), we made $\alpha_1 + [60 - 1/3(h_d(x))] = \rho$, and we take ι, the degree of the zodiac which corresponds to ρ, from table 7 (Goldstein, *ibid.*, 162). If $\iota = x$, $x = \lambda_2$; if not, the procedure is repeated starting from a value slightly different from x until identity between ι and x is reached. The same process is used to find λ_3, excepting that 2/3 and 120 replace, respectively, to 1/3 and 60. For λ_{11}, we take from table 4 the degree of the zodiac, x, which corresponds to $\alpha_{10} + 1/3(\alpha_{10} - \alpha_1)$, and, after finding $h_d(x)$ in Table 5, we search again in Table 4 the degree, j, which corresponds to $\alpha_{10} + [h_d(x)/3]$; if $j = x$, $x = \lambda_{11}$, iterating as above if it is not the case. The same is to be made for λ_{12}, except that 2/3 replaces 1/3 (compare with equations in North [1986], 23). Column II of Table C-3 displays values for λ_1 to λ_6 in Levi's example in sentences [127]-[38], which I have computed only to minutes using Levi's procedure, and column III the same cusps computed using North's program for the *hour lines (fixed boundaries) method* (see above section 4.3, comments to sentences [114]-[115]):

TABLE C-3		
I	II	III
λ_1	0; 0	0; 0, 0
λ_2	43; 4	43; 4,29
λ_3	69;19	69;20,10
λ_4	90; 0	90; 0, 0
λ_5	110;40	110;39,50
λ_6	136;55	136;55,31

In sentences [236]-[74] the text contains a computation of the solar eclipse of 26 June 1321 (using Levi's first solar model) in greater detail than that which appears in chapter 80 (Goldstein [1979],109-110, and 123-124). With the exception of the lunar anomaly and the true solar longitude (90;38° and Cancer 11;39°, respectively), at least 14 of the 30 values computed in the text are slightly different in the Hebrew version of this chapter (lunar parallax in longitude, relative lunar velocity, apparent lunar velocity in the 6th hour before noon, remainder of the lunar parallax in longitude, lunar parallax in longitude at the end of the 7th hour before noon, apparent lunar velocity in the 7th hour before noon, time of eclipse-middle, lunar longitude at eclipse-middle, argument of latitude, lunar latitude, lunar parallax in latitude, apparent lunar latitude, begining and end of the eclipse), both sets being internally consistent and derived from different values for the lunar parallax in longitude (0;40,19° East in the Hebrew version, and 0;40,8° here) and for the relative lunar velocity (0;30,20°$^{/h}$ and 0;30,31°$^{/h}$, respectively).

3. Text

[0] In 99° adunabimus aliqua que fecimus circa inuentionem loci solis et lune et eclipsium eorundem et aliquorum eorum ad petitionem aliquorum christianorum nobilium; quod capitulum in 5 dictiones diuiditur. Id est, in prima dictione ponemus tabulas quas fecimus ad sciendum coniunctiones et oppositiones medias et ueras et uerum locum solis in quolibet tempore et 4 domorum principalium principium prime, 4^e, 7^e et 10^e. In 2^a docebimus ex dictis quomodo habuimus scientiam que cuiuscumque coniunctionis uel oppositionis uolemus habere. In 3^a

declarabimus quomodo de coniunctione uel oppositione media deueniemus ad ueram. In 4ª docebimus inuenire uerum locum solis et quatuor domorum principalium initia in tempore sempiterno. In 5ª dabimus doctrinam intelligentibus inueniendi faciliter eclipses solares et lunares et uerum locum lune in omni tempore mediantibus aliquibus tabulis quas adiunximus predicte doctrine.

Capitulum 99

[1] Et quia ad requisitionem multorum christianorum nobilium laborauimus ad ordinandum computum coniunctionum et oppositionum mediarumque uerarum facili modo ualde, intantum quod etiam infans paruus possit practicari faciliter, [2] nobis apparuit quod hic esset iste computus conscribendus, quia in eo est profactus mirabilis ad inueniendum faciliter tempus coniunctionis et oppositionis cuiuslibet tam in preterito quam in presenti quam etiam in futuro. [3] Et quia postquam fuit iste computus ordinatus fuit nobis ex aliquibus eclipsibus solaribus et lunaribus reuelatum quod est necessarium addere equationi solis finalis circa 0;9° et quod est necessarium ponere latitudinem aliquam in diametris sperarum motus diuersitatis in principio dicti motus, ut recitabimus in sequenti capitulo, si Deo placuerit, [4] computum coniunctionum et oppositionum secundum hanc equationem solis sic correctam ordinabimus in hoc loco et latitudinis antedicte.

[5] Et sermo noster circa predicta in 5 dictiones diuiditur. [6] In prima declarabimus tabulas quas fecimus ad inueniendum coniunctiones et oppositiones medias atque ueras et uerum locum solis tempore sempiterno et 4 principalium domorum principia, scilicet prime, quarte, septime et decime. [7] In secunda declarabitur quomodo inueniatur queuis coniunctio et oppositio media. [8] In tertia quamlibet ueram inueniri docebimus. [9] In quarta docebimus inueniri uerum locum solis tempore coniunctionis et oppositionis uel quouis alio tempore et gradum ascendentem in oriçonte Aurayce, que distat ab oriente 9;46ʰ, et gradum medij celi in quolibet oriçonte. [10] In quinta docebimus scire faciliter computus eclipsium solariumque lunarium ex aliquibus tabulis ordinatis ad istud. [11] Et ibi etiam ordinauimus tabulas ad sciendum uerum locum lune in quolibet tempore secundum dispositionem predictam.

Dictio 1

[12] Et sciendum quod radix computus tabularum dictarum est post annum 1320 incarnationis Christi in oriçonte Aurayce. [13] Et in prima tabula sunt ordinate medie coniunctiones et oppositiones in eodem oriçonte usque ad 8 annos solares continentes circa 99 coniunctiones. [14] Et sciendum quod ex certa scientia coniunctiones sunt scripte atromento nigro, oppositiones uero scripte sunt rubeo. [15] Et dicta tabula est diuisa in 5 spatia. [16] In primo est scriptus mensis solaris. [17] In secundo dies mensis illius in quo est coniunctio uel oppositio. [18] In tertio est scripta feria in qua est coniunctio uel oppositio. [19] In quarto est scriptus

numerus horarum equalium post meridiem. [20] In quinto numerus minutorum. [21] Et post quamlibet coniunctionum uel oppositionum est scriptus locus lune motus diuersitatis in tempore illo in spatijs tribus. [22] In primo est scriptus numerus signorum, in secundo numerus graduum, in tertio numerus minutorum. [23] Et principium anni supponimus in omnibus nostris computibus in prima die mensis martij. [24] Et numerus annorum est scriptus in ista tabula iuxta lineam principij anni.

[25] In secunda tabula sunt 69 linee, quarum quelibet seruit tempori totius tabule prime, scilicet 8 annis solaribus, uni diei, 12 horis, 41 minutis et circa 13 secunda, que tabula in spatia quinque diuiditur. [26] In primo est scriptus numerus linearum de uno usque in 69, [27] in secundo numerus dierum addendorum numero dierum tabule prime uel subtrahendorum ab eo, [28] in tertio numerus feriarum addendarum uel subtrahendarum, in quarto hore, in quinto minuta; [29] que tabula comprehendit 552 annos, 105 dies, 11 horas, 24 minuta. [30] Et post quamlibet lineam est scriptus locus lune motus diuersitatis modo quo supra. [31] Tabula tertia habet 40 lineas, quarum quelibet tempori totius tabule secunde deseruit; ideo comprehendit 22091 annos, 201 dies; [32] que tabula in spatia 5 diuiditur modo quo supra. [33] Et ibidem est scriptus locus lune motus diuersitatis modo quo supra. [34] Tabula uero quarta habet 28 lineas, quarum quelibet deseruit toti tempori tabule tertie et per consequens comprehendit 618563 annos, 156 dies; [35] que solum in 3 diuiditur spatia, quia non sunt ibi hore neque minuta. [36] Et ibidem est scriptus locus lune motus diuersitatis modo quo supra. [37] Et post totum tempus contentum in tabulis istis reincipitur ab earum principio, set post duas reuolutiones superfluum dierum est 313 dies propter bisextum.

[38] Prima autem aliarum tabularum ordinatarum ad hoc continet equationem in horis et minutis prouenientem excentricitatis de causa in die anni et parte diei in quo est coniunctio uel oppositio media. [39] Et in latitudine tabule istius scripti sunt menses anni in 12 spatijs, et in longitudine dicte tabule dies mensium sunt descripti, [40] equationem cuiuslibet diei in directo eius sub quolibet mense inueniens. [41] Et in secunda tabula sunt descripti menses et dies modo eodem, [42] et continet equationem in minutis hore prouenientem ex dierum naturalium inequalitate in tempori anni in quo est coniunctio uel oppositio media. [43] Tabula tertia continet quandam equationem prouenientem propter locum lune motus diuersitatis, [44] et in latitudine tabule scripta sunt signa, in longitudine uero gradus. [45] Quarta tabula continet quandam equationem prouenientem in tempore anni in quo est coniunctio uel oppositio media propter excentricitatem et propter inequalitatem dierum et propter locum lune motus diuersitatis simul. [46] Et in latitudine tabule sunt scripta signa motus diuersitatis et gradus de 15 in 15, et in longitudine tabule sunt ordinati menses anni et dies de 6 diebus in 6 dies gradatim, excepto quod in principio martij est gradus dierum 4 et in fine mensium 31 dierum est gradus 7 dierum, excepto fine martij ubi est gradus 6 dierum sicut

in alijs. [47] Quinta tabula est ordinata ad sciendum uerum locum solis in qualibet die anni; [48] in latitudine tabule sunt scripti menses anni per ordinem, et in longitudine eius dies mensium. [49] Et in sexta tabula sunt scripte ascensiones signorum in spera recta incipiendo a capite Capricorni. [50] Et hec tabula est facta et locata in secunda distinctione 62 capituli huius primi tractatus huius quinti libri. [51] In septima tabula sunt scripte ascensiones omnes signorum in oriçonte Aurayce ubi altitudo poli est 44°, que incipit ab Arietis capite; que est facta et posita in quinta dictione prelibati capituli. [52] Has duas tabulas si quis uellet posset in hunc locum transferre, ut sibi facilius in practicando occurrerent.

Dictio 2

[53] Si queris coniunctionem uel oppositionem mediam inuenies eam ordinatam in tabula prima usque ad 8 annos solares. [54] Et si tempus in quo queris est post annum Christi 1328, subtrahas ab annis Christi quos habes 1320 annos et tantum plus quantum est a principio martij ad primam coniunctionem, si queris coniunctionem, uel primam oppositionem si queris oppositionem. [55] Et si residuum est minus tempore contento in secunda tabula, diuidatur per 8 annos, 36 horas, 41 minuta, et numerus quotiens in secunda tabula est intrandum. [56] Verbi gratia, si numerus quotiens esset 10, cum eo intrabis in 10 lineam; [57] qua seruata residuo annorum et mensium quod superauit ex diuisione predicta, quod debet esse minus 8 annis, 36 horis, 41 minutis, addas tempus quod est a principio martij ad primam coniunctionem, si queris coniunctionem, uel primam oppositionem, si queris oppositionem, et cum eo intrabis annum et mensem cuius coniunctionem uel oppositionem tu queris; [58] cuius introitus linea seruata, quod est in eius directo coniungatur cum eo quod est in directo predicte linee decime prius seruate, dies cum diebus, feria cum feria, et quodlibet cum quolibet sui generis. [59] Et si hore sunt 24 uel ultra, de 24 facias diem unam, qua[m adda]s diebus mensis et feriis. [60] Et si nichil remanet facta diuisione predicta, intrabis lineam prime coniunctionis prime tabule, si queris coniunctionem, uel prime oppositionis, si queris oppositionem, et inuento in eius directo utaris ut prius. [61] Et sic habebis mensem et diem coniunctionis uel oppositionis quam queris.

[62] Et si tempus coniunctionis uel oppositionis quam queris est ante tempus nostre radicis, uideas quantum antecedit horam prime coniunctionis prime tabule, si queris coniunctionem, uel oppositionis, si queris oppositionem, et illud totum diuidas per 8 annos, 36 horas, 41 minuta. [63] Et si diuisio est punctalis cum numero quotiens tabulam secundam intrabis modo quo supra, et inuento in eius directo utaris ut prius. [64] Si uero diuisio non est punctalis, addas numero quotiens unitatem et tunc cum eo tabulam secundam intrabis. [65] Et residuum diuisionis subtrahas ab annis 8, horis 36, minutis 41, et subtractionis residuo addas tempus a principio martij ad primam coniunctionem uel oppositionem. et utaris modo quo supra. [66] Et inuenies lineam prime tabule quam tu queris, et tunc subtrahas ab inuento in ea illud quod est in directo linee seruate de tabula secunda.

[67] Et residuum erit tempus coniunctionis uel oppositionis quam queris. [68] Et hinc est notum faciliter qualiter istis tabulis sit utendum, si tempus distans a nostra radice tempore tabule secunde sit maius, quia tunc tempus coniunctionis uel oppositionis in tribus tabulis est querendum. [69] Et si istud tempus est post radicem, coniungatur totum quod est in lineis omnium aliarum tabularum ab eo quod est in linea tabule prime, et sic deuenies in tempus coniunctionis uel oppositionis quam queris. [70] Et eodem modo inuenies locum lune motus diuersitatis addendo addenda, et subtrahendo subtrahenda.

[71] Et sciendum quod non expedit computare modo predicto annos scriptos in tabulis, set solum computabis mensis et dies, ad sciendum in quo die anni ueniet computus quem tu queris. [72] Set computes largo modo reuolutionem prime tabule solum 8 annos, et reuolutionem secunde solum 552, et reuolutionem tertie solum 22080 annos, et tunc opportebit te computare annos inuentos in lineis seruatis modo quo supra. [73] Et tunc scies quo anno et die tuus computus finietur et hinc scies faciliter tempus coniunctionis uel oppositionis. [74] Quam et qualitercumque tu computes, expedit quod attendas ad annos que sunt in tertia tabula et in quarta; [75] quia si anni dictarum tabularum coniuncti sunt 3 uel 4 supra reuolutionem 4 annorum et in nulla earum per se sunt 3 uel 4 supra reuolutionem 4 annorum, tunc subtrahes de diebus anni et de ferijs unam diem propter diem bisexti computatum in anno quarto.

Dictio 3

[76] Et postquam tempus coniunctionis uel oppositionis medie tu sciueris ac etiam lune locum motus diuersitatis, in uere coniunctionis uel oppositionis notitiam cum eodem loco et tempore poteris faciliter deuenire. [77] Ad quod faciendum, illud quod de horis et minutis in primis duabus tabulis equationis inuenies in directo illius diei accipies, et illud sub horis et minutis coniunctionis et oppositionis medie scribes. [78] Deinde queres locum lune motus diuersitatis in tabula tertia, et quod inuenies in directo signi et gradus scribes cum alijs. [79] Item queres in tabula quarta diem mensis in quo es in longitudine tabule et locum lune motus diuersitatis in latitudine, et equationem quam in directo utriusque loci inuenies scribes ut supra. [80] Quibus omnibus agregatis a resultante subtrahes unam diem et residuum coniunctionis est tempus oppositionis uel coniunctionis uere in certitudine magis possibili.

[81] Et si non inuenitur in tabulis tempus medie coniunctionis uel oppositionis punctaliter, accipies equationem propinquiorem precedentem et propinquiorem sequentem, et de diferentia earum secundum proportionem. [82] Verbi gratia, si tempus coniunctionis uel oppositionis medie mensis martij est 3 dies, 8 hore post meridiem, accipies equationem tertie diei et quarte, et de diferentia earum talem proportionem qualem habent 8 ad 24. [83] Et acceptum addes equationi tertie diei, si est addendum, uel subtrahes, si est subtrahendum. [84] Et eodem modo facies si non inuenies locum lune motus diuersitatis punctaliter.

[85] Et est sciendum quod dies anni secunde tabule transeunt per unam diem in 119 annis et tribus quartis anni. [86] Et ideo quando transierint 119 anni et tres quarte anni post radicem, reduces equationem scriptam in secunda die martij in primam et scriptam in tertia in secundam, et sic de singulis mensibus et diebus. [87] Et oppositum horum esset in tempore elapso ante radicem. [88] Et similiter in quibuslibet 119 annis et tribus quartis anni erit fiendum. [89] Item dies prime tabule et quarte transeunt per unam diem in 43 annis. [90] Et ideo transactis 43 annis reduces equationem secunde diei ad primam, et sic de singulis, et eodem modo erit in quibuslibet 43 annis fiendum. [91] Et oppositum esset fiendum in tempore elapso ante radicem. [92] Et causa horum uirtualiter continetur in uerbis nostris superius.

Dictio 4

[93] Si scire uolueris uerum locum solis tempore coniunctionis uel oppositionis uel quouis alio tempore, hoc inuenies in tabula quinta modo nunc dicendo, [94] et ex hoc scies uerum locum lune tempore coniunctionis uel oppositionis uere. [95] Sciendum est autem primo quod ista tabula fuit ordinata pro uero loco solis in qualibet meridie anni Christi 1321 in oriçonte Aurayce. [96] Ideo uolens scire uerum locum solis post istam radicem subtrahas ab annis Christi 1320 et serua residuum; [97] a quo subtrahas omnes quaternarios possibiles subtrahi, ita quod remaneant 4 uel minus, et illud erit annus in quo eris de reuolutione 4 annorum. [98] Et si es in primo anno reuolutionis predicte, scriptum in tabula est uerus solis locus in qualibet meridie anni illius. [99] Si es in secundo, erit 6 horis post meridiem. [100] Si in tertio, erit 12 horis post meridiem. [101] Si uero in quarto, erit 6 horis ante meridiem. [102] Et sciendum pro quolibet anno completo post radicem inuento in tabula addenda sunt $0;0,30°$, et sic usque ad 43 annos post radicem uteris. [103] Quibus elapsis descendes per unam diem in tabula; [104] intendo dicere quod illud quod est scriptum in secunda die martij reducas in primam, et ita de singulis. [105] Et tunc subtrahas ab inuento in tabula $0;37,54°$. [106] Et utaris in 43 annis sequentibus sicut usus fuisti in annis reuolutionis predicte. [107] Verbi gratia, si sunt transacti 4 anni post reuolutionem primam 43 annorum, descendes in tabulam per unam diem et non subtrahes de inuento in ea nisi $0;35,54°$, quia debes addere $0;2°$ propter 4 annos completos. [108] Et hoc modo te habeas in qualibet reuolutione 4 annorum, ita quod in quolibet anno sequenti subtrahas $0;0,30°$ minus quam in precedenti usque ad 43 annos, et tunc descendens in tabulam per unam diem procedas ut prius usque ad alios 43 annos, et sic de singulis in infinitum. [109] Et oppositum predictorum omnium est fiendum annis ante radicem; [110] intendo dicere quod illud quod erat addendum erit subtrahendum et illud quod erat subtrahendum erit addendum, et ubi descendebatur per unam diem in tabula ascendetur.

[111] Et uero loco solis scito queras in oriçonte recto ascensiones gradus çodiaci in quo est sol in tabula sexta. [112] Et si coniunctio uel oppositio est in meridie,

erit gradus in quo est sol gradus meridiei, qui est principium decime domus. [113] Si uero coniunctio uel oppositio est post meridiem, multiplices hore et minuta que sunt post meridiem per 15 et proueniens addas ascensionibus gradus in quo est sol, et serua. [114] Et gradus cuius ascensiones sunt equales seruato est principium 10 domus. [115] Quo scito et ascensionibus eius, queras ascensiones eis equales in 7 tabula. [116] Et gradus çodiaci eis respondens erit gradus ascendens siue principium prime domus in oriçonte in quo sunt ordinate. [117] Et hijs duabus domibus scitis leue est scire alias duas cauiculas, quia gradus oppositi principijs primarum sunt principia secundarum. [118] Et in notitiam principiorum aliarum domorum potes cum astrolabio faciliter deuenire.

[119] Quo non obstante, quia est ualde difficile inuenire punctaliter principia aliarum domorum, hic intelligentibus damus doctrinam qua possint ea punctaliter inuenire sine alicuius aminiculo instrumenti. [120] Que doctrina est ista. [121] Queras in ascensionibus oriçontis recti diferentiam inter ascensiones gradus ascendentis et principij quarte domus, et accipias tertiam eius partem; [122] quam addas ascensionibus gradus ascendentis oriçontis predicti et uideas cui gradui çodiaci in dicto oriçonte facta additione respondeant, et serua. [123] Deinde queras in tuo oriçonte semiarcum diei gradus seruati, cuius tertiam partem subtrahas de 60. [124] Et residuum addas ascensionibus gradus ascendentis in tuo oriçonte et uideas cui gradui zodiaci in eodem oriçonte respondeant iste ascensiones cum addito. [125] Et si gradus cui respondent est minus seruato, addas ei semigradum uel unum gradum uel maius quousque acceptum per suum semiarcum modo predicto additum ascensionibus ascendentis in tuo oriçonte consentiat ut sue ascensiones punctales dicto gradui respondenti facta additione. [126] Et si gradus respondens est plus seruato, subtrahas a respondente paulatim quousque eodem modo idem fiat.

[127] Verbi gratia, sit principium Arietis gradus ascendens, [128] et per consequens principium Cancri est principium quarte domus. [129] Et est notum quod ascensiones a principio Arietis in oriçonte recto usque ad principium Cancri sunt 90°; quorum pars tertia est 30°, quibus additis ascensionibus gradus ascendentis, que sunt 90°, fiunt 120°. [130] Et est notum quod gradus respondens ascensionibus istis in oriçonte recto est 2;12° Tauri, et iste est gradus seruatus. [131] Et est notum quod semiarcus diei gradus seruati est 102° et circa 0;6°; cuius tertia pars est 34;2°; [132] qua subtracta a 60° [remanent 25;58°; quibus additis ascensionibus gradus ascendentis] in oriçonte Auraice, que sunt 0, erunt 25;58°. [133] Et est notum quod gradus respondens ascensionibus istis in oriçonte Auraice est 14° et circa 0;59° Tauri, et hoc est multo plus gradu seruato. [134] Ideo subtrahamus ab hoc gradu respondente secundo paulatim modo predicto. [135] Et inueniemus quod si iste gradus respondens ponatur 13;4° Tauri, ei consentient ascensiones accepte modo predicto; [136] quia tertia pars semiarcus diei 13;4° Tauri est 35;18° tantum, quibus subtractis a 60° remanent 24;42°; [137] quibus additis ascensionibus gradus ascendentis in oriçonte Aurayce, 0, erunt 24;42°, quibus respondent 13;4° Tauri in oriçonte predicto. [138] Quare est notum quod istud est principium domus

secunde. [139] Et eodem modo inuenies principium tertie domus excepto dum-taxat quod, ubi accipiebas tertiam partem, duas tertias partes accipias, et ubi tertiam partem semiarcus diei subtrahebas de 60°, duas tertias subtrahas de 120°.

[140] Principium 11 et 12 domus inuenies modo dicendo. [141] Nam tertiam partem diferentie inter ascensiones principij 10 domus et ascensiones gradus ascen-dentis in oriçonte recto addas ascensionibus principij 10 domus, et uideas quis gradus çodiaci eis in oriçonte recto respondeat et serua. [142] Deinde tertiam partem semiarcus diei gradus seruati addas ascensionibus principij 10 domus et uideas quis gradus çodiaci eis in recto oriçonte respondeat. [143] Et si gradus respondens est met gradus seruatus, ipse est principium domus 11; [144] si uero gradus respondens est plus uel minus seruato, a gradu respondente paulatim subtrahas aut addas et procedas modo quo supra. [145] Et eodem modo inuenies principium 12 domus saluo quod ubi accipiebas tertiam partem duas tertias partes accipias. [146] Et sciendum quod inuento principio 11 domus debent inueniri inter ipsum principium et gradum ascendentem in tuo oriçonte ascensiones 4 horarum inequalium diei primi gradus dicte domus 11, [147] et inuento principio 12 domus debent inueniri inter ipsum principium et gradum ascendentem ascensiones 2 horarum inequalium diei primi gradus dicte 12 domus. [148] Et sciendum quod querendo tres domus sub terra inuenimus nos horis inequalibus gradus oppositi, quia respectu inferioris emisperij polus antarticus est superior. [149] Et ideo sunt ascensiones signorum meridionalium ibi tales quales ascensiones septentrionalium signorum sunt nobis. [150] Hijs autem 6 domibus scitis, 6 relique erunt scite, quia gradus oppositi harum principijs sunt principia reliquarum.

[151] Si autem tuus oriçon ab orizonte Aurayce in longitudine est diuersus, leue est tabulas istas ad tuum reducere subtrahendo diuersitatem longitudinis inter eos, si tua est maior, uel eam addendo, si tua est minor, dato quod longitudinem computes ab oriente, ut facimus. [152] Si autem eam ab occidente tu computas, econtrario facias. [153] Si uero latitudo esset diuersa, expediet facere tabulas ascensionum oriçontis illius sicut fecimus in oriçonte Aurayce utendo doctrina data superius. [154] Et ex illis tabulis deuenies in notitiam domorum 12 modo nunc tradito. [155] Poteris autem habere scientiam diuersitatis longitudinis inter tuum oriçontem et oriçontem Aurayce ex sequentibus tabulis, quibus tempus uere coni-unctionis et oppositionis cognoscitur, primo cognito tempore medij alicuius eclipsis lunaris, quia diuersitas temporis medij illius eclipsis inter tuum oriçontem et istum est longitudinis predicte diuersitas.

[156] Quoniam autem est ualde difficile inuenire domos 12 modo predicto, ordinauimus hic supra promissum de gratia speciali 12 tabulas ad eas inueniendum facilime in oriçonte Aurayce, quarum quelibet in latitudine in 6 spatia est diuisa et in longitudine in lineas 30 diuiditur. [157] Tabula prima attribuitur Arieti isto modo: quia in prima linea primi spatij supponitur primus gradus Arietis gradus ascendens, et in eius directo in secundo spatio ponitur principium domus secunde, et in tertio tertie, et sic de singulis usque ad sextam. [158] In secunda uero linee

primi spatij supponitur secundus gradus Arietis gradus ascendens, et in eius directo in secundo spatio scribitur principium domus secunde, et cetera ut prius, et sic de singulis gradibus Arietis in singulis lineis primi spatij. [159] Et in eorum directo in sequentibus spatijs ponuntur principia aliarum domorum usque ad sextam. [160] Secunda tabula eodem modo attribuitur Tauro, tertia Geminis, et sic de singulis signis et tabulis. [161] Hijs autem 6 domibus scitis, 6 relique erunt scite, ut dictum est supra. [162] Has autem duodecim tabulas nunquam scripsimus seorsum in hijs que fecimus ad petitionem aliquorum christianorum nobilium, que sunt adunata in isto capitulo, ut promisimus antea. [163] Ideo nunc paulo superius diximus quod has tabulas hic ordinauimus supra promissum de gratia speciali ad faciliorem domorum inuentionem.

Dictio 5

[164] Tabule quas expedit nos habere impromptu ad inueniendum eclipses solis et lune sunt iste. [165] Quarum prima et secunda sunt ordinate ad inueniendum locum capitis draconis ad annos, menses, dies, horas et minuta horarum que sunt facte superius ad radicem annorum Christi 1300. [166] Si cui uero esset cure hic eas locare, ponat eas ad radicem suppositam in isto capitulo. [167] Tertia est ordinata ad inueniendum latitudinem lune ad meridiem et septentrionem secundum suam distantiam a capite draconis. [168] Quarta est eclipsis lune sole stante in auge et sephel. [169] Quinta est eclipsis solis ipso stante in auge et sephel. [170] Sexta est minutorum proporcionalium ad equandum computus eclipsium solariumque lunarium. [171] Septima est ad inueniendum uerum cursum lune in una hora secundum suam distantiam a motu diuersitatis principio et uerum cursum solis in una hora secundum distantiam suam ab auge, et est superius posita; [172] set quia postea ei addidimus ad sciendum numerum horarum medie diei in oriçonte Aurayce secundum distantiam solis a principio Cancri, ideo iterum hic eam locauimus. [173] Octaua est ad inueniendum diuersitatem aspectus lune in longitudine et latitudine ipsa stante in principio cuiuslibet signi in qualibet hora ante meridiem et post. [174] Nona est ad inueniendum diuersitatem aspectus lune respectu spere solis tempore coniunctionis uel oppositionis secundum distantiam suam a cenith. [175] Et est superius posita ubi declaratur quomodo possimus nos ea iuuare ad sciendum diuersitatem lune in temporibus omnibus. [176] Omnes tabule precedentes sunt superius posite in locis diuersis, set si quis uellet posset hic eas locare. [177] Decima est ad sciendum uerum cursum lune qualibet die post mediam coniunctionem uel oppositionem usque ad 14 dies secundum locum motus diuersitatis, et est gradualis de 10° in 10°.

[178] Si id quod queris non est in tabulis istis punctaliter quoad tempus uel locum, queras in eis duo loca propinquiora ei quod queris, precedentem uidelicet et sequentem, et de diferentia eorum secundum proportionem accipias. [179] Verbi gratia, si uis scire diuersitatem aspectus lune ipsa stante in 12° Arietis una hora ante meridiem, queras hoc in tabula luna stante in principio Arietis et quod inuenies in

directo unius hore ante meridiem scribas seorsum, et eodem modo facias luna stante in principio Tauri. [180] Et de diferentia inter ista duo loca talem proportionem accipias qualem habent 12° ad totum signum, qui 12° sunt 2 quinarij signi unius; [181] quod acceptum addas diuersitati inuente luna stante in principio Arietis, si erit addendum, uel subtrahas, si fuerit subtrahendum.

[182] Et scietur ex istis tabulis recitatis latitudo lune tempore coniunctionis uel oppositionis modo dicendo. [183] Scias primo locum medium capitis draconis illo tempore ex tabulis ad hoc ordinatis, cui uerum locum lune in eodem tempore addas. [184] Et proueniens est argumentum latitudinis lune, cum quo intrabis tabulam latitudinis lune. [185] Et in directo argumenti inuenies latitudinem lune septentrionalem, si argumentum est minus 180°, uel meridionalem, si est plus.

[186] Scies autem uerum locum lune illo tempore scito uero loco solis, quia luna est simul cum sole, si est coniunctio, uel in gradu et minuto opposito, si est oppositio. [187] Et scita latitudine lune tempore oppositionis, poteris scire utrum luna eclipsetur et, si sic, quantum, et alias conditiones que sunt ordinate in tabula eclipsis lune modo statim dicendo. [188] Nam si latitudo lune est 0;56,31° uel plus, non est possibilis eclipsis lunaris; si uero minus, est possibilis. [189] Et si est minus 0;55,33°, necessario erit eclipsis. [190] Et scito quod possibilis est eclipsis, intres cum latitudo lune illo tempore in duas tabulas eclipsis lune et scribas quod inuenies in directo latitudinis in utraque tabula, scilicet quantitatem digitorum eclipsatorum et omnium temporum in qualibet tabula descriptorum. [191] Et de diferentia omnium in utraque tabula inuentorum talem proportionem accipies qualem habent minuta proportionalia que sunt in directo distantie tunc solis ab auge ad 60; [192] quod acceptum addas inuentis in tabula eclipsis lune sole stante in auge, si est eis addendum, uel subtrahas, si est subtrahendum ab eis. [193] Et proueniens est quantitas eclipsis lune illo tempore, et quantitas uelocitatis lune in quolibet tempore descripto in tabula. [194] Et scies distantiam solis ab auge scito loco augis tempore radicis istius capituli et scita quantitate motus istius. [195] Et iam sciuimus per experientias longas et certas quod locus eius erat tempore dicte radicis in Cancro 2° et circa 0;38°. [196] Et secundum quod inuenimus per experientias nostras eis addendo experientias omnium antiquorum perfectius loquentium in ista scientia, aux mouetur per 1° in 43 annis solaribus, 221 diebus et circa 22 horas.

[197] Et sciendum quod si luna tunc est ante caput draconis uel caudam, medium eclipsis erat tanto tempore post oppositionem ueram in quanto tempore fieret uelocitas descripta in tertio spatio tabule, et econtrario si luna esset post caput uel caudam. [198] Scito autem hoc toto, scies omnia tempora eclipsis modo dicendo. [199] Scribas locum lune motus diuersitatis tempore oppositionis medie. [200] Deinde multiplica diferentiam horarum et minutorum que sunt inter oppositionem ueram et mediam per 0;32,39°, et proueniens addas loco lune motus diuersitatis, si oppositio uera est post mediam, uel subtrahas, si est ante. [201] Et proueniens, facta additione uel subtractione, erit locus lune motus diuersitatis

tempore oppositionis, ex quo scies uerum cursum lune in una hora ex tabula ad hoc ordinata. [202] Et ex distantia solis ab auge scies uerum cursum solis in una hora in eodem tempore, quem subtrahas de uero cursu lune in illo tempore, et residuum erit uelocitas cursus lune in una hora, quam serua. [203] Et per seruatum diuidas proueniens in spatio tertio tabule eclipsis lune facta additione uel subtractione supradicta, et numerus quotiens erit tempus inter oppositionem ueram et medium eclipsis, [204] et medium eclipsis erit scitum scito tempore oppositionis uere. [205] Item diuidas per seruatum prouentum simili modo in spatio quarto, cuius diuisionis quotiens erit tempus a principio eclipsis usque ad medium uel a medio usque ad finem. [206] Item diuidas per seruatum proueniens similiter in spatio quinto, cuius diuisionis quotiens erit tempus a puncto totalis eclipsis usque ad medium uel a medio usque ad illuminationis principium. [207] Et istud sufficit ad sciendum omnia tempora eclipsis lunaris. [208] Et scies per horas medie diei illius in tuo oriçonte si totum tempus uel aliqua pars temporis eclipsis ueniet infra terminos noctis, quia solum illud in tuo oriçonte uidebitur.

[209] Ad eclipsim uero solis sciendum, primo scias quantitatem diuersitatis aspectus lune in longitudine tempore coniunctionis uere in oriçonte Aurayce ex tabula ad hoc ordinata; [210] set si es in alio oriçonte expedit quod facias per eo tabulam similem huic si uelis hoc faciliter. [211] Quo scito uideas quanta est diferentia diuersitatis aspectus in longitudine inter horam precedentem inmediate horam uere coniunctionis et ipsam horam uere coniunctionis, si luna est ante medium celi, uel inter horam uere coniunctionis et horam inmediate sequentem, si luna est post medium celi; [212] quam diferentiam subtrahas a uelocitate lune in una hora in illo tempore, si diuersitas aspectus est maior in dicta hora precedenti uel sequenti quam in hora uere coniunctionis, uel addas, si minor. [213] Et proueniens erit uelocitas uisa in illa hora. [214] Et si una diuersitas est ex parte orientis, alia ex parte occidentis, quod accidit in horis in quarum aliqua parte est nulla diuersitas aspectus in longitudine, ambe diuersitates simul sunt diferentia diuersitatis aspectus in longitudine inter unam horam et aliam, que est subtrahenda a uelocitate lune ut supra. [215] Et residuum est uelocitas uisa in luna in illa hora. [216] Et scita uelocitate uisa in luna in illa hora, accipias de ea talem proportionem qualem habes de hora, si luna est ante medium celi, uel qualis restat de hora, si luna est post medium celi. [217] Et si acceptum secundum proportionem est equalem diuersitati aspectus in longitudine tempore uere coniunctionis, coniunctio uisa erit in principio hore illius, si luna est ante medium celi, uel in fine ipsius, si luna est post medium celi. [218] Et si diuersitas aspectus in longitudine tempore uere coniunctionis est minor dicto accepto, diuidatur dicta diuersitas aspectus per uelocitatem uisam in luna in illa hora; [219] cuius diuisionis quotiens erit tempus quo coniunctio uisa est ante ueram, si luna est ante medium celi, uel post eam, si luna est post medium celi. [220] Voco autem medium celi gradum çodiaci distantem ab ascendente 90° punctaliter, quia tunc nulla diuersitas est in longitudine, et ante est diuersitas longitudinis ex parte orientis, et post ex parte occidentis. [221]

Et si diuersitas aspectus in longitudine est maior predicto accepto, subtrahas ab ea dictum acceptum et serua residuum. [222] Deinde queras uelocitatem uisam in luna, ut supra, in tota hora que est ante horam uere coniunctionis, si luna est ante medium celi, uel in tota hora sequenti, si luna est post medium celi. [223] Et utaris hac uelocitate uisa et seruato modo quo supra, et hoc modo uenies ad minutum hore in quo est coniunctio uisa; [224] quo scito scies uerum locum lune tempore coniunctionis uise modo dicendo.

[225] Scias primo locum coniunctionis uere modo quo supra. [226] Deinde multiplica uerum cursum lune in hora una in illo tempore per medium tempus inter coniunctionem uisam et ueram, et proueniens subtrahas de uero loco lune tempore coniunctionis uere, si uisa est ante ueram, uel ei addas, si est post. [227] Et proueniens facta subtractione uel additione erit uerus locus lune tempore coniunctionis uise. [228] Et ex hoc scies modo quo supra latitudinem ueram lune tempore illo ad septentrionem uel meridiem. [229] Deinde queras in tabula diuersitatis aspectus in oriçonte Aurayce diuersitatem aspectus lune in latitudine ex parte meridiei tempore coniunctionis uise, quam addas uere latitudini, si [uera] est ex parte meridiei, uel subtrahas minorem de maiori, si uera est ex parte septentrionis. [230] Et proueniens erit latitudo uisa in luna ex parte latitudinis maioris. [231] Cum qua latitudine intres ambas tabulas eclipsis solaris et procedas modo declarato in eclipsi lunari, et deuenies in quantitatem digitorum eclipsatorum et in quantitatem uelocitatis uise in luna a principio eclipsis usque ad medium, uel a medio usque ad finem.

[232] Et scies tempus principij eclipsis et tempus finis modo dicendo. [233] Diuidas uelocitatem uisam in luna a principio eclipsis usque ad medium per uelocitatem uisam in luna in tota hora una ante coniunctionem uisam. [234] Et si prima uelocitas est plus una hora, diuidas illud plus per uelocitatem uisam in luna in tota alia hora ante horam predictam, cuius diuisionis quotiens ducet te in tempus principij eclipsis. [235] Et eodem modo utaris uelocitate uisa in luna a medio eclipsis usque ad finem per uelocitatem uisam in luna in horis post coniunctionem uisam.

[236] Verbi gratia, uolumus scire eclipsim solarem anni Christi 1321. [237] Et scimus quod coniunctio media fuit 25 die mensis junij 9;26^h post meridiem. [238] Et coniunctio uera fuit 26 die eiusdem 5;20^h ante equalem meridiem, que sunt 5;33^h ante meridiem uisam. [239] Et tunc erat uerus locus solis in Cancro 11° et circa 0;39°, et ibi erat etiam luna. [240] Et locus lune motus diuersitatis tunc erat in 90° et circa 0;38°. [241] Et tunc erat diuersitas aspectus in longitudine ex parte orientis 0;40,8°. [242] Et diuersitas aspectus in longitudine in fine sexte hore ante meridiem 0;41,2°, et diuersitas aspectus in longitudine in fine quinte hore erat 0;39,29°. [243] Et sic erat in fine [sexte] hore maior quam in fine quinte 0;1,33°; [244] quibus subtractis a uelocitate lune in una hora in illo tempore, que est 0;30,31°, remanet uelocitas lune uisa in illa sexta hora 0;28,58°; [245] per que multiplicetur transitum de sexta hora ante meridiem, que est 0;27^h que era[n]t

transacta tempore coniunctionis uere, cuius multiplicationis proueniens est 0;13,2°; [246] quibus subtractis a diuersitate aspectus lune in longitudine tempore coniunctionis remanent 0;27,6°. [247] Et queramus modo predicto uelocitatem uisam in luna in tota hora septima ante meridiem. [248] Et inueniemus diuersitatem aspectus in longitudine in fine hore septime minorem quam in fine sexte 0;0,53°. [249] Sequitur quod uelocitas uisa lune in ista hora septima sit 0;31,24°; [250] per que diuidantur 0;27,6° que remanserant, cuius diuisionis quotiens est circa 0;51,48^h. [251] Sequitur quod coniunctio uisa fuit ante meridiem 6;51^h et circa 0;0,48^h. [252] Et tunc erat uerus locus lune, ut est declaratum ex cursu lune in una hora, in Cancro 10;59° et circa 0;0,25°; [253] cui uero loco lune addito medio loco capitis draconis fuit argumentum latitudinis lune 173;27,3°.

[254] Sequitur quod uera latitudo lune septentrionalis tunc esset 0;30,45°. [255] Et diuersitas aspectus in latitudine erat tunc meridionalis 0;34,14°. [256] Et remanet latitudo lune uisa meridionalis 0;3,29°. [257] Sequitur quod eclipsis fuit 10 digitorum et circa 0;30^d. [258] Sequitur quod uelocitas lune uisa a principio eclipsis usque ad medium fuit 0;27° et circa 0;0,36°. [259] Et iam habebamus de septima hora 0;8,12^h; que multiplicando per uelocitatem lune uisam in tota septima hora, que est, ut declaratum est supra, 0;31,24°, proueniens est 0;4,17°; [260] quibus subtractis a 0;27,36°, remanent 0;23,19°. [261] Et quia non inuenimus in tabula diuersitatis aspectus in directo principij Leonis nisi 7;23^h, deueniemus in scientiam uelocitatis lune uise in una hora ante septima hora ex diuersitate aspectus 0;23^h ibi inuentorum de octaua hora. [262] Et iam inuenimus in illa tabula 7;23^h ante meridiem diuersitatem aspectus in longitudine 0;38,39°, et dicta diuersitas erat maior 0;1,21° in fine septime hore quam in fine septime hore et 0;23^h, ut declaratum est supra; [263] quo posito sequitur secundum proportionem quod maioritas septime hore respectu totius octaue sit 0;3° et circa 0;0,31°; [264] quibus additis uere uelocitatis lune, erit uelocitas lune uisa in octaua hora 0;34° et circa 0;0,1°; [265] per que diuisis 0;23,19° que remanserant, est quotiens 0;41^h. [266] Sequitur quod 7;41^h fuit principium eclipsis ante meridiem. [267] Et si querimus qua hora fuit eclipsis completa, nos scimus quod uelocitas lune uisa a medio eclipsis usque in finem fuit 0;27,36°, et scimus quod uelocitas lune uisa in tota hora septima ante meridiem fuit 0;31,24°; [268] per que multiplicando 0;51,48^h que remanebant de septima hora a tempore coniunctionis uise, est proueniens 0;27,7°; [269] quibus subtractis de 0;27,36°, remanent 0;0,29°, que diuidendo per uelocitatem lune uisam in tota hora sexta, que erat 0;28,58°, erit quotiens 0;1^h. [270] Sequitur quod transacto 0;1^h de hora sexta fuit completa eclipsis. [271] Et hoc sufficit ad cognoscendum sine magna dificultate omnia tempora istius eclipsis.

[272] Et sciendum quod ista experientia respondebat prime doctrine punctaliter. [273] Set quia uidimus esse necesse ponere aliqualem latitudinem diametrorum a principio motus diuersitatis, ordinauimus istas equationes secundum doctrinam quam trademus in sequenti capitulo, secundum quam uidetur quod sit necesse subtrahere de medio loco lune circa 0;2°. [274] Et tunc posset concordare cum eo

quod de ista eclipsi per experientiam uidimus, quia non est nobis dubium quod eius principium non fuit post solis ortum minus tribus minutis.

[275] Si per tabulas istas scire uolueris uerum locum lune in quolibet tempore post aliqualem coniunctionem uel oppositionem mediam, scias primo mediam coniunctionem uel oppositionem et ueram, locum lune motus diuersitatis, diferentiam inter coniunctionem uel oppositionem ueram et mediam, locum lune uerum in tempore coniunctionis uel oppositionis uere, in omnibus datis superius. [276] Quibus omnibus scitis queras locum lune motus diuersitatis in tempore medio inter coniunctionem uel oppositionem ueram et mediam hoc modo, [277] quia medietatem diferentie inter coniunctionem uel oppositionem ueram et mediam multiplicabis per 0;32,39° et proueniens erit locus quesitus; [278] quod proueniens addes loco lune motus diuersitatis tempore coniunctionis uel oppositionis medie, si media est ante ueram, uel subtrahes, si est post eam. [279] Ex quo facta additione uel subtractione scies cursum lune in una hora, quem multiplices per horas et minuta diferentie inter coniunctionem uel oppositionem ueram et mediam, et serua proueniens. [280] Et si media est post ueram, iunge seruatum cum loco lune tempore coniunctionis uel oppositionis uere, et si est ante, subtrahe. [281] Et proueniens est locus lune uerus tempore coniunctionis uel oppositionis medie. [282] Cum quo et cum loco lune motus diuersitatis tempore coniunctionis uel oppositionis uere scies uerum lune locum et cursum in qualibet die completa post coniunctionem uel oppositionem mediam usque ad 14 dies hoc modo. [283] Quia queres in latitudine tabule numerum loci lune motus diuersitatis et in longitudine numerum dierum preteritorum post coniunctionem uel oppositionem mediam, [284] et quod in directo numeri utriusque inuenies, addas loco lune numero tempore coniunctionis uel oppositionis medie. [285] Et proueniens erit locus lune uerus tempore in quo queris. [286] Et si non inuenies in latitudine tabule illum numerum motus diuersitatis quem tu habes punctaliter, queras duo loca proximiora illi et accipias de diferentia ipsorum secundum proportionem. [287] Et simile facias si habes aliquas horas post diem completam, quia accipias de diferentia precedentis et sequentis diei secundum proportionem horarum quas habes ad 24. [288] Et acceptum addes inuento in directo precedentis diei et proueniens erit uerus locus quem queris.

---

1. mediarum et *corr. L.* 2. profactus *corr. L s.l. ex* perfectus; quam₁ *om. B.* 3. et₁: atque *B*; diametro *B.* 6. medias + in *B*; prime... decime: angulorum *B.* 7. queuis: quelibet *B.* 9. in quarta... inueniri *om. B.*; ascendentem: conscendentem *AB.* 10. *post* docebimus *A scr. et exp.* inuenire; -que: et *corr. L*, atque *B.* 14. atramento *BL.* 17. uel: et *BL.* 22. numerus₃ *om. B.* 28. uel subtrahendarum *om. B.* 34. 618963 *ABL.* 38. prouenientis *B*; de *om. B.* 39. spatijs + descripti *B.* 40. equationem cuiuslibet... mense inueniens *AB add. mg.* 45. equalitatem *AB.* 46. tabule₂ *om. B.* 51. omnes *om. B.* 55. si *om B*; numerus quotiens: cum numero quotienti *B.* 57. residuum *B*; quod₂ *om. B.* 59. qua[m adda]s: *AL* quas (*L corr. ex* quam), *B om.* 62. de tempore ante radicem *add. mg. AB*; est: et *ABL.* 66. quam tu: quantum *AL.* 69. totum *om. B.* 75. per se *A mg.* 76. postquam: post quod *B*; in uere: *A corr. ex* inuenitur, *a.m. corr. L ex* inueni, *B* inuenire. 78. deinde: et inde *B.* 80. *post* residuum *A scr. et exp.* coniunctionis; tempus *A mg.* 82. dies: dierum *L*; equationes *B.* 83. acceptis

B; post acceptum *A scr. et exp.* habes, *B add.* hōs. 86 et₁ *om. L;* ideo... quarte anni *a.m. mg. L;* radicem: *ABL* radices. 87. oppositis horis *L;* harum *B.* 88. erit: est *B.* 93. nunc + inuenies *ABL.* 95. primo + modo *B.* 97. quaternarios *corr. L a.m. ex* quaternationes. 103. descendens *ABL.* 107. descendens *L.* 108. *post* annorum *A scr. et exp.* usque ad 43 annos. 110. descendebant *L.* 112. *post* decime *A scr. et exp.* motus 113. est *A mg., om. B.* 116. çodiaci *mg A.* 117. principijs: primis *B.* 119. punctaliter *om. B;* demus *L;* adminiculo *BL.* 123. et deinde *B;* tertiam partem subtrahas: etiam partem seruabas *B.* 132. [remanent... ascendentis]: respondentis *ABL.* 132-3. aurayce *L.* 139. accipiebat *ABL.* 144. *A* respondentem. 145. ubi: *ABL* ibi. 146. 11: 12 *B.* 148. antarcticus *corr. a.m. s.l. L.* 149. ibi: *ABL* ubi. 150. reliquorum *L.* 151. auraice *B;* oriente *corr. s.l. a.m. L ex* orizonte. 152. tu *om. L.* 153. auraice *B.* 155. poteris: positis *B;* auraice *B;* ex: et *B.* 156. speciali: spirituali *ABL;* facillime *BL;* auraice *B.* 157. et₂ *om. B.* 163. inuentionem + Erant hic folia alba per huiusmodi tabulis faciendis *B.* 165. et *om. B;* cristi *B.* 169. Quinta... sephel *L mg.* 170. compuctus *A,* computos *corr. a.m. s.l. L.* 171. motus *L.* 172. auraice *B.* 173. *post* inueniendum *L. scr. et del.* aspectus lune; diuersitatem *A mg., B om.* 174. respectu *om. B;* tempore: et tempus *B.* 177. de *om. B.* 179. diuersitatem *om. B;* inueneris *L.* 180. de *om. B.* 183. medium: medij *B.* 184. cum... lune *L mg.* 185. uel + in *B.* 186. coniunctio: conueniens *B.* 187. quante *B;* et₃: ad *L.* 188. possibilis *corr. a.m. s.l. L ex* polus. 190. scilicet: si *B.* 191. tunc *om. B,* tuus *L.* 194. loco: locus *B.* 200. et₁ *om. B;* ueram *B corr. mg. ex* primam. 201. cursus *B.* 204. et medium eclipsis *A mg.;* scito *om. B.* 206. inluminationis *B.* 208. scias *B;* horam mediam *B;* illius: illud *B;* termino *B.* 209. *AB add. mg.* ad sciendum eclipsim solis; sciendam *B;* quantitates *B;* auraice *B.* 211. diferentia *A mg., a.m. add. s.l. L;* diuersitatis *corr. a.m. s.l. L ex* diuersitas; precedentem + et *L.* 212. differentias *B.* 216. accipies *L;* qualem: *ABL* quales (*L corr. s.l. ex* qualem); de hora *om. B.* 220. ascendente: ascendentem *AL,* excedente *B;* orientis *corr. a.m. s.l. L ex* orizontis. 222. uel *om. B.* 224. uise: uerso *B.* 225. primo: post *B.* 226. subtrahes *L.* 229. auraice *B.* 230. uisa *om. B.* 231. cum: in *B.* 236. 1321 *corr. s.l. a.m. L ex* 1531. 248. in fine *mg. A,* uel in fine *mg. B.* 250. 0;51,48ʰ: 0;52,20ʰ *ABL.* 251. 6;51ʰ et circa 0;0,48ʰ: *ABL* 6 h. 5 m. et circa 88 se. 261. Et quia... octaua hora: *om. B.* 262. hore₂ *om. B.* 263-5. et circa 0;0,31°... est quotiens 0;41: *om. B.* 269. 0;28,58°: 0;29,58° *B.* 272. ista *om. B.* 273. esse: circa *B;* equationes: rationes *B.* 275. ueram₁: uerum *BL; post* diuersitatis *B scr. et del.* tempore medio inter coniunctionem et oppositionem ueram et mediam multiplicabis per 0;32,39° et proueniens erit locus quesitus, quod proueniens addes loco lune motus diuersitatis tempore coniunctionis uel oppositionis ucre in omnibus datis superius, quibus omnibus scitis queras locum lune motus diuersitatis tempore medio. 276-7. hoc modo... ueram et mediam *AB mg.* 277. diferentie: datur *B.* 278. subtrahes *B.* 279. minutos *B,* minutas *L.* 280-1. tempore... lune uerus *mg. a.m. L.* 282. motus *rep. AB.* 284. numero: uero *L.* 286. tu: cum *B;* queris *B.* 288. acceptis *BL;* queris + spatium quatuor foliorum *B.*

Acknowledgements. – I am grateful to Prof. Dr. P. Kunitzsch (Munich) and to Dr. M. Zonta (Pavia) for checking the Arabic and Hebrew versions of *Almagest,* IV.3, respectively, and to Dr. M.C. Hernández (Sevilla) and Prof. D. Mériz (Pittsburgh) for their comments and suggestions on the translation of the Provençal text. I wish to thank the authorities of the Universiteitsbibliotheek, Leiden, for permission to reproduce Plates 1 and 2, and Dr. A. Th. Bouwman (Department of Western Manuscripts) for additional information on Ms. Scal. 46*. I am greatly indebted to Prof. Bernard R. Goldstein (Pittsburgh) for his helpful and detailed suggestions throughout the preparation of this article, for providing me with translations of Levi's unpublished Hebrew texts and with the value of the mean synodic month in Ibn Yūnus' *al-Zīj al-Ḥākimī,* and also for deciphering and translating the Hebrew sentence in f. 67ᵛ of Ms. Scal. 46.

References

Aaboe, A. (1955): "On the Babilonian Origin of some Hipparchian Parameters", *Centaurus*, 4, 122-125.

d'Alverny, M.-T. (1989): "Les traductions à deux interprètes, d'arabe en langue vernaculaire et de langue vernaculaire en latin", in G. Contamine (ed.), *Traduction et traducteurs au Moyen Âge*, Paris, C.N.R.S., 193-206.

Anglade, J. (1921a): *Grammaire de l'ancien provençal ou ancienne langue d'oc. Phonétique et morphologie*. Paris, Euvrad-Pichat (repr. Klincksieck, 1977).

Anglade, J. (1921b): *Histoire sommaire de la littérature méridionale au Moyen Âge (Des origines à la fin du XVe siècle)*. Paris (repr. Slatkine, Genève, 1973).

Barker, P. and Ariew, R. (eds) (1991): *Revolution and Continuity: Essays in the History and Philosophy of Early Modern Science*. Washington D.C., Catholic University of America Press.

al-Bīrūnī (1954-1956): *Al-Qānūn al-Mas'ūdī*. Hyderabad. 3 vols.

Brunel, C. (1926): *Les plus anciennes chartes en langue provençale: recueil des pièces originales antérieures au XIIIe siècle*. Paris, A. Picard.

Brunel, C. (1935): *Bibliographie des manuscrits littéraires en ancien provençal*. Paris, Droz (repr. Slatkine, Genève, 1973).

Byvanck, A. W. (1931): "Les principaux manuscrits à peintures conservés dans les collections publiques du Royaume des Pays-Bas", *Bulletin de la Société Française de Reproductions de manuscrits à peintures*.

Carlebach, J. (1910): *Lewi ben Gerson als Mathematiker*. Berlin, L. Lamm.

Chabás, J. (1991): "The Astronomical Tables of Jacob ben David Bonjorn", *Archive for History of Exact Sciences*, 42, 279-313.

Chabás, J. (1992): *L'Astronomia de Jacob ben David Bonjorn*. Barcelona, Institut d'Estudis Catalans.

Coyne, G. V.; Hoskin, M. A.; Pedersen, O. (eds) (1983): *Gregorian Reform of the Calendar. Proceedings of the Vatican Conference to Commemorate its 400th Anniversary, 1582-1982*. Roma, Pontificia Academia Scientiarum.

Curtze, M. (1898): "Die Abhandlungen des Levi ben Gerson über Trigonometrie und den Jacobstab", *Bibliotheca mathematica*, 12, 97-112.

Curtze, M. (1901): "Die Dunkelkammer", *Himmel und Erde*, 13, 225-236.

Dahan, G. (ed.) (1991): *Gersonide en son temps. Science et philosophie médiévales*. Louvain-Paris, E. Peeters.

Deprez, E. (1889): "Une tentative de réforme du calendrier sous Clement VI: Jean de Murs et la chronique de Jean de Venette", *École française de Rome, Mélanges d'archéologie et d'histoire*, 19, 131-143.

Delambre, J. B. J. (1819): *Histoire de l'Astronomie du Moyen Âge*. Paris, Courcier (Johnson Repr. Co., 1965).

Duhem, P. (1909): *Un fragment inédit de l'Opus tertium de Roger Bacon, précédé d'une étude de ce fragment*. Ad Claras Aquas (Quaracchi) prope Florentiam ex typographia Collegii S. Bonaventurae.

Freudenthal, G. (ed.) (1992): *Studies on Gersonides – A Fourteenth-Century Jewish Philosopher-Scientist*. Leiden, E. J. Brill.

Gabbey, A. (1991): "Innovation and Continuity in the History of Astronomy: The Case of the Rotating Moon", in Barker and Ariew (1991), 95-129.

Goldstein, B. R. (1967): *Ibn al-Muthannā's Commentary on the Astronomical Tables of al-Khwārizmī*. New Haven, Yale University Press.

Goldstein, B. R.; Swerdlow, N. (1970): "Planetary Distances and Sizes in an Anonymous Arabic Treatise Preserved in Bodleian Ms. Marsh 621", *Centaurus*, 15, 135-170.

Goldstein, B. R. (1971): *Al-Bitrūjī: On the Principles of Astronomy*. New Haven, Yale University Press.

Goldstein, B. R. (1972): "Levi ben Gerson's Lunar Model", *Centaurus*, 16, 257-284.

Goldstein, B. R. (1974a): *The Astronomical Tables of Levi ben Gerson*. Hamden, CT, Archon Books.

Goldstein, B. R. (1974b): "Levi ben Gerson's Preliminary Lunar Model", *Centaurus*, 18, 275-288.

Goldstein, B. R. (1975): "Levi ben Gerson's Analysis of Precession", *Journal for the History of Astronomy*, 6, 31-41.

Goldstein, B. R. (1976): "Astronomical and Astrological Themes in the Philosophical Works of Levi ben Gerson", *Archives internationales d'histoire des sciences*, 26, 221-224.

Goldstein, B. R. (1979): "Medieval Observations of Solar and Lunar Eclipses", *Archives internationales d'histoire des sciences*, 29, 101-156.

Goldstein, B. R. (1985): *The Astronomy of Levi ben Gerson (1288-1344)*. Berlin-New York, Springer.

Goldstein, B. R. (1986): "Levi ben Gerson's Theory of Planetary Distances", *Centaurus*, 29, 272-313.

Goldstein, B. R. (1988): "A New Set of Fourteenth Planetary Observations", *Proceedings of the American Philosophical Society*, 132, 371-399.

Goldstein, B. R. and Pingree, D. (1990): *Levi ben Gerson's Prognostication for the Conjunction of 1345*. Transactions of the American Philosophical Society. Vol. 80, Part 6.

Guillemain, B. (1966): *La Cour Pontificale d'Avignon, 1309-1376. Étude d'une société*. Paris, Éditions E. de Boccard.

King, D. A. and Saliba, G. (eds) (1987): *From Deferent to Equant. A Volume of Studies in the History of Science in the Ancient and Medieval Near East in Honor of E. S. Kennedy*. Annals of the New York Academy of Sciences, 500. New York, The New York Academy of Sciences.

Lay, J. (1991): *L'abrégé de l'*Almageste, *attribué à Averroès, dans la version hébraïque*. École Pratique des Hautes Études, Section des Sciences Religieuses, thèse soutenue le 6 Mars 1991. Thèse de doctorat (nouveau régime).

Mancha, J. L. (1989): "Egidius of Baisiu's Theory of Pinhole Images", *Archive for History of Exact Sciences*, 40, 1-35.

Mancha, J. L. (1992a): "Astronomical Use of Pinhole Images in William of Saint-Cloud's *Almanach planetarum* (1292)", *Archive for History of Exact Sciences*, 43, 275-298.

Mancha, J. L. (1992b): "The Latin Translation of Levi ben Gerson's *Astronomy*", in Freudenthal (1992), 21-46.

Mancha, J. L. (1993): "La determinación de la distancia del sol en la *Astronomía* de Levi ben Gerson", *Fragmentos de Filosofía*, 3, 97-127.

Mancha, J. L. (1997): "Levi ben Gerson's Astronomical Work: Chronology and Christian Context", *Science in Context*, 10, 471-493.

Mancha, J. L. (1998): "Heuristic Reasoning: Approximation Procedures in Levi ben Gerson's *Astronomy*", *Archive for History of Exact Sciences*, 52, 13-50.

Millás Vallicrosa, J. M. (1947): *Abraham Ibn Ezra. El libro de los fundamentos de las tablas astronómicas*. Madrid, C.S.I.C.

Millás Vallicrosa, J. M. (1959): *La obra Séfer Hešbón mahlekot ha-kokabim (Libro del cálculo de los movimientos de los astros) de R. Abraham Bar Hiyya ha-Bargeloni*. Madrid, C.S.I.C.

Molhuysen, P. C. (1910): *Bibliothecae Universitatis Leidensis Codices Manuscripti. II. Codices Scaligerani praeter orientales*. Lugduni-Batavorum.

Munk, S. (1856-1866): *Le guide des égarés. Traité de Théologie et de Philosophie par Moïse ben Maimoun dit Maïmonide*. Paris, G.-P. Maisonneuve & Larose (repr. 1970).

Nallino, C. A. (1899-1907): *Al-Battānī sive Albatenii Opus Astronomicum*. Milano, Pubblicazioni del Reale Osservatorio di Brera in Milano, XL.

Neugebauer, O. (1962): *The Astronomical Tables of al-Khwārizmī*. Hist.-Filos. Skr. Dan. Vid. Selsk., *4*, no. 2. København.

Neugebauer, O. (1967): *The Code of Maimonides*. Book Three, Treatise Eight. Sanctification of the New Moon. Tr. by S. Gandz, with supp. and intr. by J. Obermann, and astronomical commentary by O. Neugebauer. New Haven, Yale University Press.

Neugebauer, O. (1975): *A History of Ancient Mathematical Astronomy*. Berlin-New York, Springer.

Neugebauer, O. (1987): "The Chronological System of Abu Shaker (A.H. 654)", in King and Saliba (1985), 279-293.

North, J. D. (1983): "The Western Calendar – 'Intolerabilis, Horribilis et Derisibilis'; Four Centuries of Discontent", in Coyne, Hoskin, and Pedersen (1983), 75-113.

North, J. D. (1986): *Horoscopes and History*. London, The Warburg Institute. (Warburg Institute, Surveys and Texts, XIII.)

Pedersen, O. (1983): "The Ecclesiastical Calendar and the Life of the Church", in Coyne, Hoskin, and Pedersen (1983), 17-74.

Pines, S. (1967): "Scholasticism after Thomas Aquinas and the Teachings of Hasdai Crescas and his Predecessors", *Proceedings of the Israel Academy of Sciences and Humanities*, vol. I, no. 10.

Renan, E.; Neubauer, A. (1893): "Les écrivains juifs français du XIVe siècle", *Histoire littéraire de la France*, XXXI, 351-789. Paris, Imprimerie Nationale.

Roberts, V. (1957): "The Solar and Lunar Theory of Ibn al-Shāṭir", *Isis*, *48*, 428-432.

Sabra, A. I. (1979): "Ibn al-Haytham's Treatise: Solution of Difficulties Concerning the Movement of *Iltifāf*", *Journal for the History of Arabic Science*, *3*, 388-422.

Saliba, G. (1979): "The First non-Ptolemaic Astronomy at the Maragha School", *Isis*, *70*, 571-579.

Saliba, G. (1980): "Ibn Sīnā and Abū ʿUbayd al-Jūzjānī: The Problem of the Ptolemaic Equant", *Journal for the History of Arabic Science*, *4*, 376-403.

Shatzmiller, J. (1991): "Gersonide et la société juive de son temps", in Dahan (1991), 33-43.

Sesiano, J. (1984): "Une arithmétique médievale en langue provençale", *Centaurus*, *27*, 26-75.

Steinschneider, M. (1899): "Devarim ʿAtiqim", *Mimisrach Umimaarabh*, *4*, 40-43.

Swerdlow, N.; Neugebauer, O. (1984): *Mathematical Astronomy in Copernicus's De Revolutionibus*. Berlin-New York, Springer.

Thorndike, L. (1934): *A History of Magic and Experimental Science*. Vol. III. New York, Columbia University Press.

Toomer, G. J. (1984): *Ptolemy's* Almagest. London, Duckworth.

Touati, C. (1973): *La pensée philosophique et théologique de Gersonide*. Paris, Les Éditions de Minuit.

Weil-Guény, A. M. (1992): "Gersonide en son temps: un tableau chronologique", in Freudenthal (1992), 355-365.

Zinner, E. (1925): *Verzeichnis der astronomischen Handschriften des deutschen Kulturgebietes*. München, C. H. Beck.

Zonta, M. (1993): "La tradizione ebraica dell'*Almagesto* di Tolomeo", *Henoch*, *15*, 325-350.

ADDENDA

pp. 304–6: A detailed exposition of Levi's objections to Ptolemy can be found in J.L. Mancha
and Gad Freudenthal, "Levi ben Gershom's criticism of Ptolemy's astronomy. Critical editions
of the Hebrew and Latin versions and an English annotated translation of chapter 43 of the
Astronomy", *Aleph* 5 (2005), pp. 35–167.

pp. 350, 352: The following items must be added to the list of references: K. Chemla and S.
Pahaut, "Remarques sur les ouvrages mathématiques de Gersonide", in G. Freudenthal (ed.),
1992, pp. 149–91, and C. Sirat, "La tradition manuscrite des *Guerres du Seigneur*", in G.
Dahan (ed.), 1991, pp. 301–28.

Right Ascensions and Hippopedes: Homocentric Models in Levi ben Gerson's *Astronomy* I. First Anomaly

1. Introduction

Thirty years ago it was usually assumed that Eudoxus' model for planetary motion, transmitted to us through some passages in Aristotle's *Metaphysics*, XII.8, and Simplicius' commentary on Aristotle' *De caelo*, was fully reconstructed only in 1874 by Schiaparelli (who demonstrated that two concentric spheres rotating with constant but opposite angular velocity about two inclined axes generate an hippopede, i.e., an eight-shaped curve resulting from the intersection of the sphere and a cylinder, or double cone, which touches internally the sphere at the cross-over point of the hippopede, where the double cone has its vertex).[1] From this assumption it followed, as a corollary, that how a cinematic model of this kind could represent planetary motion was probably completely forgotten during the Middle Ages.

However, we know now that the principles underlying the homocentric models were understood and utilized by Islamic astronomers at least as early as Ibrāhīm b. Sinān (908–946) (Ragep 1993, p. 452). Although Latin texts from the 14th century show traces of Ibn al-Haytham's homocentric device to account for the oscillatory motion of Ptolemy's epicyclic diameters (Mancha

1990), no text discussing the detailed astronomical applications of homocentric spheres is known to us from the Latin West, excepting very late works by Regiomontanus (*ca.* 1465), Amico (1536), and Fracastoro (1538) (Swerdlow 1972, 1997; di Bono 1990, 1995). Yet, in the *Astronomy* of Levi ben Gerson (1288–1344), two entire chapters, 22 and 32, deal with the ability of homocentric models to account for the motion in longitude and anomaly, respectively, of the planets.[2] Levi gives us a lengthy discussion of different arrangements of homocentric spheres and finally rejects them due to their inadequacy to represent observations and Ptolemy's planetary equations.

The object of this paper is to present an edition, with translation and commentary, of chapter 22 of the Latin version of Levi's *Astronomy*.

For the historiography of medieval astronomy and homocentric theory, two points which follow from the analysis of the text deserve some attention. The first one concerns the fair and simple approximation used by Levi to calculate the width of the eight-shaped curve produced by an Eudoxan couple (which he does not describe as resulting from the intersection of a cylinder or double cone and the sphere), namely Ptolemy's procedure for the computation of right ascensions – never suggested, as far as I know, by modern scholars. The second one, that the use of this right ascensions-method to deal with the geometry of Simplicius' hippopede was not original in Levi's text,[3] but probably an standard procedure in the Middle Ages: it was already known in Islamic Spain in the XIIth century, as it is attested in the chapter devoted to the motion of the fixed stars in al-Biṭrūjī's *Kitāb fī al-hay'a*.[4]

2. Latin Text[5]

Capitulum 22

[0] In 22° inquiremus illud quod sequitur de diuersitate ad motum longitudinis propter motum polorum et dicemus proprietates que ad istam positionem sequuntur.

[1] Diuersitatem que potest prouenire in planeta ex motu polorum possumus ymaginari ponendo tres motus quorum quilibet sit motui planete equalis, quorum unus sit ab occidente in orientem, secundus ad meridiem et ad septentrionem sit tertius. [2] Et polus unus sit in ecliptica, polus uero alterius distet a predicto in quantitate equationis maioris planete in motu

longitudinis, et isti poli uertuntur circa polos primos in circumferentia distantie ab eis. [3] Et declarabitur inferius quod motus istius dispositionis concordaret cum motu diuersitatis quantum possibile est, ut patebit inferius.

[4] Dico quod ex ista dispositione sequitur maior equatio in motu longitudinis in quantitate arcus magne circumferentie que est inter unum polum et reliquum. [5] Ad cuius probationem sit AB arcus ecliptice in quo est motus longitudinis planete, et punctus C sit polus spere, in qua non est planeta, qui est in ecliptica, et punctus D sit polus secunde spere, in qua est planeta in medio duorum polorum dicte secunde spere, qui polus D uertitur circa polum C. [6] Et ponatur quod arcus magne circumferentie qui transiret de puncto D ad punctum C faceret angulum rectum super circumferentiam AB. [7] Et signemus circa C circumferentiam DEFA, et hec circumferentia intersecet arcum AB in duobus punctis E, A, et protrahamus magne circumferentie arcum DCF, qui intersecat circumferentiam DEFA in duobus punctis D, F, et est recta super circumferentiam arcus AB, ut supra. [8] Et ponatur quod quantitas arcus EB sit 90° et quantitas arcus CEG sit 90° et quantitas arcus ACEH sit 90°. [9] Et protrahamus super G arcum stantem super eclipticam ad angulum rectum, hoc est, arcum GI, cuius quantitas sit 90°, et hec est magna circumferentia in qua de septentrione ad meridiem uel econtrario mouetur planeta.

[10] Et ponatur quod planeta sit in puncto G quando polus sue spere est in puncto D, quia tunc non cognoscitur diuersitas in planeta propter latitudinem polorum, quia sicut distantia de puncto C ad punctum G est 90°, ita distantia de puncto D ad punctum G est 90°. [11] Sequitur quod punctus G est polus circumferentie DCF et ideo arcus de puncto G ad punctum D est 90°, et quia distantia stelle a polo est 90°, sequitur quod planeta sit in puncto G. [12] Et eadem ratione ostenditur [quod] quando polus spere erit in puncto F, planeta erit etiam in puncto G, hoc est dicere in puncto ecliptice qui distat a puncto C 90°. [13] Ideo puncti G et C girauerint in isto tempore in motu longitudinis 180°, quia tempus in quo completur reuolutio poli est equale tempori in quo completur motus longitudinis. [14] Ideo quando polus D peruenerit ad punctum E, inuenietur planeta longe a puncto C plus quam 90° in quantitate arcus puncti C ad punctum E; ideo inuenietur tunc planeta in puncto B, quia planeta semper distat a polo spere sue 90°, et tunc planete equatio addenda erit ut arcus GB, qui est equalis arcui CE. [15] Ideo quando polus D peruenerit ad punctum A, inuenietur planeta distare a puncto C minus 90° in quantitate arcus CA; ideo tunc planeta erit in puncto H, qui

distat a puncto A 90°, et erit planete equatio subtrahenda arcus GH, qui est equalis arcui CA.

[16] Dico quod dispositio ista ponit equationem maiorem ad 90° a principio motus longitudinis et ad 270° ab eodem principio, quia oportet quod principium motus longitudinis ponatur in puncto D uel F, quia ibi est nulla equatio. [17] Quia circumferentia ADEF diuiditur in 4 partes equales in punctis A, D, E, F, in duas partes equales in punctis E, A, quia [linea ACE] transit per eius polum, et etiam quia duo arcus DC, CE sunt equales duobus arcubus DC, CA, et angulus DCE est equalis angulo DCA, quia uterque est rectus, sequitur quod linea que ueniret de puncto D ad punctum E esset equalis linee que ueniret de puncto D ad punctum A. [18] Sequitur quod arcus DE sit equalis arcui DA, quia eorum sunt corde equales et quilibet arcuum est minor 180°, et eodem modo demonstraretur quod arcus EF esset equalis arcui DE; sequitur quod quilibet arcus est equalis cuilibet. [19] Sequitur quod quando polus D peruenit ad punctum A motus est 90° et quando ad punctum E 270°.

[20] Dico ergo quod equatio maior est quando polus D est in puncto A uel E. [21] Quia si dicatur quod non, ponatur quod esset inter punctum D et punctum A et sit in puncto K, et ponatur quod equatio sit in puncto K maior quam arcus GH et sit GHL, et protrahamus magne circumferentie arcum KL. [22] Et est notum quod arcus KL essct 90°, quod est impossibile. [23] Ad cuius impossibilis probationem protrahamus super punctum

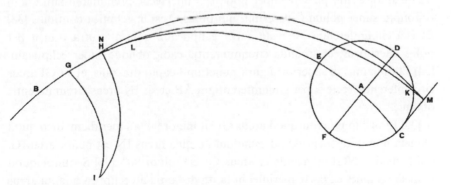

Fig. 1. This figure is repeated three and two times in mss. *A* and *B*, respectively, always incorrecly because it was difficult for the copyists to get with compass that arcs HM and LKM intersect at M. In the third figure in ms. *A* and the second one in ms. *B* letters A and C are interchanged. Mss *L* and *M* lack figures.

A magne circumferentie arcum AM facientem angulum rectum super circumferentiam AB, et est notum quod iste arcus tangit circumferentiam DA non intersecando ipsam. [24] Et protrahamus arcum LK magne circumferentie quousque intersecet arcum AM in puncto M, et protrahamus magne circumferentie arcum HM. [25] Et est notum quod arcus HM est equalis arcui HA, qui est 90°, et angulus MAH est rectus et arcus MH est maior arcu MKL, ut demonstrat Theodosius. [26] Sequitur quod arcus MKL est minor 90°; set arcus KL erat 90°; sequitur quod pars sit maior toto, quod est impossibile. [27] Unde sequitur quod equatio que prouenit quando polus D est in puncto K, est minor equatione proueniente quando polus D est in puncto A. [28] Et isto modo probaretur quod maior equatio in ista dispositione est in puncto A uel E quam in quocumque alio puncto, quod uolebamus probare.

[29] Item dispositio ista distinguitur quia quantitas diametri planete uidetur in quolibet loco in eadem mensura. Et in istis duabus proprietatibus seu distinctionibus concordat dispositio spere concentrice cuius est motus excentricus.

[30] Set proprietas propria istius dispositionis est quod planeta habet latitudinem ad meridiem et septentrionem bis in reuolutione eadem. [31] Et maior latitudo uisa in ista dispositione est ad 45° capitis draconis et caude, ante et post; uocamus autem caput draconis unum duorum locorum in quibus una circumferentiarum quas describit planeta inter septentrionem et meridiem intersecat aliam, et locum oppositum caudam. [32] Verbi gratia, circumferentia quam describit planeta super polum D intersecat circumferentiam quam suo motu super polum C describit, qui polus C est in ecliptica continue. [33] Et ista circumferentia est recta super eclipticam, quia alterutra transit per polos alterutrius, et secunda circumferentia cadit oblique super eclipticam. [34] Et ponamus in superiori figura punctum I caput draconis, et sit GI arcus circumferentie recte super circumferentiam AB et sit IN arcus circumferentie oblique.

[35] Et primo ponatur quod arcus GI sit minor 90° ad meridiem: dico quod planeta non potest uenisse ad punctum N, quia arcus IN est maior arcu IG, quia angulus NGI est rectus et arcus GI est minor 90°. [36] Sequitur quod planeta remanet ex parte meridiei in tanta distantia ab ecliptica quanta arcus IN est maior arcu IG. [37] Et ponatur nunc quod arcus GI sit 90°: est notum quod arcus IN est etiam 90° et tunc planeta in ecliptica est reuersus. [38] Et si bene queratur per modum qui in libro isto ponetur, inuenietur quod latitu-

do planete crescit quousque arcus GI sit 45°, et inde decrescit quousque dictus arcus sit 90°, et tunc planeta est in ecliptica, ut dictum est supra. [39] Et ponatur nunc quod arcus GI sit maior 90°: dico quod tunc planeta transibit eclipticam ex parte septentrionis, quia tunc arcus IN est minor arcu IG, quia angulus NGI est rectus et arcus GI est maior 90°. [40] Et hinc est notum quod si arcus GI fuerit 180°, punctus G erit cauda draconis et planeta erit reuersus in eclipticam in dicto puncto G. [41] Et erit notum, ut supra, quod quando arcus GI erit 135°, erit maior latitudo septentrionalis. [42] Et isto modo ostendetur in alijs duabus quartis quia in una quarta erit latitudo meridionalis et septentrionalis in alia, quia ponetur quod punctus qui est cauda fuerit motus a loco puncti G ad punctum I, et ponendo arcum GI minorem 90° erit latitudo meridionalis, et ponendo ipsum maiorem 90° erit latitudo septentrionalis.

[43] Sequitur quod in ista dispositione planeta habeat latitudinem ad meridiem et septentrionem bis in reuolutione eadem. [44] Et est notum ex hijs que sunt demonstrata in Almagesti, que demonstrabimus infra, quod differentia inter arcum GI et IN est multo minor quam distantia inter polum D et polum C, quia ponendo inter polum çodiaci et polum spere recte distantia 23;33° non est diferentia dictorum arcuum maior 2;29°.

[45] Et hinc est notum quod si ponatur ecliptica circumferentia GI et remaneant poli C et D ut supra, ex isto motu sequitur latitudo planete ab ecliptica in quantitate arcus GH et arcus GB, qui sunt equales arcubus CA et CE. [46] Et ista latitudo erit in una parte reuolutionis meridionalis et septentrionalis in alia. [47] Et in longitudine sequetur equatio modica et multo minor arcu CA et esset addenda et subtrahenda bis in reuolutione eadem. [48] Et hic est completa declaratio eius quod sequitur ex ista dispositione de equatione et de latitudine in planeta ex motu polorum.

[49] Posset etiam componi ista dispositio cum qualibet predictarum; set nolumus prolixe loqui de ista materia in hoc loco, tum quia consequens talis compositionis non est ignotum illi qui uidit nostra precedentia uerba, tum quia una pars istius compositionis habet locum in dispositionibus motus diuersitatis, tum etiam quia manifestum est quod consequentia istius dispositionis non concordant cum hijs que de equatione in motu planetarum in longitudine nos uidemus.

0 positionem: compositionem *LM*. 2 unus: unius *B*; polus uero: et polus *B*. 4 maior *om. B*. 5 polorum: punctorum *ABLM*. 7 circumferentie *add. in L*

alia manu mg.; sit 90° et quantitas arcus ACEH *add. in L a. m. mg.* 9 rectum hoc est arcum: scilicet *B*; hec: hoc *LM*; econtrario: econuerso *LM* (*L corr. a. m. ex* econtra). 13 puncti: punctum *ABLM*; girauerint: girauerunt *B*. 18 est: *L corr. ex* esset, *M* esset. 25 Theodetius *A*, Theondetius *LM*. 31 et locum *rep. L.* 32 continuo B. 33 recta: *L a.m. corr. ex* uera; alterutrius: *L a.m. corr. ex* altitudinis. 35 meridiem: *ABLM* septentrionem. 38 et$_1$ *om. B*; per: *L a.m. corr. ex* quod. 42 in$_3$: *L a.m. corr. ex* et. 44 notum + quod B. 45 sequeretur LM. 47 sequeretur M. 54 predictarum: dictarum B.

3. *Translation*

Chapter 22

[0] In the 22th chapter we will investigate the anomaly for the motion in longitude that follows from the motion of the poles and we will explain the consequences which follow from this configuration [of orbs].

[1] We can conceive the anomaly that for a planet follows from the motion of the poles by assuming three motions, all of them equal to the [mean] motion of the planet, of which the first takes place from west to east, the second to the south, and the third to the north. [2] A pole [i.e., the pole of the second sphere] is placed on the ecliptic, the pole of the other [sphere, namely the third one,] is removed from it by a distance equal to the amount of the maximum correction in longitude for the planet, and these poles are moved around the first poles [i.e., the poles of the first sphere] along the circumference equidistant from them. [3] We shall establish later that the motion [which results] from this configuration agrees as far as possible with the [observed] motion in anomaly, as it will be evident below.

[4] I assert that from this configuration it follows that the maximum correction for the motion in longitude is equal to the arc of the great circle passing through the second and the third pole. [5] To prove it, let AB be the the arc of the ecliptic along which the motion in longitude of the planet takes place, let point C, on the ecliptic, be the pole of the [second] sphere which does not carry the planet, and point D the pole of the second sphere [in fact, the third one], which is moved around pole C; the planet is situated on the great circle of this sphere equidistant from its poles [i.e., on its equator]. [6] Let the plane of the great circle passing through points D and C be perpendicular to the circle AB. [7] We trace around C circumference DEFA, which intersects arc

AB at two points, E and A, and we draw arc DCF of a great circle, perpendicular to the arc of circumference AB, as said above, and intersecting circumference DEFA at points D and F. [8] Let arc EB, as well as arcs CEG and ACEH, be 90°. [9] We draw passing through G an arc [of a great circle] perpendicular to the ecliptic, namely arc GI, whose amount is 90°; on this great circle is moved the planet from north to south or contrariwise.

[10] Let the planet be at point G when the pole of its sphere is at point D, for at this place no anomaly follows for the planet from the inclination of the poles, since just as the distance between points C and G is 90°, so the distance from point D to point G is 90°. [11] It follows that point G is the pole of circumference DCF and arc GD is 90°; it also follows that the planet is at G, since the distance of the planet from its pole is 90°. [12] For the same reason, when the pole of the [third] sphere is at point F, the planet will also be at point G, that is, on a point of the ecliptic situated at 90° from point C. [13] Thus, points G and C will have rotated during this time 180° with the motion in longitude, for the period of the revolution of the pole [of the third sphere around the pole of the second] is equal to the period of the motion in longitude. [14] So, when pole D reaches point E, the distance of the planet to point C will be greater than 90° in the amount of arc CE; consequently, as the distance of the planet from the pole of its sphere is always 90°, the planet will then be at B and the positive correction for the planet will be arc GB, equal to arc CE. [15] Therefore, when pole D reaches point A, the distance of the planet from point C will be lesser than 90° by the amount of arc CA; in consequence, the planet will then be at point H, at 90° from point A, and the negative correction will be arc GH, equal to arc CA.

[16] I assert that in this configuration [of orbs] the maximum correction occurs at 90° and 270° from the beginning of the motion in longitude, which must be placed at point D or point F, since there is no correction at these points. [17] Since circumference ADEF is divided in four equal parts at points A, D, E, and F, for line ACE passes through its pole, and arcs DC and CE are equal to arcs DC and CA, and angle DCE is equal to angle DCA, for both are right angles, it follows that line from point D to point E is equal to line from point D to point A. [18] It also follows that arc DE is equal to arc DA, since their chords are equal and they are smaller than 180°, and in the same way it can be demonstrated that arc EF is equal to arc DE; thus, all these arcs are equal. [19] Therefore, when pole D reaches point A it has been moved 90°, and 270° when it reaches point E.

[20] Thus, I assert that the maximum correction occurs when pole D is at points A or E. [21] To prove it, consider it has been denied and the maximum correction occurs between points D and A, v.g. at point K; let the correction at point K be arc GHL, greater than arc GH, and let us draw arc KL of a great circle. [22] It is known that arc KL should be 90°, which is impossible. [23] To prove that it is impossible, we draw through point A arc AM of a great circle perpendicular to circle AB; it is know that this arc will be tangent to circumference DA, without intersecting it. [24] Let draw arc LK of a great circle intersecting arc AM at point M, and draw also arc HM of a great circle. [25] It is known that arc HM is equal to arc HA, which amounts to 90°, and MAH is a right angle; as Theodosius demonstrates [in its *Sphaerics*], arc MH is greater than arc MKL. [26] It follows that arc MKL is lesser than 90°; however, [we assumed] arc KL to be 90°; consequently, it follows that the part is greater than the whole, which is impossible. [27] Therefore, the correction which results when pole D is at K is smaller than the correction which results when pole D is at A. [28] In this way we can prove for this configuration [of orbs] that [when the pole of the planet is] at A or E the correction is greater than at any other point [of the circumference ADEF].

[29] This configuration [of orbs] can also be distinguished from others because the apparent diameter of the planet does not vary at the different places [of its motion in longitude], and in these two properties or characteristics it agrees with the configuration that assumes a concentric sphere whose [uniform] motion takes place around an eccentric point.

[30] However, a special characteristic of this configuration is that the planet has twice a southern and northern latitude for each revolution. [31] The greatest apparent latitude in this configuration takes place about 45° before and behind the ascending and descending node; we call 'ascending node' one of the two points where the circumferences described by the planet between north and south intersect, and 'descending node' the opposite. [32] The circumference described by the planet with the motion of the sphere whose pole is D intersects the circumference described by the planet with its motion around pole C, which always remains on the ecliptic. [33] This circumference intersects the ecliptic at right angles, as each one passes through the poles of the other, and the second circumference intersects the ecliptic at oblique angles. [34] In the previous figure, let point I be the ascending node, GI the arc of the circumference perpendicular to circumference AB, and IN the arc of the oblique circumference.

[35] Consider first that arc GI is lesser than 90° measured to the south: I affirm that the planet cannot reach point N, as arc IN is greater than arc IG, since angle NGI is right and arc GI is lesser than 90°. [36] It follows that the planet remains to the south [of the ecliptic] distant from it by an amount equal to the difference between arcs IN and IG. [37] Let now arc GI be 90°: it is known that arc IN is also 90°, and then the planet is again on the ecliptic. [38] So, if the matter is well investigated according to the method that we will explain in this book, it shall be found that the latitude of the planet increases till arc GI reaches 45°, and behind this point it decreases till the aforementioned arc is 90°, and then the planet is on the ecliptic, as above said. [39] Let now arc GI be greater than 90°: I assert that the planet will then cross the ecliptic to the north, since arc IN is then lesser than arc IG, as angle NGI is a right angle and arc GI is greater than 90°. [40] Therefore, when arc GI is 180°, point G will be the descending node, and then the planet will return again to the ecliptic at the mentioned point G. [41] As it was above shown, the greatest northern latitude will occur when arc GI is 135°. [42] In the same way it can be demonstrated for the two remaining quadrants that in one of them the latitude will be to the south, and in the other to the north, as then the descending node will be moved from the position of point G to point I, and when arc GI is lesser than 90° the latitude will be to the south, and when the same arc is greater than 90° the latitude will be to the north.

[43] It follows that in this configuration [of orbs] the planet will exhibit northern and southern latitudes twice for each revolution. [44] It is however known, from the propositions proved in the *Almagest*, which we will also prove later, that the difference between arcs GI and IN is very much smaller than the distance between pole D and pole C, since if we assume the distance between the pole of the zodiac and the pole of the celestial equator is 23;33°, the difference between the aforementioned arcs is not greater than 2;29°.

[45] It is hence known that if [in our previous figures] we consider circumference GI to be the ecliptic, with poles C and D as above, from this motion will follow a [variation in] latitude of the planet from the ecliptic in the amount of arc GH and arc GB, which are equal to arcs CA and CE. [46] This latitude will be to the south in a part of the [zodiacal] revolution [of the planet], and to the north in the remaining one. [47] It will also follow a small correction in longitude, very much smaller than arc CA, to be added and subtracted twice for each revolution. [48] And so we have completed the explanation of the correction [in

longitude] and the latitude which follow from the motion of the poles in this configuration.

[49] This configuration can also be combined with any of the preceding ones; but we do not wish to speak largely here of this matter not only because the consequences of such configuration are not unknown to the reader of our preceding words, but because a part of this matter has its [proper] place in [the chapters where we will discuss] the configurations for the motion in anomaly, and because it is evident that the consequences of this configuration do not agree with what we observe about the correction in longitude in the motion of the planets.

4. Commentary

In the Latin version of Levi's *Astronomy*, the expression "motion of the poles" denotes arrangements of solid homocentric spheres and, more specifically, Eudoxan couples.[6]

A model from which a correction to the mean motion in longitude results is briefly described in the first sentences (§§ 1–2). Levi does not mention, however, a central point of his cosmology, established later in chapter 29 of the *Astronomy*, namely an orb is carried by the orb placed immediately below it (Goldstein 1986). Eudoxus' order of spheres is thus inverted, and the sphere of diurnal motion, not mentioned by Levi, is the innermost of the set, not the outermost. The first sphere, whose poles are the poles of the ecliptic, is moved eastward with the mean motion in longitude of the planet; the second one, placed above the first, has its poles on the ecliptic circle, and the axis of the third and upper one is inclined an angle i with respect to that of the second. The second and third spheres, carried along the ecliptic by the first one, are moved southward and northward, respectively, also with the mean motion in longitude of the planet. So, Levi's three spheres are identical to spheres 2, 3, and 4 in Schiaparelli's interpretation of Aristotle's description of Eudoxus' planetary model (*Metaphysics*, 1073b 17–1074a 15), except that they are applied to explain the first anomaly, not direct and retrograde motion – i.e., they play the same role that Callippus' two additional spheres for the Sun according also to Schiaparelli's interpretation (Schiaparelli 1877, pp. 82–84 and pp. 100–101 (Simplicius' text)).

In Fig. 2, let circle ACEHGB be the equator of the first sphere (i.e., the

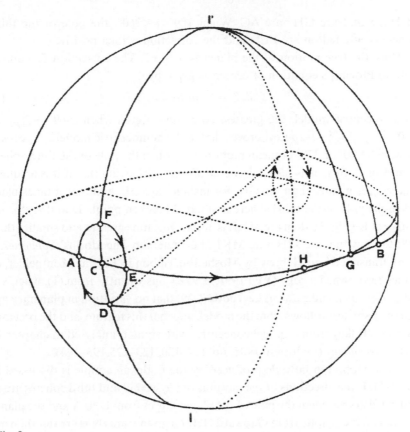

Fig. 2.

ecliptic) with axis II′, and C the pole of the second sphere; the pole of the
third one rotates around C on circle ADEF, whose radius i is equal to the
maximum equation of center. The planet is placed on the equator of the third
sphere, so that the distance from the planet to its pole is always 90°. The
correction in longitude derived from the model is explained in §§ 10–19. At
the initial position, when κ, the mean motion, is 0°, the pole of the third
sphere is at D (or F) and the planet at G; as the correction is negative for
$0°<\kappa<180°$, the motion of the second sphere carries the pole of the third one
from D (or F) towards A and the planet from G towards I′ (or I); finally, the
third sphere moves the planet in the opposite direction through the same
angle; thus, when $\kappa=90°$, the pole of the third sphere is at A and the planet

at H, so that arc GH=arc AC. When $180° < \kappa < 360°$, the pole of the third sphere is situated on arc FED and the correction is then positive.

Thus, the true motion of the planet is $\mu = \kappa + \xi$. The correction ξ, 'equivalent' to Ptolemy's equation of center, is given by

$$sin\ \xi = -sin\ i\ sin\ \kappa \tag{1}$$

In an eccentric model, the greatest correction occurs when $\kappa = 90° + \xi_{max}$ or $270° - \xi_{max}$. In §§ 20–28 Levi proves that in the homocentric model ξ_{max} occurs however at 90° or 270° of mean motion, i. e., when the pole of the third sphere is at A or E, by means of a *reductio ad absurdum* showing that if it is assumed that ξ_{max} is greater than arc GH, for instance arc GL, and that it takes place when the pole of the third sphere is situated between points D and A, for instance at K (Fig. 3), then arc MKL is at the same time greater and smaller than 90°; to show that arc MH>arc MKL, Levi refers to Theodosius' *Sphaerics*, a work translated into Hebrew by Moshe Ibn Tibbon in 1271 at Montpellier, of which Levi owned a copy (Weil 1991, p. 93). Consequently, from (1), when $\kappa = 90°$, $sin\ \xi = sin\ i$, and $\xi_{max} = i$. Levi adds (§ 29) that no variation in planetary apparent diameters follows from the model, and that this feature, and the previous one, also follow from a model concentric with equant, analysed in chapter 20 of the *Astronomy* (Goldstein 1985, pp. 117–120, 123–125,193–194).

The deviation in latitude produced by the Eudoxan couple is discussed in §§ 30–44. If the directions of the motions of the second and third spheres are as indicated above, when the pole of the planet moves from D to A and the planet recedes on the ecliptic from G toward H, for a given κ arc IN is greater than arc GI (Fig. 4); therefore, the planet cannot reach point N and it remains to the south of the ecliptic; its latitude increases until $\kappa = 45°$ and then decreases until $\kappa = 90°$ and the planet is again on the ecliptic, for the pole of the third sphere is then at A and arc IN=arc GI=90°; when the pole of the third sphere moves from A to F and the planet from H toward G, arc IN will be smaller than arc GI, and the planet will cross the ecliptic to the north, etc. It is clear that this sinusoidal curve along the ecliptic becomes an eight-shaped curve if the first sphere does not move.

In Schiaparelli's reconstruction the latitude β of the planet derives from

$$sin\ \beta = -sin^2\ \tfrac{1}{2}\ i\ sin\ 2\kappa\ , \tag{2}$$

where $sin^2\ \tfrac{1}{2}\ i = \tfrac{1}{2}(1 - cos\ i) = r$, the radius of the cylinder touching internally the sphere; so, β reaches its maximum when $2\kappa = 90°$, and consequently $sin\ \beta_{max} = r$.

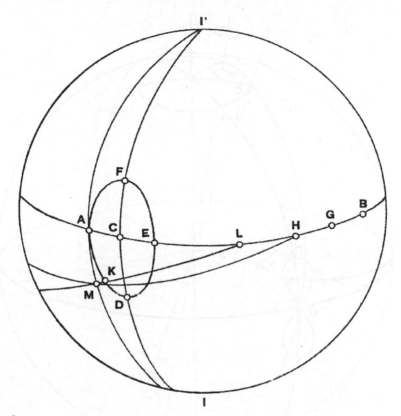

Fig. 3.

According to Levi, the greatest latitude is reached when $\kappa=45°$ (§ 31), which is correct, and the latitude of the planet is equal to the difference between arcs IN and IG (§ 36; i.e., arc PN in Fig. 4), which is literally wrong, for the arcs measuring latitudes are perpendicular to the ecliptic and arc IN, as Levi himself explicitly asserts in § 37, does not intersect the ecliptic at right angles except when $\kappa=90°$. Levi suggests in § 44 that β can be computed from (or approximated by) Ptolemy's procedure to calculate the rising time at *sphaera recta*, α, which corresponds to an ecliptic arc λ, namely

$$tan\ \alpha = cos\ \varepsilon\ tan\ \lambda \tag{3}$$

where ε is the obliquity of the ecliptic, and whose numerical results are tabulated in the *Almagest* at the end of I, 16, and in II, 8 (Toomer 1984, pp. 74

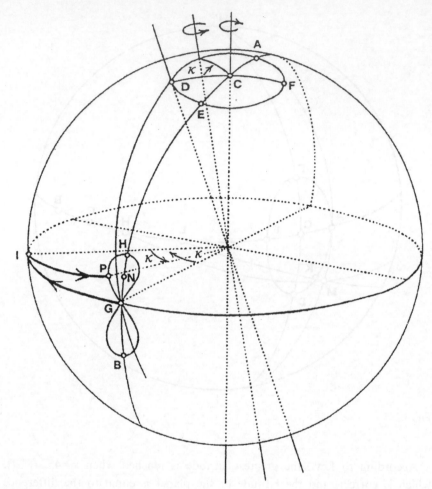

Fig. 4.

and 100–103). If we made arc IG=κ=α, arc IN=λ, and i=ε, arc PN in the model, which is equal to β according to Levi, is given by

$$\beta = -tan^{-1} (tan \, \kappa / cos \, i) - \kappa \qquad (4)$$

Now, whereas it is true that the greatest latitude on the hippopede is reached when 2κ=90° (that is, when the motion of the pole of the third sphere around the pole of the second one is 45°), it is not true that the greatest value for arc PN computed from (4) always occurs for $\kappa=45°$, as it depends on angle i (indeed, for $i=16°$, the maximum deviation occurs when $\kappa=44°$, and the value of κ corresponding to the maximum deviation decreases for increasing values of i, so that for $i=47°$, for instance, arc PN reaches its maximum when $\kappa=40°$). It is also obvious that the eight-figure generated by (1) and (4) is symmetric only about 90°, whereas Schiaparelli's hippopede has a double symmetry, about 45° and 90°.

For small eccentricities and computed to the seconds, there is no difference between the values of β computed from (2) and (4), as when $i=2;23°$, as required for Venus, or the maximum difference is insignificant, $0;0,2°$, for instance, when $i=6;32°$, as required for Saturn. For greater eccentricities, the values of β deriving from (2) are very well approximated by (4), as it can be seen in Tables 1 and 2, where column (a) lists the values derived from (4), column (b) those derived from (2), and column (c) the difference (a)−(b), computed taking angle i equal to $11;25,16°$ (Ptolemy's maximum equation of center for Mars, computed from tan $(\xi_{max}/2)=e/60$, with e=6) and $23;33°$ (Levi's value for the obliquity of the ecliptic[7]), respectively.

From a more general view, Levi's reasoning shows that it is possible to 'see' the behaviour of the Eudoxan couple and to deal with the curve on which the planet moves (with sufficiently accurate results) without reconstructing it as the intersection of the sphere with a cylinder. It can be easily done with the three great circles diagram used for the computation of right ascensions (Neugebauer 1975, p. 1214, figure 24): we only need to rotate the horizon

	Table 1			Table 2		
κ	(a)	(b)	(c)	(a)	(b)	(c)
10	0;11,52	0;11,38	0;0,14	0;53,16	0;48,58	0; 4,18
20	0;22,16	0;21,53	0;0,23	1;39,18	1;32, 2	0; 7,16
30	0;29,55	0;29,29	0;0;26	2;12,11	2; 4, 1	0; 8,10
40	0;33,55	0;33,31	0;0,24	2;28, 8	2;21, 2	0; 7, 6
44	0;34,22	0;34, 1	0;0,21	2;29,25	2;23, 7	0; 6,18
45	0;34,23	0;34, 2	0;0,21	2;29,17	2;23,12	0; 6, 5
50	0;33,48	0;33,31	0;0,17	2;25,55	2;21, 2	0; 4,53
60	0;29,37	0;29,29	0;0, 8	2; 6,34	2; 4, 1	0; 2,33
70	0;21,56	0;21,53	0;0, 3	1;32,54	1;32, 2	0; 0,52
80	0;11,38	0;11,38	0;0, 0	0;49, 5	0;48,58	0; 0, 7

circle around its diameter perpendicular to the equator (which is equivalent to the rotation of the pole of sphere 3 around the pole of sphere 2), and to compute at n-degrees intervals the difference between arcs λ and α.

Levi asserts in §§ 45–47 that if the model is rotated 90° (that is, if circle IG in Fig. 4 is the ecliptic and point C its pole), it can be used to produce oscillations in latitude which reach their northern and southern maximum values – GH and GB, respectively, given now by formula (1) – only one time each revolution of κ. The position in longitude of the planet will be consequently disrupted by an amount equal to the width of the eight-figure, producing very small positive and negative corrections twice each revolution in longitude, which can be computed with formula (4).

Finally, Levi repeats in § 49 that the main features of the model (maximum equation at 90°, double deviation in latitude, and no variation in distance and apparent diameter) disagree with planetary equations and observations, and postpones the discussion of other models to the chapter dealing with the motion in anomaly.

REFERENCES

Di Bono, M.
1990: *Le Sfere Omocentriche di Giovanni Battista Amico nell'Astronomia del Cinquecento*, Genoa.
1995: "Copernicus, Amico, Fracastoro and Ṭūsī's Device: Observations on the Use and Transmission of a Model", *Journal for the History of Astronomy* 26, pp. 133–154.
Dreyer, J. L. E.
1906: *History of Planetary Systems from Thales to Kepler*, Cambridge (repr. as *A History of Astronomy from Thales to Kepler*, New York, 1953).
Freudenthal, G.
1992: "Sauver son âme ou sauver les phénomènes: sotériologie, épistémologie et astronomie chez Gersonide", in G. Freudenthal (ed.), *Studies on Gersonides – A Fourteenth-Century Jewish Philosopher-Scientist*, Leiden, pp. 317–352.
Genequand, C.
1986: *Ibn Rushd's Metaphysics. A Translation with Introduction of Ibn Rushd's Commentary on Aristotle's Metaphysics, Book Lām*, Leiden.
Goldstein, B. R.
1974: *The Astronomical Tables of Levi ben Gerson*, Hamden, Conn.
1985: *The Astronomy of Levi ben Gerson (1288–1344)*, Berlin-New York.
1986: "Preliminary Remarks on Levi ben Gerson's Cosmology", in D. Novak and N. Samuelson (eds.), *Creation and the End of Days: Judaism and Scientific Cosmology*, Lanham, MD, pp. 261–76.
1997: "Saving the Phenomena: The Background to Ptolemy's Planetary Theory", *Journal for the History of Astronomy* 28, pp. 1–12.

Heath, T. L.
  1913: *Aristarchus of Samos*, Oxford (repr. New York, 1981).
Mancha, J. L.
  1990: "Ibn al-Haytham's Homocentric Epicycles in Latin Astronomical Texts of the XIVth and XVth Centuries", *Centaurus* 33, pp. 70–89.
  1998: "The Provençal Version of Levi ben Gerson's Tables for Eclipses", *Archives Internationales d'Histoire des Sciences* 48 (141), pp. 269–352.
Mendell, H.
  1998: "Reflections on Eudoxus, Callipus, and their Curves: Hippopedes and Callippopedes", *Centaurus* 40, pp. 177–275.
Millás, J. M.
  1947: *El libro de los fundamentos de las Tablas astronómicas de R. Abraham Ibn 'Ezra*, Madrid-Barcelona.
Munster, S.
  1546: *Sphaera mundi autore Rabbi Abrahamo Hispano filio R. Haijae ...*, Basileae.
Nallino, C. A.
  1903: *Al-Battānī sive Albatenii Opus astronomicum*, Milan, vol. 1.
Neugebauer, O.
  1953: "On the 'Hippopede' of Eudoxus", *Scripta Mathematica* 19, pp. 225–229.
  1975: *A History of Ancient Mathematical Astronomy*, Berlin-Heidelberg-New York.
North, J.
  1994: "The Hippopede", in A. von Gotstedter (ed.), *Ad Radices*, Stuttgart, pp. 143–154.
Ragep, F. J.
  1993: *Naṣīr al-Dīn al-Ṭūsī 's Memoir on Astronomy*, Berlin-New York.
Schiaparelli, G.
  1877: "Le sfere omocentriche di Eudosso, di Callippo e di Aristotele", *Memorie del Reale Istituto Lombardo, Classe di scienze matematiche e naturali, vol. XIII* (repr. in *Scritti sulla storia della Astronomia antica*, Bologna, 1925–27, II, pp. 1–112).
Swerdlow, N.
  1972: "Aristotelian Planetary Theory in the Renaissance: Giovanni Battista Amico's Homocentric Spheres", *Journal for the History of Astyronomy* 3, 1972, pp. 36–48.
  1997: "Regiomontanus's Concentric-sphere Models for the Sun and the Moon", *Journal for the History of Astronomy* 30, pp. 1–23.
Toomer, G. J.
  1984: *Ptolemy's Almagest*. London-New York.
Weil, G. E.
  1991: *La bibliothèque de Gersonide d'après son catalogue autographe*, Louvain-Paris.
Yavetz, I.
  1998: "On the Homocentric Spheres of Eudoxus", *Archive for History of Exact Sciences* 52, pp. 221–278.
  2001:
    "A New Role for the Hippopede of Eudoxus", *Archive for History of Exact Sciences* 56, pp. 69–93.

NOTES

1. On Schiaparelli's reconstruction of Eudoxus' system see Schiaparelli 1877, Dreyer, 1906, pp. 87–107, Heath 1913, pp. 190–211, and Neugebauer 1953 and 1975, pp. 677–685.

North 1994 shows that the intersection of a sphere with a parabolic surface also produces Schiaparelli's hippopede. Although the historical plausibility of Schiaparelli's reconstruction has been recently discussed (Goldstein 1997; see also Mendell 1998, pp. 264–266; Yavetz 1998 and 2001), I shall use, for convenience, the expression 'Eudoxan couple' to denote a model with two concentric spheres moved with equal and opposite motions around inclined axes.

2. Levi's purpose in the *Astronomy*, in agreement with his rationalistic views on knowledge, was to find the true arrangement of orbs that produces in the heavens the observed motions of the planets. Thus, a preliminary step of his method was to analyse different geometrical models in order to decide which of them can represent the phenomena, and this is the object of a long section in his book (chapters 20–26, and 32–38), placed immediately after the initial chapters on trigonometry and theory and use of observational instruments. On chapter 20, see Goldstein 1985, pp. 114–129 and 192–197.

3. Levi makes no claim of originality and the only source on homocentric astronomy mentioned in chapter 32 is Averroes' faulty text of Aristotle's *Metaphysics*: after discussing the ability of a model with two Eudoxan couples to represent the variation in the equation of anomaly due to the eccentric, Levi quotes *Metaphysics*, 1074a 1–14, according to the Hebrew translation of the Arabic text given by Averroes in his long commentary, where the backward motion of the counteracting (*anelíttousai*) spheres introduced by Aristotle in Callipus' system was wrongly translated as a spiral (*lawlab*) motion (see Genequand 1986, pp. 55 and 181). In my opinion, this proves that, despite his long relation with a Christian circle, Moerbeke's translations of Aristotle and Simplicius were not known by Levi, for Simplicius' commentary on *De caelo* was apparently never rendered into Arabic and Hebrew.

4. See my forthcoming "On al-Biṭrūjī's Theory of the Motion of the Fixed Stars", where I show that to account for the only two motions of the fixed stars admitted by al-Biṭrūjī (an apparent motion of access and recess, and another one, true and visible to the eye, in declination), he indeed uses an Eudoxan couple. Textual evidence suggests that Levi knew al-Biṭrūjī's work in an abridged and revised version from which the chapter on the fixed stars was at least partially lacking.

5. Copies of the Latin version of Levi's *Astronomy* are preserved in at least four manuscripts: Vat. Lat. 3098 (*A*), Vat. Lat. 3380 (*B*), Lyon, Bibliothèque Municipale 326 (*L*), and Milano, Ambros. 327 (*M*). On the *stemma* of the manuscripts, see Mancha 1998, pp. 333–334. The text of the chapter is found in *A* 16rb–vb, *B* 104vb–105vb, *L* 51r–53r, and *M* 52v–54v; sentence [0] is taken from the table of contents of the work: *A* 1rb, *B* 71va, *L* 2v, and *M* 2v. Paragraphing, punctuation, and capitalization are due to the editor. Insertions are enclosed in square brackets, as well as sentence numbers added for ease of reference. I have retained the spelling of *A*, with only one exception indicated in the critical apparatus. Anachronically, sexagesimal numbers appear in the Latin text in modern notation.

6. For instance, in chapter 25: «Set quomodo poterit fieri iste motus, ita quod non appareat in planeta de isto motu nisi motus centri epicicli, dicemus ponendo duos motus equales unum contrarium alteri, sicut dictum est supra de motu polorum» (*A*, f. 28vb), and in chapter 61: "Et ex isto manifeste apparet quod motus tardus uisus in stellis fixis est semper equalis, et quod iste motus earum non est ex motu polorum ut posuerunt aliqui antiquorum, quia si sic esset, impossibile esset quod in istis duobus temporibus uideremus istos motus equales" (*A*, f. 48ra–b). With identical meaning, the expression can be found in many medieval texts, and it probably derives from the Latin versions of al-

Biṭrūjī and Averroes. In a passage in Abraham ibn Ezra's *Liber rationibus tabularum*, the expression seems to allude to an homocentric model to account for changes in stellar declinations (and trepidation ?): "Antiqui omnes et Hermes et indi et doctores ymaginum omnes in hoc consentiunt, quod in circulo firmamenti duo motus sunt, ascendendi in septentrionem et descendendi in austrum; inter hos tamen est aliqua discordia, nam magistri ymaginum dicunt eos motus esse polorum, indi vero duorum circulorum qui sunt in capite arietis et libre" (Millás 1947, p. 77; see also a passage by Abraham bar Ḥiyya, Munster 1546, p. 196). At Levi's times, Qalonymos ben Qalonymos uses it in a letter to Joseph Ibn Kaspi dated 1318: «T'a-t-on dit et as-tu entendu dire qu'un système astronomique a eté trouvé, ne postulant ni épicycle, ni excentrique, ni un mouvement des pôles, de sorte que les principes naturels soient respectés?» (Freudenthal 1992, p. 338).

7. The value given by Levi in § 44, namely 2;29° (accurately computed, 2;29,17°) is also found in his table for rising times at *sphaera recta* (47;29° for Aqu 15°; see Goldstein 1974, p. 160). The same result for β_{max} (2;29,17° instead of 2;23,12°) is obtained using formula (4) in Neugebauer 1975, p. 680, which is incorrect, as, if arc XPAQ in Neugebauer's figure 30 (*ibid.*, p. 1359) represents the surface of the sphere, point P cannot represent the planet, for the planet describes the hippopede on the surface of the sphere.

Ibn al-Haytham's Homocentric Epicycles in Latin Astronomical Texts of the XIVth and XVth Centuries

1. Introduction

In his *Maqāla fī ḥarakat al-iltifāf (Treatise on the Movement of Iltifāf)* and *Ḥall shukūk ḥarakat al-iltifāf (Solution of Difficulties Concerning the Movement of Iltifāf)*[1], Ibn al-Haytham (965–ca. 1040 AD) proposed an Eudoxan device of two concentric spheres with different poles to account for each one of the oscillatory motions attributed by Ptolemy to the diameters of the planetary epicycles in the theory of the latitudes (*Almagest,* XIII, 2). Ibn al-Haytham's purpose was to provide a physically acceptable alternative to the Ptolemaic small circles producing the inclination (*egklisis, mayl*) of the apogee-perigee diameter of the epicycle for the superior and inferior planets and the slant (*loxōsis, inḥirāf*) of a second diameter laying between the mean longitudes of the epicycle and perpendicular to the first for the inferior ones. Since the motion attributed by Ptolemy to the epicyclic diameters must be produced by a solid spherical body, Ibn al-Haytham encloses the epicycle within a concentric sphere, whose poles are located at a distance from the apogee and the perigee of the epicycle equal to the Ptolemaic maximum inclination of the epicyclic diameter, which rotates with the same motion as that of the endpoints of the diameter of the epicycle on the small circles of the Ptolemaic model.

In figure 1 [Ragep (1982), 1:290a], let A and B be respectively the apogee and perigee of the epicycle. Points K and L are the poles of the first concentric sphere, and the distances KA and LB are equal to the maximum oscillation caused by the inclination. The motion of the enclosing sphere around KL carries the epicycle and, therefore, points A and B trace out the same circular path as in the Ptolemaic model. But, as every other point on the epicycle, with the exception of the points of the axis KL, also performs a complete revolution, in order to avoid this displacement Ibn al-Haytham postulates another enclosing sphere between the first one and the epicycle, whose motion is the same as the first one but in the opposite direction. The poles M and N of this second sphere are located directly above the points A and B and they also perform circular paths with the motion of the sphere KL (in such a way that for any position of A caused by the motion of the sphere KL, M is always located directly above A). When the sphere KL rotates through the angle a, as in figure 2 [Ragep (1982), 1:291a)], carrying the epicyclic apogee from A_1 to A_2, it carries also the point S (i.e.,

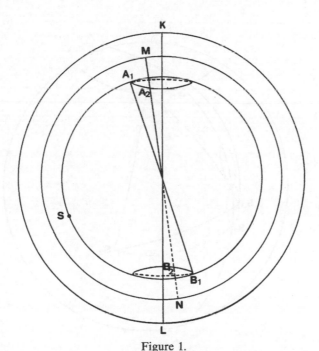

Figure 1.

the endpoint of the epicyclic diameter lying between the mean distances) from S_1 to S_2 along a path parallel to that of point A from A_1 to A_2. The motion of the second concentric sphere, whose pole M is then located directly above point A_2, leaves A stationary at position A_2 but moves the point S from S_2 to S_3 through the same angle a. As position S_3 does not coincide with position S_1, while the point A describes a complete revolution around K, the point S will perform a narrow hippopede with respect to the great circle A_1S_1 (cf. also Neugebauer (1970), pp. 182–3). Obviously, this procedure requires two additional spheres for the epicycle of each of the five planets and two more for Venus and Mercury.

But Ibn al-Haytham tried only to provide a physical mechanism of solid spheres for Ptolemy's assumptions in the *Almagest*, and his configuration reproduces two objectionable features of the Ptolemaic model: the non-uniform motion of the small circles (which is identical to the motion of the epicyclic center on the deferent) and the disruption in the longitude theory. The disruption

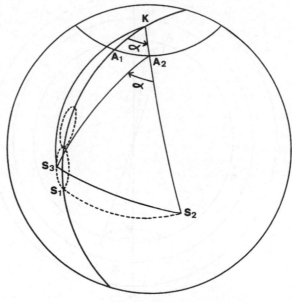

Figure 2.

caused by the inclination *(egklisis, mayl)* reaches its maximum, being then equal to the latitudinal inclination, when the planet is at the epicyclic apogee or perigee and the center of the epicycle is at the nodes (for the superior planets) or at the apogee or perigee of the deferent (for the inferior ones). Two centuries later Naṣīr al-Dīn al-Ṭūsī (1201–1274 AD) in *al-Tadhkira fi ᶜilm al hay'a* (1261), after criticizing Ibn al-Haytham's arrangement and suggesting some modifications on it, proposed his own solution using a spherical version of his "couple"[2]. Furthermore, al-Ṭūsī proved that the inclination *(prosneusis, muḥādhāt)* of the diameter of the lunar epicycle towards a point diametrically opposed to the center of the eccentric, may be considered as an oscillatory motion similar to that performed by the diameters of the planetary epicycles in the theory of the latitudes and solved with the same technique[3].

The object of this paper is to present some passages drawn from three astronomical Latin texts of the XIVth and XVth centuries where, in the context of a discussion on Ptolemy's equant, lunar model and latitude theory, physical arrangements of two concentric spheres enclosing the epicycle (or the eccentric) identical to that of Ibn al-Haytham are unequivocally described, although without parameters, in order to provide alternatives to the Ptolemaic mathematical procedures.

2. Homocentric Epicycles in Latin Astronomical Texts

1. The first text is the *Tractatus de reprobatione eccentricorum et epiciclorum (A Treatise Concerning the Refutation of the Eccentrics and the Epicycles)* by Henry of Hesse (ca. 1325–1397), whose non-Ptolemaic solar model, concentric with the equant, is well known. Written in Paris in 1364, *De reprobatione* is probably the first of Hesse's extant works and the only one in his vast production directly connected with astronomy. The aim of the work is to prove that the eccentrics and epicycles of Ptolemaic astronomy cannot be interpreted as physical mechanisms actually existing in the heavens, but as mere mathematical hypotheses justified by their predictive function. The criticism of Ptolemaic cosmology in *De reprobatione* includes a detailed refutation of Ptolemy's theory of planetary

distances and sizes, and objections concerning the irregular motion of the epicycle caused by the equant and the disruption of the predicted longitude positions due to the latitude theory; Hesse's other arguments, however, are only relevant as criticism of the incorrect exposition of the Ptolemaic astronomy contained in the *Theorica planetarum communis*[4].

The first mention of an arrangement of homocentric spheres enclosing the epicycle occurs in the section V of *De reprobatione*, at the end of the second argument against epicycles, which deals with the problem of the equant. Hesse's argument begins in the form of a *reductio ad absurdum*. When the center of the epicycle is located at the apogee of the eccentric, the epicyclic mean apogee is defined as the point of the epicycle farthest from the earth and determined by the line passing through the center of the earth, the center of the eccentric, and the center of the epicycle. This line is perpendicular to the tangent to the eccentric which passes through its apogee. The epicyclic mean apogee corresponds in fact, according to Hesse, to an invariable point situated in the interior concavity of the deferent sphere, which is the place with respect to which the motion of the epicycle is computed[5]. The epicycle is carried by the eccentric deferent, and this deferent is a solid sphere which rotates around an axis passing through its center. Consequently, the line joining the epicyclic mean apogee and the center of the epicycle must always be perpendicular to the tangent to the eccentric for any position of the epicycle on the eccentric. When the center of the epicycle is not in the apsidal line the epicyclic mean apogee points out to the center of the equant. From this it follows that the line joining the mean apogee of the epicycle and its center will be perpendicular to the tangent to the eccentric, although this line does not end at the center of the deferent; which is absurd. Therefore, when the center of the epicycle is not in the apsidal line, it is evident that the line which determines the epicyclic mean apogee will not end at this invariable point located in the interior concavity of the deferent sphere.

Moreover, to attribute to the epicycle an oscillatory motion in order to explain the alignment of its mean apogee with the equant point is in fact equivalent to attributing to it a non-uniform motion, because the same sphere would be moved by turns in opposite directions; which is impossible. The use of two concentric spheres enclosing the epicycle, Hesse concludes, does not prevent the epicyclic mean apogee from being displaced with respect to the aforementioned point in the interior concavity of the deferent sphere. Note Hesse's claim about the identity of the oscillatory motion of the diameter of the lunar epicycle towards the *prosneusis* point and that performed by the diameters of the planetary epicycles in latitude theory. The translation of the full argument is as follows[6]:

Secondly, concerning to the principal argument, if the epicycles existed, as it is generally stated in the *Theories* [*of the planets*], it would follow that a line not passing through the center [of a circle] would be perpendicular to that circle; which is impossible, because only a line passing through the center can be perpendicular. The antecedent is proved because the mean apogee will be considered invariable in the circumference where the motion of the epicycle around its own center is measured; because, if not so, the distance of the planet from the [epicyclic] true apogee could never be found with complete certainty in the tables of mean anomalies, since in these tables the uniform motion of the epicycle is computed from this invariable point, as is obvious for anyone who examines the figure. And it is as if the beginning of Aries was changed sometimes onwards, sometimes backwards: using the tables of mean motions, the true distance of the planet from Aries could never be found.

When the epicycle is at the apogee of the eccentric, the diameter of the epicycle between the mean apogee and the opposite point is perpendicular to the eccentric, because if it is prolonged it goes through the center of the eccentric. Then let this mean apogee be called *A*, which is necessarily invariable in that circumference [see Fig. 1]. Therefore, the line that comes from the mean apogee down to the center of the epicycle will necessarily be perpendicular to the eccentric, and it will occur anywhere the epicycle is, since it describes the eccentric. So let the epicycle be moved a certain distance from the apogee, i.e., to the point *B* corresponding to the mean longitude. And let a line be drawn pointing out to the mean apogee, that everybody draws from the center of the equant through the center of the epicycle, excepting the case of the moon. This line, either ends at *A* or at some point distant from *A*. It cannot be the second [possibility], because at the point *A* of the said circumference the mean apogee was placed when the epicycle was at the apogee of the eccentric. If so, the mean apogee would have been displaced to another point; which is opposed to what they maintain. Therefore, the first [possibility] must be accepted. Consequently, since this straight line begins at a point that is not the center of the eccentric, and since, as it has been said, the line between *A* and the center of the epicycle is perpendicular to the eccentric, the conclusion follows clearly.

And there is no use in saying that the epicycle has a certain movement of inclination by means of which its diameter between the mean apogee and the opposite point is always directed or inclined towards the center of the equant or to another point as in the case of the moon; because then it follows that the epicycle is moved non-uniformly in longitude; which is opposed to the tables of mean anomalies. And it would follow too a displacement of the point from which those tables begin to compute the motion, and this with respect to the inmobile circumference in which the velocity of the epicycle is measured, as the previous reasoning obviously shows.

The first is evident because the motion of inclination would be sometimes opposite to the motion of the epicycle in longitude. As in the case of the moon: when the epicycle is at the apogee of the eccentric, the diameter [between the mean apogee and the opposite point] is perpendicular to the eccentric; but when the epicycle comes down [on the eccentric] eastwards, the diameter deviates continuously from its perpendicularity westwards, till the epicycle reaches the mean longitudes, when that inclination will be the greatest, as can be proved[7]. From this point on, [the diameter] approaches the perpendicularity eastwards till the epicycle is at the perigee of the eccentric, where the diameter will be again perpendicular to the eccentric. And, however, while the epicycle of the moon was moved continuously westwards [with its motion in anomaly], the epicycle was also inclined from the perpendicularity either west or eastwards.

Therefore, the revolution of the epicycle became quicker in the first quarter of the

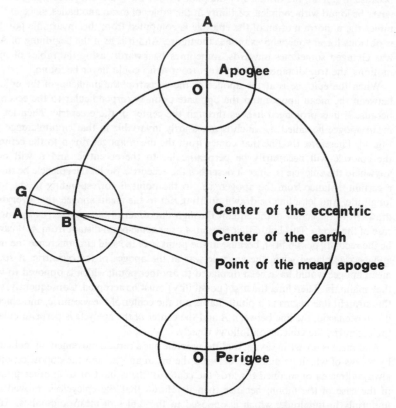

Figure 3: Paris, Bibl. Nat. 16401, f. 59r. *B* is missing in the MS.

eccentric and slower in the second one. And, moreover, since the epicycle of the moon has no motion in latitude, it follows from the inclination that the same point would be moved at the same time on the same circumference in opposite directions; which seems evident from the aforementioned hypothesis and is impossible and opposed to one of the initial assumptions.

And it is useless also to imagine an epicycle immediately enclosing another which would move the epicycle carrying the planet with a motion of inclination or with a motion like that of the latitude in the other planets; because it is equally necessary for the motion to be computed in the tables beginning at a point of this concave surface that, according to everybody's statements, encloses immediately the epicycle. Consequently, the same can be concluded as before[8].

Hesse's second discussion of the subject occurs in the eleventh argument against epicycles, in the same section of *De reprobatione*, while discussing the problem of the epicycle motion in latitude. However, Hesse now mentions only one additional sphere enclosing the epicycle, and this probably as the result of carelessness, as Hesse has written in the passage just quoted that the arrangement of two concentric spheres is valid for the inclination of the epicycle of the moon as well as for the oscillatory motion in latitude of the epicyclic diameter for the remaining planets.

There is, however, another interpretation of this passage which cannot be excluded. According to it, the Ptolemaic small circle producing the inclination of the apogee-perigee diameter of the epicycle is transferred to the surface of a concentric sphere surrounding the epicycle. The perigee of the epicycle is then attached to the endpoint of the radius of the small circle on the surface of the enclosing sphere, the radius of this small circle being equal to the maximum inclination of the epicyclic diameter. The perigee of the epicycle is thus moved just as the point Aries 0° of the eighth sphere in the trepidatory model attributed in the Latin West to Thābit b. Qurra (d. 901)[9]. Such a proposal seems to be very similar to the one Peurbach (1423–61) made in his *Theoricae novae planetarum* (1454; first printed ed., Nüremberg, ca. 1474) to explain the *deflection* of the eccentrics (*egklisis tou ekkentrou, mayl falak al-cārij al-markaz*; called by Peurbach *deviatio*) for the inferior planets (see below). Be that as it may, Hesse rejects this additional concentric sphere because the maximum distance of a planet would not be equal to the minimum distance of the immediately superior one.

Eleventhly, if the epicycles existed, the same simple body would be moved immediately and by itself with different motions; which is false, because the same cause [*potentia*] cannot produce at the same time different effects on the same body.

The antecedent is proved as everybody has established that the epicycles are moved eastwards with a uniform motion around their poles, and they also assert that the epicycles have a motion in latitude with which they are moved in the following way: the northernmost point of the epicycle sometimes is nearer the ecliptic and sometimes farther from it. And this cannot occur only because of the first motion that acts upon the epicycle by itself by an intrinsic principle. Thus, either this second motion acts upon the epicycle also by itself or by accident. If the first possibility occurs, either what we wanted follows or that two motors are appropriate in a direct manner to the same orb. But it may be that somebody could suppose another epicycle enclosing the epicycle carrying the body of the planet, which would be moved by itself with this motion in latitude and thus would influence the epicycle carrying the planet; but, if it were so, the maximum distance of Mercury from the earth could not be in any way equal to the minimum distance of Venus; which is contrary to al-Farghānī and to the position they maintain[10].

2. The second text is the *Tractatus de reprobationibus epiciclorum et eccentricorum*, written by *magister* Julmann in 1377[11]. Like Hesse's work, the first part of Julmann's *De reprobationibus* contains a criticism of Ptolemaic cosmology, now based on arguments extracted from Aristotelian physics and from optical theory. It is followed by a refutation of Ptolemy's theory of planetary distances and sizes taken mostly from Hesse's work, which Julmann quotes. In the second part of this work, Julmann proposes a version of the homocentric theory (five orbs for the three superior planets, six for Venus, Mercury and the Moon, and three for the Sun). Moreover, Julmann claims that the Sun has a motion in latitude, bringing forward as authorities Aristotle's *Metaphysics* (XII, 8; 1073 b) and an incorrect statement of the *Theorica planetarum communis*[12].

Concerning the subject of this article, in the context of a discussion about the number of orbs required by Ptolemaic cosmology, Julmann asserts that the proposers of eccentrics and epicycles must add two concentric spheres for the epicycle of the superior planets to explain the inclination of the apogee-perigee epicyclic diameter and four concentric spheres for the inferior ones (two for the inclination and two others for the slant). For Venus and Mercury, Julmann adds, it is necessary to suppose two other concentric spheres now enclosing the eccentric to explain the deflection of the eccentric planes of these planets. The passage reads as follows:

Once these things have been treated, there follows a remark on the latitudes of the planets. The proposers of the eccentrics and the epicycles claimed that Saturn, Jupiter and Mars have a double motion in latitude; this is true, but it cannot be explained with eccentrics and epicycles unless you agree to add two other epicycles for any of these planets; because, as they assert, each of them is moved to the south and to the north with a motion of reflection and declination of the epicycle, as is evident in [the book of] al-Farghānī and in the *Theory of the planets*. That the epicycles are moved [in this way] is proved because it is said [in the *Theory of the planets*]: "when the epicyclic center is at the nodes, the epicyclic plane lies on the eccentric"[13]. But the epicyclic plane cannot reach this position from its inclination without a motion; consequently, the epicycle is necessarily moved with two motions, one from west to east, another from north to south and vice versa. It follows also from the third assumption[14], that one of these motions acts upon the epicycle by accident, and this accidental motion corresponds to another body upon which, according to the fourth assumption[15], that motion acts by itself, and this body is also moved by the eccentric from west to east.

Then let us draw in this epicycle a great circle, and this circle will be the second epicycle. And as this second epicycle is not moved around, but reflected, because otherwise the planet would go out from the zodiac, it is necessary to suppose another or other epicycles that reflect the other two, given that it is impossible that they reflect by themselves. Consequently, they need to add at least two epicycles. So, they either wanted to avoid a great number of eccentric orbs or were not able to find a way to save appearances with concentric orbs; however, defeated by natural reason, they needed to suppose this great number of orbs. They fall thus in the hollow where during a long time they had reasons to fall.

In the same way, they need to suppose three epicycles for Venus and Mercury as in the case for the superior planets. But, what of it? For any of these planets there is established another motion in latitude which comes from the involution [i.e., the slant], which acts in the following way: when the epicyclic center of any of them is situated in the eccentric apogee, then the epicycle of Venus begins to move to the north, and the epicycle of Mercury to the south; but when the epicyclic center of any of them is situated in the eccentric perigee, then the epicycles are moved in the opposite direction. So, the epicycles of Venus and Mercury are moved with three motions: one of which is from west to east across the zenith, the second from south to north and vice versa also across the zenith, the third towards north and vice versa by the east. The first motion is called motion in longitude; the second, motion of inclination; the third, motion of involution. For the first two motions of Venus and Mercury they must suppose two epicycles besides the epicycle carrying the body of the planet, as is the case for the three superior planets. And for the third motion, it is equally necessary that they suppose two other orbs for the epicycle, and consequently circles for the epicycle, one carrying it on, another bringing it back. Therefore, it is evident from the aforesaid and from the assumptions made previously that they need to suppose five epicycles for the motions of Venus and Mercury. And they suppose two eccentrics. There are thus seven orbs for each of the inferior planets and five orbs for each of the superior ones.

For Venus and Mercury it is necessary to add two more orbs, one of which carries the eccentric perigee to the south in the direction opposite to the other which carries the eccentric apogee to the north, given that they assert that the eccentric apogee of Venus and Mercury are moved so, although one contrarily to the other. Consequently, each of the inferior planets has nine orbs and each of the superior ones five[16].

3. The third text is the *Commentariolum super Theoricas novas planetarum* by Albert of Brudzewo (1445–1495), written around 1483 as a course on Peurbach's *Theory of the planets* read at Cracow University, and published twice (Milan, 1495; Cracow, 1900). In Peurbach's aforementioned work there are two brief passages on the subject of enclosing epicycles and eccentrics with concentric orbs. The first occurs in the chapter entitled *De declinatione et latitudine*, while dealing with the deflection of the eccentrics of the inferior planets. After describing the orbs corresponding to the models for these planets, Peurbach writes:

Thus, because of the aforementioned deflections, it seems necessary to add to the enumerated orbs another orb concentric to the earth enclosing all the others, in such a way that the aforementioned deflections were produced by the motion of trepidation of this orb[17].

Later, in the section on the slant of the epicycles, Peurbach mentions again concentric orbs surrounding now the epicycles of the inferior planets (but it is difficult to decide whether Peurbach refers here to Ibn al-Haytham's device of two concentric orbs for each motion of the epicycles in latitude or rather to his own proposal of a concentric orb moved by a trepidatory motion):

Because of the aforementioned inclinations and slants of the epicycles, some people postulate small orbs inside of which the epicycles are located, and which produce these inclinations and slants with their motion[18].

There is no commentary on these passages in Brudzewo's *Commentariolum*; but, on discussing the motion of the lunar epicycle, considered by Peurbach as irregular, Brudzewo mentions an arrangement of two concentric spheres enclosing the lunar epicycle to explain the motion of its mean apogee, called by Brudzewo *motus declinationis et reflexionis* (or *reclinationis*, or *inclinationis*). After quoting the beginning of chapter 5, book V, of the *Almagest*, Brudzewo writes:

> Ptolemy seems to point out with these words that [the motion of] declination and reclination do not derive only from computation, but from a motion actually observed in the moon. But the moon, because of this motion of declination and reflection[19], gets certain appearances after [the beginning of] their elongation from the Sun, i.e., that it appears concave or sliced, and this around the fifth day after conjunction. It also appears swollen or gibbous, and this around the tenth day after conjunction. So, to explain this apparent motion of the moon, some imagine an epicycle for the moon, that has another epicycle inside, which moves the epicycle carrying the moon with this motion of declination and reflection; which does not seem to be inappropriate[20].

3. Concluding Remarks

There is no reason to suppose that this Eudoxan device was created by the authors of the texts presented above. This possibility is explicitly excluded in Peurbach's *Theoricae* and in Brudzewo's *Commentariolum*, where the arrangement of two concentric spheres is attributed to some unknown author(s). It has been suggested that there is a link between Ibn al-Haytham's criticism of Ptolemaic cosmology and Copernicus, not directly but through some intermediary connected with the transmission to the West of Maragha planetary theory or perhaps with a western Aristotelian tradition of objections to Ptolemy[21]. It is possible that the texts by Hesse, Julmann, and Brudzewo translated here were the traces of such a critical tradition related to homocentric theory, insofar as the Arabic works mentioned in the introduction were apparently never translated into Latin during the Middle Ages. In any case, it is unlikely that the two homocentric spheres theme was original in Hesse's text: there is enough internal evidence in *De reprobatione* to suggest that most of the arguments contained in the work were taken by Hesse from yet unknown sources. Be that as it may, it is evident that this tradition drew upon the capabilities of the two homocentric spheres device and of its usefulness to provide a physical mechanism for the different oscillatory motions of the Ptolemaic epicyclic diameters required by the equant and latitude theory.

Acknowledgments

I wish to express my gratitude to Prof. B. R. Goldstein for his commentaries on this article and to Dr. G. Rosińska for providing me with photographs of MS Knihovna Metropolitní Kapitoly 1272. Microfilms of manuscripts were kindly supplied by the Directors of Bibliothèque Nationale, Paris; Stiftsbibliothek, Melk; Bayerische Staatsbibliothek, Munich; Nationalbibliothek, Vienna.

NOTES

1. The *Treatise on the Movement of Iltifāf* is now believed to be lost, but the *Solution of Difficulties Concerning the Movement of Iltifāf*, written by Ibn al-Haytham as a reply to a criticism to the *Treatise,* is preserved and has been edited by Sabra (1979). Abū ʿUbayd al-Jūzjānī, a younger contemporary of Ibn al-Haytham and Ibn Sīnā's student and biographer, wrote a work entitled *Kayffiyat tarkīb al-aflāk* where he tried also to apply homocentric spheres with different poles to solve the problems of the equant and the components of latitude. The *Kayffiyat* is apparently lost, but a compendium of the book by the author himself is preserved and has been edited by Saliba (1980).
2. Cf. the edition of al-Ṭūsī's *Tadhkira* by Ragep (1982), esp. 1, pp. 285–98, and 2, pp. 107–15, for the details of Ibn al-Haytham's arrangement and al-Ṭūsī's criticism and solution. On the subject of the latitude theory in al-Ṭūsī's *Tahrir al-majisṭī (A Redaction of the Almagest)*, see also Saliba (1987).
3. In fact, al-Ṭūsī proves that the spherical version of his "couple" may also be applied to three similar problems involving an oscillatory motion of a point along an arc of a sphere: the oscillation of the inclined planes of the eccentrics of Venus and Mercury, the change in the obliquity of the ecliptic between a maximum and a minimum value, and trepidation. On Ibn al-Haytham's criticism of the Ptolemaic equant in *al-Shukūk ʿalā Baṭlamyūs (Doubts concerning Ptolemy)*, cf. Sabra (1978).
4. On Hesse's solar model, cf. Kren (1968). Hesse's *De reprobatione* is also the source of Copernicus' arguments against the Ptolemaic lunar model in the *Commentariolus*, through Regiomontanus' *Epitome in Almagestum* (V, 22). One of the extant copies of *De reprobatione* is by Regiomontanus himself (Wien, Nat. Bibl. MS lat. 5203, 110r-117v), who again used Hesse's arguments in his correspondence with G. Bianchini (d. *ca.* 1470), the Italian astronomer [cf. Curtze (1902), pp. 218, 265 and 266] and his *Disputationes contra Cremonensia in planetarum theoricas deliramenta* (1464, published in 1475).
5. Although obviously the deferent sphere is moved around its own center, Hesse seems to consider its interior concavity immobile with respect to the motion of the epicycle. Thus, this interior concavity is the place where the epicycle is moved; cf. Aristotle, *Physics*, IV, 4(212 a).
6. There are nine extant copies of *De reprobatione:* Praga, Knihovna Metropolitní Kapitoly

1272, 45r–54r; Melk, Stiftsbibl. 51, 210ra–218va (incomplete); Paris, Bibl. Nat. 16401, 55r–67v; Roma, Vat. Lat. 4082, 87ra–97rb, and Barberini 350, 61ra–65vb (incomplete); Princeton Lib., Garret 95, 297–340; Oxford, Bodleian Lib., Bodley 300, 53ra–64vb; Wien, Nat. Bibl. 5203, 100r–117v, and Utrecht, Univ. Bibl. 725, 218r–246r. The copies of Oxford and Princeton MSS, both probably from the first half of the XVth century [cf. North (1976), 2, pp. 37–39] and entitled *Tractatus contra theoricam planetarum,* do not mention the author and seem to be a first version of the text. I am preparing a critical edition, with translation and commentary, of this work and the following quotations are based on a collation of the nine copies; but I will give the location of the quoted passages in the three oldest ones: Praga, Melk, and Paris, cited from now on as *K, M,* and *P,* respectively.

7. Hesse's text is wrong on this point, insofar as the maximum inclination does not occur at mean distances from the earth of the epicycle center. In the lunar model, the maximum inclination occurs when the line between the epicycle center and the *prosneusis* point is perpendicular to the apsidal line [cf. Sabra (1978), 126; Ragep (1982), 2, pp. 107–8], and then the double elongation is more than 90°. When the center of the epicycle is at mean distances, the double elongation is slightly less than 90° (exactly, 84; 2°).

8. *"Secundo ad principale. Si essent epicicli sicut communiter inuenitur in theoricis, sequeretur lineam preter centrum transeuntem circulo esse perpendicularem; quod est impossibile cum solum sit talis linea perpendicularis per centrum progrediens. Antecedens probatur nam media aux ponatur inuariabilis in circumferentia qua mensuratur uelocitas epicicli circa proprium centrum; quia alias, cum ab ea computetur motus epicicli uniformis in tabulis mediorum argumentorum, nunquam inueniretur in eisdem tabulis ueraciter distantia ab auge uera planete, sicut patet cuilibet intuenti figuram. Et est sicut si caput Arietis uariaretur quandoque ante et post: per tabulas equalium motuum non utique inueniretur uera distantia planete ab Ariete.*

Sed cum epiciclus est in auge eccentrici notum est quod diameter epicicli inter eiusdem augem mediam et oppositum est perpendicularis super eccentricum, quia in continuum ducta per eius centrum transiret. Vocetur ergo illa aux A, que necessario inuariabilis erit in dicta circumferentia. Igitur necessario linea descendens ab ea ad centrum epicicli est super eccentricum perpendicularis et ubicumque fuerit epiciclus, quia ipse describit eccentricum. Veniat ergo epiciclus ad aliquam distantiam ab auge, scilicet ad punctum B longitudinis medie. Et trahatur linea ostendens augem mediam, quam omnes trahunt a centro equantis preterquam in luna per centrum epicicli; que linea uel terminatur ad A uel in punctum distantem ab A. Non secundum, quia in A puncto dicte circumferentie signabatur aux media cum epiciclus esset in auge et sic iam esset uariata ad alium punctum; quod est contra eorum positionem predictam. Ergo dabitur primum. Igitur cum talis linea recta incipiat a puncto preter centrum eccentrici et cum linea inter A et centrum epicicli sit eccentrico perpendicularis, ut dictum est, clarissime sequitur propositum.

Et nihil ualet dicere epiciclum habere quemdam motum flexionis per quem eius diameter inter augem mediam et eius oppositum semper dirigatur uel flectatur ad centrum equantis uel ad alium punctum ut in luna; quia ex hoc sequitur epiciclum difformiter moueri in longitudine, quod est contra tabulas mediorum argumentorum. Et etiam sequeretur uariatio puncti a quo ille tabule incipiunt motum, et hoc in illa circumferentia quiescente in qua uelocitas epicicli mensuratur, sicut euidenter argumentum iam factum demonstrat.

84

Primum istorum patet quia ille motus flexionis quandoque esset in contrarium motus epicicli in longitudine. Sicut in luna: quando epiciclus est in dicta auge eccentrici diameter est eccentrico perpendicularis; sed eo descendente uersus orientem illa diameter continue declinat ab illa perpendicularitate uersus occidentem, donec ueniat epiciclus in mediam longitudinem ubi maxime declinat, ut demonstrabile est. A quo situ reclinat ad dictam perpendicularitatem uersus orientem usque ad oppositum augis, ubi illa diameter iterum est perpendicularis eccentrico. Et tamen semper epiciclus lune mouebatur interim uersus occidentem, siue flecteretur epiciclus a perpendicularitate uersus occidentem uel orientem. Ergo necessario reuolutio epicicli ibi uelocitabatur et hic retardabatur. Et quod plus est, cum epiciclus lune non habeat aliquem motum in latitudine, sequitur ex illa flexione eundem punctum super eadem circumferentia simul uersus oppositos terminos moueri, quod manifeste uidetur ex dicta ymaginatione; quod est impossibile et contra quoddam suppositum in principium.

Nec ualet hic ymaginari unum epiciclum inmediate includentem alium qui moueret epiciclum deferentem astrum motu flexionis uel motu secundum latitudinem in alijs planetis; quia eque bene oportet motum incipi in tabulis ab aliquo puncto illius superficiei concaue, que epiciclum inmediate locat sicut omnes dicunt. Igitur concluderetur idem quod prius" (K, f. 47vb–48ra; M, f. 211vb–212ra; P, f. 59r–v).

9. The treatise *On the motion of the eighth sphere* is not extant in the original Arabic, though it is preserved in Latin translation. This Latin version has been published by Millás Vallicrosa (1950), pp. 496 ff., and translated into English, with commentary, by Neugebauer (1962), pp. 290–99. Ragep (1982), 1, pp. 219–29, has presented evidence for believing that *De motu octavae sphaerae* is not by Thābit, but originated in Spain.

10. *"Undecimo. Si essent tales epicicli, idem corpus simplex eque inmediate et per se moueretur diuersis motibus; quod est falsum, quia eadem potentia secundum idem non potest in eodem secundum idem sui diuersos effectus producere simul.*

Probatur antecedens quia epicicli ponuntur ab omnibus moueri uersus orientem motu uniformi super suis polis, et cum hoc ponuntur habere motum in latitudine quo sic mouentur: quod punctus epicicli altissimus quandoque est ecliptice propinquior et quandoque remotior. Et hoc non potest contingere ex solo primo motu qui inest epiciclo per se primo ab intrinseco principio per appropriationem. Vel ergo ille alter motus inest ei etiam per se uel per accidens. Si primum, sequitur propositum uel quod duo motores sint appropriati eidem orbi inmediate.

Sed forte aliquis poneret alterum epiciclum super epiciclum deferentem corpus astri, qui per se moueretur illo motu latitudinis et eum influeret epiciclo deferente planetam; quod si esset, nullo modo maxima distantia Mercurij a terra esset equalis minime distantie Veneris ab eadem; quod est contra Alfraganum et eorum positionem"(K, f. 48vb; M, f. 212vb; P, f. 60v).

11. This work is apparently preserved in only two copies: *Codex Latinus Monacensis* 26667, ff. 109va–116ra, and Wien, Nat. Bibl. MS lat. 5292, ff. 180r–197r. The text was composed between 21th June 1377 (an observation of the lunar eclipse of this date is mentioned) and 26th September of the same year. No other work by Julmann is known. The following quotations are based on a collation of these two copies and editorial insertions are enclosed in square brackets.

12. *"Nullam autem latitudinem in sole ponunt in eorum scriptis et uera experientia ostendetur*

esse falsum (...) Supponatur secundo etiam quod eleuatio in meridie solis quando est in auge eccentrici sui sit maior quam eleuatio ecliptice super orizontem in duobus gradibus; hoc scibitur in fine Theorice *et experientia docet"* (CLM 26667, f. 113vb; Wien, Nat. Bibl. 5292, f. 191r). For the passage of the *Theorica communis,* see the Latin text in Carmody (1942), pp. 46–47, and Pedersen's English translation [Grant (1974), p. 464].

13. *"Inclinatur autem epiciclus ab ecentrico ita quod semper erit planetam inter eclipticam et centrum epicicli nisi cum centrum epicicli sit in capite uel cauda draconis, tunc enim epiciclus est directus in ecentrico; ..."* [Carmody (1942), VII: 97, p. 42; cf. also Grant (1974), p. 462: 102]. It must be noted that in this chapter the author of the *Theorica communis* presents to the reader a non-Ptolemaic latitude theory taken from al-Khwār-izmī. For Ptolemy (*Almagest,* XIII, 2), when the center of the epicycle of the superior planets is situated at the nodes, the apogee-perigee epicyclic diameter lies on the eccentric plane, but the epicyclic plane lies on the plane of the ecliptic.

14. The third assumption made by Julmann at the beginning of the text reads: *"Nulli corpori simplici insunt plures motus simplices per se" (CLM* 26667, f. 109va; Wien, Nat. Bibl. 5292, f. 180r).

15. *"Quarta [suppositio]: omnis motus simplex, qui per accidens alicui corpori inest, alteri corpori inest per se" (CLM* 26667, f. 109va; Wien, Nat. Bibl. 5292, f. 180r).

16. *"Ex dictis sequitur determinatio circa latitudinem planetarum uenientes. Dixerunt enim Saturnum, Jouem et Martem duplices habere latitudines; quod quidem uerum, sed non poterit saluare per eccentricos et epiciclos nisi uoluerit superaddere cuilibet eorum duos epiciclos; quoniam, ut dicunt, quilibet eorum mouetur ad meridiem et ad septentrionem secundum motum reflexionis et declinationis epicicli, ut patet in Alfragano et in theorica communi planetarum. Sed quod epicicli moueantur patet quia dicitur: quando centrum epicicli est in nodis, tunc epiciclus est directus. Sed non potest a declinatione uenire ad rectitudinem nisi per motum; ergo necessario epiciclus mouetur duobus motibus, uno scilicet ab occidente in oriens, altero a septentrione in meridiem et econtra. Sequitur ulterius ex tertia suppositione quod unus ex illis motibus sibi insit per accidens, et illi motui aliud corpus, quarta suppositione correspondente, apparet cui insit per se, quod etiam notatum est duci ab orbe eccentrico ab occidente in oriens.*

Signetur igitur in ipso circulus magnus et erit epiciclus secundus ex descriptione epicicli. Et quia ille epiciclus non circumducitur totaliter, sed reflectitur, alias enim planeta exiret zodiacum, necesse est poni alium uel alios qui reflectant priores duos epiciclos, quoniam ex se reuolui impossibile est. Quare oportet eos ad minus superaddere duos epiciclos. Voluerunt autem euitare multitudinem orbium eccentricorum uel non poterant inuenire modum saluandi apparentias per concentricos, quam tamen multitudinem oportet eos ponere naturali ratione uictos. Cadunt igitur ad foueam ad quam incidendi rationem longo tempore habuerunt. Similiter in Venere et Mercurio oportet eos ponere epiciclos tres sicut in superioribus. Sed qui? Unicuique eorum ponitur alia latitudo que prouenit ex inuolutione que taliter fit: quando centrum epicicli alicuius eorum est in auge eccentrici, incipit tunc epiciclus moueri uersus septentrionem in Venere, in Mercurio autem uersus meridiem; sed in opposito augis eccentrici cum fuerit centrum epicicli alicuius eorum, tunc mouebitur motu contrario priori motu. Mouentur igitur epicicli Veneris et Mercurij tribus motibus, quorum unus est ab occidente in oriens per cenith regionum, secundus a meridie in septentrionem et econtra etiam per cenith regionum, tertius a meridie in septentrionem et econtra per oriens. Primus uocatur motus longitudinis, secundus declinationis, tertius motus inuolutionis. Ad primos duos motus eorum oportet eos ponere duos epiciclos preter primum qui defert corpus planete sicut in tribus superioribus planetis. Et pro tertio motu similiter est necesse quod ponant duos orbes epicicli, et per consequens circulos epicicli, unum qui ducat, alium qui reducat. Quare manifestum est ex iam dictis et ex suppositionibus quod oportet eos ponere ad saluandum motus Veneris et Mercurij 5 epiciclos. Et ponunt duos eccentricos; erunt ergo 7 orbes in quolibet eorum et 5 orbes in quolibet trium superiorum. Adhuc in Venere et Mercurio necessarium est duo esse orbes, quorum unus faceret eccentrici [oppositum augis moueri ad meridiem] uersus alium qui reducat augem ad septentrionem, quoniam dicunt auges eccentricorum Veneris et Mercurij sic moueri, licet diuersimode; quare unusquisque· ex illis duobus habet 9 orbes et unusquisque ex superioribus 5" (CLM 26667, f. 113va–b; Wien, Nat. Bibl. 5292, f. 190r–v).

17. *"Propter dictas autem deuiationes, orbibus praenumeratis alium mundo concentricum praedictos omnes includentem superaddi uidetur oportere: ad cuius motum trepidationis praedictae deuiationes accidant"* [Peurbach (1490), f. iiiv]. This passage is paraphrased by Joannes Baptista Capuanus in his commentary to Peurbach's *Theoricae: "Quarto infert quod cum epicycli deferens habeat motum proprium in longitudine çodiaci: motus hic in latitudine ei non erit proprius siquidem corpora singula singulis feruntur motibus: erit igitur aliud corpus mundo concentricum ambiens totam spheram cuiusque eorum Veneris et Mercurij quod motu proprio hoc deuiationis motu feratur et rapiat orbes eorum"* [Capuanus (1518), f. 242va].

18. *"Propter dictas epiclorum inclinationes atque reflexiones orbes parui epicyclos intra se locantes a quibusdam ponuntur ad quorum motum eaedem contingunt"* [Peurbach (1490), f. iiii]. The passage is also discussed by Erasmus Reinhold (1511–1553) in his commentary to Peurbach's *Theoricae* (first ed., Wittenberg, 1542), but it is interpreted as if Peurbach's *orbes parui epicyclos intra se locantes* were identical to Ptolemy's small circles: *"Ptolemaeus, postquam obseruationibus didicit tales fieri epicyclorum* egkliseis kaí loxoseis, *vt etiam ostendat qua ratione tales in coelo motus existere queant, circellos, seu, vt ipse vocat,* kukliskous, *ipsis epicyclis apposuit. Qua de re consulant studiosi ipsum Ptolemaeum lib. ultimo, cap. 2 & Theonis diligentissimi commentarios in eiusdem librum. Non enim paucis ea ratio explicari potest, quanquam praeter nostram spem ita creuerunt haec qualiacumque scholia, quibus discentium studia iuuare cupimus, vt iustum pene commentarium efficere videantur"* [Reinhold (1557), 156v–157r].

19. Ptolemy (*Almagest*, V, 5) obviously does not attribute like Brudzewo those lunar appearances (concave, gibbous) to the motion of inclination of the apogee-perigee epicyclic diameter; he simply asserts that "from individual observations taken at distances of the moon [from the Sun] when it is sickle-shaped or gibbous (which occur when the epicycle is between the apogee and the perigee of the eccentre), we find that the moon has a peculiar characteristic associated with the direction in which the epicycle points" [Toomer (1984), pp. 226–227].

20. *"Videtur [Ptolemaeus] in istis verbis innuere, quod declinatio et reclinatio non ex sola computatione proveniat, sed ex motu realiter apparenti in Luna. Luna enim ratione istius declinationis et reflexionis consequitur quasdam figuras post elongationem eius a Sole, videlicet quod apparet concava seu excisa, et hoc circa quintum diem fere post coniunctionem. Apparet etiam tumida vel gibbosa, et hoc circa decimum diem fere post*

coniunctionem. Propter ergo salvare istum motum apparentem in Luna, quidam imaginantur epicyclum talem in Luna, quod habeat alium intra se inclusum, qui movet epicyclum deferentem Lunam motu declinationis et reflexionis, quod non videtur esse inconveniens" [Birkenmajer (1900), pp. 67–68].

The beginning of Brudzewo's commentary on the passage of Peurbach's *Theoricae* is closely similar in wording to Hesse's argument quoted above in note 9: "Notandum circa litteram, quod Mathematici, motus astrorum praesertim aequales calculantes, supponunt eos a certo punto semper invariabili computari, ideo, quia aliter non possent devenire in certum locum astri, si terminus a quo computant motum esset variabilis. Quemcumque enim arcum motus invenirent, semper in ipsius maiore aut minore parte esset astrum, termino a quo variabili existente, *sicut si caput Arietis variaretur quandoque ante et post, non utique inveniretur vera distantia planetae ab Arietis initio,* sicut patet in figura" [Birkenmajer (1900), p. 62].

Birkenmajer includes in the critical apparatus of Brudzewo's text a note in the margin of MS L (Ossol. 759), 51r, by the copyist John of Croyba. In this note the subject of a second concentric epicycle to explain the oscillatory motion of the lunar epicyclic diameter is mistakenly related to the problem of the spots of the moon, which are considered sometimes in the Middle Ages as an evidence against the Ptolemaic model. The text reads as follows: *"Lunam quidam imaginantur habere duos epicyclos, unum maiorem, alterum minorem, in quo est eius corpus situatum, et ita epicyclus superior tantum, motu declinationis et reflexionis movetur. Et pro tanto illa macula quae in luna aspicitur, semper una et eadem apparet propter istum epicyclum: quod non esset, si talis epicyclus non esset"* [Birkenmajer (1900), p. 68].

Another marginal note in the fol. 29 of the same MS (corresponding to Peurbach's text, but displaced by Birkenmajer to the page 145 of his edition) deals with the same subject. This copyist asserts that the oscillatory motion in latitude is attributed by some astronomers to the orb(s) which enclose(s) the epicycle(s); he adds, however, that in his opinion the order of motions must be inverted: *"Nonnulli enim astronomi dicunt, quod epicycli habeant inter se alios epicyclos maiores, eos ambientes, qui istum motum latitudinis causare debent; epicyci vero eos ambientes longitudinis motum causant"* (I have substituted *majores* for *minores* in Birkenmajer's edition).

21. Swerdlow (1976), p. 117

BIBLIOGRAPHY

Birkenmajer,
    1900: *Commentariolum super Theoricas Novas Planetarum Georgii Purbachii in Studio Generali Cracoviensi per Mag. Albertum de Brudzewo diligenter corrogatum A.D. MCCCCLXXXII.* Post editionem...curavit L. A. Birkenmajer, Cracoviae typis... Universitatis Jagiellonicae.
Capuanus,
    1518: *Sphera cum commentis in hoc volumine continentis... Theorice noue planetarum*

Georgij Purbachij...cum expositione D. Joannis Baptiste Capuani de Manfredonia Canonici regularis ordinis sancti Augustini..., Octaviani Scoti, Venetiis.

Carmody,

1942: *Theorica planetarum Gerardi*, Berkeley.

Curtze, M.

1902: "Briefwechsel Regiomontan's mit Giovanni Bianchini, Jacob von Speier und Christian Roder", *Abhandlungen zur Geschichte der mathematischen Wissenschaften*, XII, pp. 185–336.

Grant, E. (ed.),

1974: *A Source Book in Medieval Science*, Cambridge, Mass., Harvard University Press.

Kren, C.

1968: "Homocentric Astronomy in the Latin West. The *De reprobatione eccentricorum et epiciclorum* of Henry of Hesse", *Isis*, 59, pp. 269–281.

Millás Vallicrosa, J. M.

1950: *Estudios sobre Azarquiel*, Madrid-Granada, C.S.I.C.

Neugebauer, O.

1962: "'On the Solar Year' and 'On The motion of the eighth sphere'", *Proceedings of the American Philosophical Society*, 106, pp. 264–299.

1970: *The Exact Sciences in Antiquity*, 2nd pr., Providence, Brown University Press.

North, J.D.

1976: *Richard of Wallingford*, 3 vols. Oxford, Clarendon Press.

Peurbach,

1490: *Theoricae nouae planetarum Georgii purbachii astronomi celebratissimi...* Venetiis mandato & expensis nobilis uiri Octauiani scoti... Anno Salutis M.cccc.lxxxx. quarto nonas octobris.

Ragep, F. J.

1982: *Cosmography in the 'Tadhkira' of Naṣīr al-Dīn al-Ṭūsī*, Ph. D. Dissertation, Harvard University.

Reinhold,

1557: *Theoricae novae planetarum Georgii Purbachii Germani ab Erasmo Reinholdo Salueldensi pluribus figuris auctae, & illustratae scholiis....* Parisiis, Apud Carolum Perier, in vico Bellovaco sub Bellerophonte.

Sabra, A. I.

1978: "An Eleventh-Century Refutation of Ptolemy's Planetary Theory", *Studia Copernicana*, XVI, Warsaw, Ossolineum, pp. 117–131.

1979: "Ibn al-Haytham's Treatise: Solutions of Difficulties Concerning the Movement of *Iltifāf*", *Journal for the History of Arabic Science*, 3, pp. 388–422.

Saliba, G.

1980: "Ibn Sīnā and Abū ʿUbayd al-Jūzjānī: The Problem of the Ptolemaic Equant", *Journal for the History of Arabic Science*, 4, pp. 376–403.

1987: "The Role of the *Almagest* Commentaries in Medieval Arabic Astronomy: A

Preliminary Survey of Ṭūsī's Redaction of Ptolemy's *Almagest"*, *Archives Internationales d'Histoire des Sciences*, 37, pp. 3–20.

Swerdlow, N.
1976: "Pseudodoxia Copernicana...", *Archives Internationales d'Histoire des Sciences*, 26, pp. 108–158.

Toomer,
1984: *Ptolemy's Almagest*. Translated and annotated by G. J. Toomer. London, Duckwort.

On IBN AL-KAMMĀD's Table for Trepidation

Communicated by J. North

1. Introduction

In an article recently published in this journal, Chabás & Goldstein (1994) offered a careful analysis of the *zīj al-Muqtabis*, a set of astronomical tables composed by Abū Jaᶜfar Ahmad ben Yīsuf Ibn al-Kammād in Islamic Spain around the end of the XIth or the beginning of the XIIth century, as preserved in a Latin translation made by John of Dumpno in 1260 in Palermo (Madrid, Biblioteca Nacional, MS 10023, ff. 1r–66r). The present paper deals only with Ibn al-Kammād's table for "the equation of access [and recess] of the first point of Aries [from the vernal point]" (*tabula directionis aduenctionis capitis arietis*), reconstructed by Chabás & Goldstein using Azarquiel's second model for trepidation, for which an alternative reconstruction is presented taking into account a hitherto unexplored and unedited anonymous Castilian translation of Ibn al-Kammād, made probably also in the XIIIth century (Segovia, Catedral, MS 115, ff. 218vb-220vb).

2. Some Remarks on the Reconstruction of Ibn al-Kammād's Table by Chabás and Goldstein

It is a standard procedure to identify medieval astronomical tables of equations by their maximum values, insofar as some of the parameters underlying the computation can easily be derived from them. Thus, the maximum value in the table for the equation of access in the *zīj al-Muqtabis* ($P_{max} = 9;59°$, corresponding to an argument of 90°) has been taken as its distinctive feature[1]. Samsó (1992, p. 323) for instance, in his comments

---

[1] Ibn al-Kammād's trepidation table is preserved in at least four different sources: the afore-mentioned manuscript of the Biblioteca Nacional in Madrid, f. 35v; a treatise on astronomical instruments by Abū l-Hasan ᶜAlī bᶜUmar al-Marrākushī (*ca.* 1262; cf. Sédillot, 1834, p. 131); the Tables of Barcelona, compiled under the patronage of the king Peter of Aragon (epoch 1320; edited by Millás, 1962, pp. 194–5, according to MS 21 of the Biblioteca Lambert Mata at Ripoll), and the tables of Juan Gil (*ca.* 1350; London, Jews College, MS Heb. 135, f. 78b). I am grateful to Prof. Bernard R. Goldstein for providing me with variant readings in this Hebrew manuscript and for his comments on a draft of this paper.

2

on Ibn al-Kammād's table, made some attempts to derive parameters from this maximum value using Azarquiel's third model for trepidation[2] based on the equations:

$$\delta = r \cdot \sin(i) \tag{1}$$

$$\sin(P) = \sin(\delta)/\sin(\varepsilon) \tag{2}$$

where i is the argument, ε the obliquity of the ecliptic, and $r = 4;7,58$. He pointed out that $P_{max} = 9;59°$ leads to an unacceptable value for the obliquity of the ecliptic, $24;33,50°$, and, with the value for the obliquity of the ecliptic used in Ibn al-Kammād's table for solar declination (f. 35v), $23;33°$, it leads to $r = 3;58,19$, close to the $3;54$ used in Azarquiel's first model; Samsó, however, did not recompute the table, claiming only that perhaps Ibn al-Kammād had not computed it according to Azarquiel's third model. Chabás & Goldstein (1994, p. 24–27) have used an identical procedure: they started with $9;59°$ and recomputed the table using Azarquiel's second model with equation

$$\sin(P) = r \cdot \sin(i)/60 \tag{3}$$

where $r = 10;24$, since arcsin $(10;24/60) = 9;58, 54° \approx 9;59°$. (Table 1 displays in column (a) the argument in degrees, in columns (b) the entries in the table, in columns (c) the values according to the reconstruction by Chabás and Goldstein [hereafter called R_1], and in columns (d) the difference between text and recomputation.) However, in my opinion, it is difficult to accept the claim that the value $10;24$ for r in equation (3) "is made plausible by the two other models for trepidation associated with Azarquiel and his followers" (Chabás & Goldstein, 1994, p. 27), because (a) agreement between computation (using equations based on Azarquiel's first and third model and assuming $P_{max} = 10;24°$) and text of Ibn al-Bannā' and Ibn al-Raqqām's trepidation table (where $P_{max} = 10;24°$) is irrelevant for deciding on the supposed use by Ibn al-Kammād of Azarquiel's second model, and (b) any one of Azarquiel's three models indeed produces very similar results once it is assumed for all of them that $P_{max} = 9;59°$ (or any other value for P_{max})[3]. In fact the agreement with Ibn al-Kammād's table is slightly better using equations [1] and [2] with $r = 3;58,19$ and $\varepsilon = 23;33°$ than using equation (3) with $r = 10;24$ or using equation

$$\sin(P) = \sin(i) \cdot \sin(P_{max}) \tag{4}$$

which corresponds to Azarquiel's first model (see Table 2).

Moreover, it is known that sometimes maximum (and minimum) values in the tables are less informative than one might expect: Goldstein (1980, 1992) himself has shown that the maximum entries in some Alfonsine lunar velocity tables (the one which appears in the 1483 edition, and those attributed to John of Genoa and John of Montfort) are probably due to errors in the calculation or some initial corruption which affected the

---

[2] On Azarquiel's trepidation models see Millás, 1943–1950, pp. 282–294; Neugebauer, 1962b, pp. 183–4; Goldstein, 1964; Samsó, 1992, pp. 228–236.

[3] The trepidation table preserved by Ibn al-Bannā' and Ibn al-Raqqām can be also reproduced using equation (3) with $r = 10;49,52$.

Table 1.

(a) arg.	(b) text 0ˢ	(c) R₁	(d) dif.	(b) text 1ˢ	(c) R₁	(d) dif.	(b) text 2ˢ	(c) R₁	(d) dif.
1	0;10	0;10,23	0	5;16	5; 7,18	9	8,52	8;43,10	9
2	0;20	0;20,47	1	5;25	5;16,12	9	8;56	8;48,12	8
3	0;31	0;31,11	0	5;34	5;25, 1	9	9; 1	8;53, 3	8
4	0;41	0;41,34	1	5;43	5;33,44	9	9; 5	8;57,45	7
5	0;53	0;51,56	1	5;52	5;42,20	10	9;10	9; 2,17	8
6	1; 3	1; 2,17	1	6; 4	5;50,51	13	9;14	9; 6,39	7
7	1;14	1;12,37	1	6;16	5;59,15	17	9;17	9;10,51	6
8	1;25	1;22,56	2	6;29	6; 7,33	21	9;21	9;14,53	6
9	1;35	1;33,13	2	6;41	6;15,44	25	9;24	9;18,45	5
10	1;45	1;43,29	2	6;43	6;23,49	19	9;28	9;22,26	6
11	1;56	1;53,43	2	6;57	6;31,46	25	9;31	9;25,57	5
12	2; 8	2; 3,55	4	7; 2	6;39,37	24	9;35	9;29,18	6
13	2;19	2;14, 4	5	7; 6	6;47,20	19	9;38	9;32,28	6
14	2;30	2;24,11	6	7;10	6;54,56	15	9;41	9;35,28	6
15	2;41	2;34,16	7	7;14	7; 2,24	12	9;45	9;38,17	7
16	2;50	2;44,18	6	7;21	7; 9,45	11	9;47	9;40,56	6
17	2;58	2;54,17	4	7;28	7;16,58	11	9;49	9;43,24	6
18	3; 6	3; 4,13	2	7;35	7;24, 3	11	9;51	9;45,41	5
19	3;15	3;14, 6	1	7;42	7;31, 0	11	9;52	9;47,47	4
20	3;25	3;23,55	1	7;48	7;37,49	10	9;55	9;49,42	5
21	3;36	3;33,40	2	7;54	7;44,29	10	9;56	9;51,27	5
22	3;47	3;43,22	4	8; 0	7;51, 1	9	9;56	9;53, 0	3
23	3;57	3;53, 0	4	8; 7	7;57,25	10	9;57	9;54,23	3
24	4; 8	4; 2,33	5	8;13	8; 3,40	9	9;58	9;55,35	2
25	4;19	4;12, 3	7	8;20	8; 9,46	10	9;59	9;56,35	2
26	4;29	4;21,28	8	8;25	8;15,43	9	9;59	9;57,25	1
27	4;39	4;30,48	8	8;31	8;21,31	9	9;59	9;58, 4	1
28	4;49	4;40, 3	9	8;37	8;27,10	10	9;59	9;58,31	0
29	4;58	4;49,13	9	8;41	8;32,39	8	9;59	9;58,48	0
30	5; 7	4;58,18	9	8;47	8;38, 0	9	9; 59	9;58,54	0

M = Madrid, MS 10023; R = MS Ripoll 21 (Millás); S = Sédillot; G = Gil (MS Jews Colege, Heb. 135).Variant readings. 1;14 RSG: 1;54 M. 2;50 MSG: 2;51 R. 2;58 RSG: 2;18 M. 3;25 MRG: 3;24 S. 4;58 RSG: 4;38 M. 5;7 MRG: 5;6 S. 6;41 MRS : 6;40 G. 6;43 R: 7;53 M, 6;53 S, 6;50 G. 6;57 RSG: 7;57 M. 7;48 MSG: 7;54 R. 7;54 MS: 7;58 R, 7;55 G. 8;7 MSG: 8;6 R. 8;20 MSG: 8;19 R. 8;25 MRG: 8;24 S. 9;52 MRS: 9;55 G. 9;56 MRS: 9;58 M. 9;57 MRS: 9;58 G. 9;59 MRS: 9;58 G. 9;59 MRS: 9;58 G. 9;59 MRS: 10;0 G.

subsequent transmission, and that they imply parameters not used in fact in the computation. In the case of the tables by John of Genoa and John of Montfort, the corruption adduced for the last entries affect all the known copies; therefore, it must have affected their common ancestor. As for Ibn al-Kammād, Chabás and Goldstein have clearly demonstrated that many tables in the manuscript which preserves John of Dumpno's translation are corrupt: see, for instance, in the table for the solar equation (pp. 8–9)

Table 2.

(a) arg.	(b) text	(c) R_1	(d) dif.	(e) r= 3;58,19	(f) dif.
5	0;53	0;52	1	0;52	1
10	1;45	1;43	2	1;44	1
15	2;41	2;34	7	2;34	7
20	3;25	3;24	1	3;24	1
25	4;19	4;12	7	4;12	7
30	5; 7	4;58	9	4;59	8
35	5;52	5;42	10	5;43	9
40	6;43*	6;24	19	6;24	19
45	7;14	7; 2	12	7; 3	11
50	7;48	7;38	10	7;38	10
55	8;20	8;10	10	8;10	10
60	8;47	8;38	9	8;38	9
65	9;10	9; 2	8	9; 2	8
70	9;28	9;22	6	9;23	5
75	9;45	9;38	7	9;38	7
80	9;55	9;50	5	9;50	5
85	9;59	9;57	2	9;57	2
90	9;59	9;59	0	9;59	0

the entries corresponding to arguments from 111° to 117°, 119° to 125°, and 131° to 135°; shifts (one place) of entries in the copying process occur in the same table for arguments from 145° to 150°, and also (one or two places) in the table for the adjusted parallax in latitude (p. 21) for arguments from 23° to 30°, 33° to 35°, and 67° to 69°. The table for lunar eclipses (p. 18), copied twice in the Madrid MS (ff. 52v and 57v), is an extreme case of deterioration: for its 36 entries, f. 57v offers 19 variant readings, and the copy edited by Millás (1962, p. 238) adds another 16. Many of these accidents are probably scribal errors; but others (those of the table for solar eclipses, for example, which seems to derive from Yahyā Ibn Abī Mansūr), are also present in Arabic MSS (p. 23).

Could the ancestor of the preserved (Arabic, Latin, and Hebrew) versions of Ibn al-Kammād's table have been affected by these kinds of accidents (corruption and shifts)? The answer to this question, in my opinion, must be positive. Entries for arguments of 7°, 17°, 29°, 40°, and 41° are clearly corrupt[4], and an inspection of the line-by-line differences suggests that at least entries for arguments from 35° to 45° and for 58°–59° are also corrupt. The irregular pattern of the table is clearly shown in Table 3, which displays the differences between successive entries at 5°-intervals in the trepidation

---

[4] Consequently, I have preferred to give alternative readings in columns (b) of Table 1; see the critical apparatus. Although the Madrid MS is the oldest of the four which preserve the table, there is no reason to prefer its readings for 7°, 17°, and 29°, to those of the other (independent and coincident) MSS.

Table 3.

(a) arg.	(b) Thabit	(c) Ibn Ish. Ibn al-B.	(d) R_1	(e) Ibn al-K.
5	0;55,52	0;54	0;52	0;53
10	0;54,44	0;54	0;51	0;52
15	0;54,40	0;53	0;51	0;56*
20	0;54, 7	0;51	0;50	0;44*
25	0;51,49	0;51	0;48	0;54*
30	0;51,18*	0;48	0;46	0;48
35	0;46,36	0;46	0;44	0;45
40	0;44, 6	0;43	0;42	0;51*
45	0;43,23	0;40	0;39	0;31*
50	0;37,25	0;37	0;36	0;34
55	0;33,48	0;33	0;32	0;32
60	0;29,56	0;30	0;28	0;27
65	0;26, 9	0;25	0;24	0;23
70	0;21,37	0;21	0;20	0;18*
75	0;17,17	0;16	0;16	0;17
80	0;12,14	0;12	0;12	0;10
85	0; 7,12	0; 7	0; 7	0; 4*
90	0; 2,47	0; 3	0; 2	0; 0*

tables found in the following sources: (b) the treatise *De motu octaue spere*, attributed in the Middle Ages to Thābit Ibn Qurra (d. 901); (c) the *zījes* of Ibn Ishāq al-Tūnisī (*f l. ca.* 1200), Ibn al-Bannā' al-Marrākushī (1256–1321), and Abū ᶜAbd Allāh Muhammad Ibn al-Raqqām (d. 1315); (d) R_1; and (e) Ibn al-Kammād's *al-Muqtabis* (I have marked with an asterisk the differences that, very likely, involve corrupt values)[5]. The same conclusion about the very poor condition of the table follows when equation (4) is used to obtain the value of P_{max} derived from the entries for P (see Table 4, in which are displayed the results for selected entries of the tables attributed to Thābit, where $P_{max} = 10;45°$ [column (b)], Ibn al-Bannā' and Ibn al-Raqqām, where $P_{max} = 10;24°$ [column (c)], and Ibn al-Kammād [column (d)]). Therefore, it seems obvious that with the textual $P_{max} = 9;59°$ a satisfactory reconstruction of the table will be impossible. In fact, the tabular entries exhibit a certain continuous pattern only for $31°–35°$ ($P_{max} \approx 10;15°$), for $52°–54°$ and $56°–58°$ ($P_{max} \approx 10;10°$), and for $69°–74°$ and $76°–81°$ ($P_{max} \approx 10;4°$).

There is also some implausibility in the claim that Ibn al-Kammād chose Azarquiel's second model for trepidation. His work, like those of the aforementioned XIIIth century astronomers, was strongly dependent on Azarquiel's authority, who in his treatise *On the fixed stars* adduced several reasons to reject his first two models and asserted that

---

[5] On the table in *De motu*, see Millás, 1943–1950, p. 508, and Neugebauer, 1962a, p. 297. On Ibn Ishāq's, Ibn al-Bannā's, and Ibn al-Raqqām's table, see Chabás & Goldstein, 1994, p. 27, and Samsó & Millás, 1994, pp. 12–13.

Table 4.

(a) arg.	(b) Thābit	(c) Ibn Ishāq Ibn al-B.	(d) Ibn al-K.
5	10;44,44	10;22,57	10;11,17
10	10;40,30	10;25,17	10; 7,44
15	10;42, 1	10;25,16	10;25,16
20	10;44,46	10;22,51	10; 2, 5
25	10;44,49	10;25, 8	10;15,33
30	10;47,52	10;24,34	10;16,28
35	10;46, 4	10;24,43	10;15,54
40	10;45, 3	10;24,18	10;29, 1
45	10;47,38	10;25,24	10;15,25
50	10;46,27	10;24, 6	10;12,16
55	10;45,35	10;23,43	10;11,27
60	10;44,58	10;24,24	10; 9,19
65	10;44,55	10;24, 1	10; 7,25
70	10;44,48	10;24, 0	10; 4,49
75	10;45, 0	10;23,28	10; 5,50
80	10;44,55	10;23,34	10; 4,16
85	10;44,41	10;23,23	10; 1,18
90	10;45, 0	10;24, 0	9;59, 0

only the third one satisfied all the appropriate requirements (Millás, 1943–1950, pp. 319–21). Moreover, Abū Muḥammad ᶜAbd al-Ḥaqq al-Ghāfiqī al-Ishbīlī, known as Ibn al-Hā'im, in his *zīj al-Kāmil fī-l-taᶜālīm*, written *ca.* 1204–1205, criticized Ibn al-Kammād's inconsistency in computing a trepidation table without taking into account the variation in the obliquity of the ecliptic (Samsó, 1992, p. 322). If Ibn al-Hā'im's criticism was correct, it is very likely that Ibn al-Kammād, as Ibn Isḥāq, Ibn al-Bannā' and Ibn al-Raqqām did later, computed his trepidation table assuming Azarquiel's third model and (inconsistently) a constant obliquity.

3. The Trepidation Model in Ibn al-Kammād's Castilian Text

The text is preserved in only one known copy of the fifteenth century (Segovia, MS 115, ff. 218vb:23–220vb:14). It was identified by Beaujouan (1968, pp. 17–18), according to whose description the codex also contains some Alfonsine texts (Book VIII of the *Libro conplido en los iudizios de las estrellas* by Aly Aben Ragel, and the *Libro de las cruzes*, both translated by Yehudah B. Moshe ha-Kohen), Castilian versions of the Tortosa almanach for 1307 (whose Latin version was edited by Millás, 1943–1950, pp. 396–400), and al-Qabīsī's *Introductorium*, and tables of true syzygies for 1431–1460. The date and translator's name of Ibn al-Kammād's text are not given. The title of the Arabic original work is not mentioned; but it is probably a chapter of one of the two *zījes* composed by Ibn al-Kammād[6], although it does not contain instructions on the use of tables, as often occurs in the canons. The syntax of the text is difficult and its meaning, sometimes, obscure.

The trepidation model is summarily described in the part of the text edited in the Appendix. A circle is drawn on the concave surface of the sphere of the equator, whose pole is the pole of that sphere, and whose radius [lit. diameter] is 23;33°. There is another circle with center on the circumference of the first one and whose diameter is 0;22°[7]. An end point of the diameter of the equator moves on this circle with an uniform motion around its center, and it bears an end point of a diameter of the sphere of the signs (i.e. a pole of the ecliptic) and also an end point of the diameter lying between the centers of the small circles whose centers are the mean first points of Aries and Libra. The diameter of these small circles, which move with uniform motion and produce the equations of access and recess, is 8°. Although no number of years is mentioned for the revolutions of the polar and equatorial epicycles, it is beyond doubt that their periods are identical[8]: when the pole of the ecliptic is situated on the mean longitude of the small circle (i.e., at mean distance between the points producing maximum and minimum obliquity), the

---

[6] The Castilian text has no title. It is entirely devoted to trepidation and its first sentence asserts: "this book on the circumference of motion composed for a centennial time was made by master Yūsuf Ibn al-Kammād..." [Este libro fizo maestre Yuçuf Benacomed sobre çircunferençia de moto sacado por tiempo seculo...]. The expression "çircunferençia de moto" seems to allude to al-Kawr ᶜalā al-dawr, translated in Chabás & Goldstein (1994, p. 3) as "the periodic rotations". However, according to the introductory section of the Latin version of al-Muqtabis(Millás, 1942, pp. 231–2), Ibn al-Kammād's largest work was al-zīj al-Amad ᶜalā al-abad, from which al-zīj al-Kawr ᶜalā al-dawr was extracted. The chapter devoted to trepidation in the Latin version of al-Muqtabis explains only how to use the table and claims that a detailed discussion of the subject can be found in al-zīj al-Amad ᶜalā al-abad ["Et nos iam narrauimus in puncto equinoctii et recessionis cius et in motu duorum capitum que sunt inter ipsos per diuersitatem toclus declinationis circuli signorum in omni tempore, et narrauimus eciam memorationem declinationis puncti capitis arietis a circulo equinoctii in meridie et septentrione ab ea, sed nos iam tractauimus super expansione eius in canone helmed hale elelbed, et si intrauimus memorationem eius in hoc canone exiebamus a termino transitus et breuitatis et tendebamus ad prolixitatem et ad multitudines"; Madrid, Biblioteca Nacional, MS 10023, f. 7va-b]; thus, it is likely that the Castilian text translates a chapter of that work.

[7] 23;33° for the radius of the first circle is probably a copyist's error; read 23;43°, as in Azarquiel's third model. The reading 0;22° is not clear, and 0;23° is also possible. Accepting 0;22°, the values for maximum and minimum obliquity are 23;54° and 23;32° (23;53° and 23;33° in Azarquiel's model where the diameter of the small circle is 0;20°). Azarquiel's mimimum obliquity was modified by some of his successors. ᶜAlī b. Khalaf determined an obliquity of 23;32,12° for 1084–85 according to Ibn Ishāq, whose zīj contains a table for solar declination based on ε = 23;32,30°. This author attributed to Azarquiel a value of 23;32,31° for 1074–75. Ibn al-Raqqām seems to have accepted for his time 23;32,40° (cf. Samsó, 1992, pp. 181, 210, 232, 422 and 425).

[8] See the Appendix, especially the sentence: "Et quando llegare el exe a la longura mediana onde ouo començado a mouerse ajuntá[n]se los dos puntos de la cabeça de Aries et de la eguación en uno onde començaron su mouimiento, et entonçe non avrá auenimiento nin arredramiento et será el comienço de los dos motos un comienço".

mean and movable first points of Aries coincide and the equation of trepidation is $0°$; when the pole of the ecliptic moves on the small circle towards the point producing the maximum obliquity[9], the equation is positive (access), increasing until it reaches its maximum when the ecliptic pole is situated at that point, and decreasing beyond it until the pole reaches the opposite mean longitude and the equation is again $0°$; the equation is negative (recess) when the pole of the ecliptic moves on the second half of the small circle, and it reaches its maximum when the pole is situated on the point determining the minimum obliquity. Thus, this section seems to confirm Ibn al-Hā'im's views in criticizing Ibn al-Kammād for deviating from Azarquiel's authority when he assumed the periods of revolution of the movable first point of Aries and the pole of the ecliptic to be identical (cf. Samsó, 1992, p. 322)[10].

Table 5 (to be compared with Table 1) displays the results of the reconstruction [R_2] of Ibn al-Kammād's trepidation table based on equations [1] and [2] with $r = 4$, as in the Castilian text, and $\varepsilon = 23;33°$, the same value used by Ibn Ishāq, Ibn al-Bannā', and Ibn al-Raqqām (there is no significant improvement using the minimum value for the obliquity required by the text, $23;32°$). As it can be seen, the approximation is clearly better than using equation (3) with $r = 10;24$, excepting obviously the last five entries of the table; see Table 6, where the first row displays differences between text and recomputation in minutes, and the other two the number of occurrences of such differences in R_1 and R_2. These differences are $\leq 2'$ for arguments from $1°$ to $11°$, $17°$ to $23°$, and $67°$ to $86°$, and $\leq 1'$ from $46°$ to $67°$ if we assume, as R_2 seems to indicate, that those entries have been shifted one place downwards. The Castilian text thus provides a very likely reconstruction of the table [11], which can be perhaps confirmed by an Arabic source still unexplored.

Appendix[12]

Por que digo que la causa de la diuersydat entre las oppiniones es de la propiadat de los mouimientos de los 2 exes del moto de los signos que se mueuen en alto et en baxo, por el qual uiene el mouimiento de la cabeça de Aries et de Libra, que es el abenimiento

---

[9] This is a tentative translation. The text uses the expression "escontra la parte de la longura más çerca" (lit. towards the place of the nearest longitude [or distance]). The opposite point is called "la longura luenga".

[10] The periods of revolution of these circles in Azarquiel's third model are, respectively, 3874 and 1850 Julian years (Millás, 1943–1950, pp. 318 and 325). The Madrid MS contains (f. 28v) a table for the mean motion of the first point of Aries identical to Azarquiel's (Chabás & Goldstein, (1994, p. 24); however, it lacks a table for the mean motion of the pole of the ecliptic.

[11] There is no way to decide if the variant reading for P_{max} in Gil's table, $10;0°$, is a trace of the lost original form suggested by R_2 (with $P_{max} \approx 10;4°$), or merely another accident.

[12] I have worked on reproductions of folios 218–20 of the Segovia MS from a poorly microfilmed copy. Only approximately half of Ibn al-Kammād's text is edited here. I have retained the spelling of the manuscript, but accentuation, punctuation, and capitalization are mine. Editorial insertions are enclosed in square brackets.

Table 5.

(a) arg.	(b) text 0ˢ	(c) R₂	(d) dif.	(b) text 1ˢ	(c) R₂	(d) dif.	(b) text 2ˢ	(c) R₁	(d) dif.
1	0;10	0;10,29	0	5;16	5; 9,43	6	8,52	8;47, 6	5
2	0;20	0;20,57	1	5;25	5;18,41	6	8;56	8;52, 9	4
3	0;31	0;31,26	0	5;34	5;27,34	6	9; 1	8;57, 2	4
4	0;41	0;41,54	1	5;43	5;36,20	7	9; 5	9; 1,46	3
5	0;53	0;52,21	1	5;52	5;45, 1	7	9;10	9; 6,19	4
6	1; 3	1; 2,47	0	6; 4	5;53,35	10	9;14	9;10,43	3
7	1;14	1;13,13	1	6;16	6; 2, 3	14	9;17	9;14,57	2
8	1;25	1;23,36	1	6;29	6;10,24	19	9;21	9;19, 0	2
9	1;35	1;33,58	1	6;41	6;18,39	22	9;24	9;22,53	1
10	1;45	1;44,19	1	6;43	6;26,47	16	9;28	9;26,36	1
11	1;56	1;54,38	1	6;57	6;34,48	22	9;31	9;30, 8	1
12	2; 8	2; 4,54	3	7; 2	6;42,42	19	9;35	9;33,30	1
13	2;19	2;15, 9	4	7; 6	6;50,28	16	9;38	9;36,42	1
14	2;30	2;25,21	5	7;10	6;58, 7	12	9;41	9;39,43	1
15	2;41	2;35,30	5	7;14	7; 5,39	8	9;45	9;42,33	2
16	2;50	2;45,37	4	7;21	7;13, 3	8	9;47	9;45,12	2
17	2;58	2;55,41	2	7;28	7;20,19	8	9;49	9;47,41	1
18	3; 6	3; 5,41	0	7;35	7;27,27	8	9;51	9;49,59	1
19	3;15	3;15,39	-1	7;42	7;34,26	8	9;52	9;52, 6	0
20	3;25	3;25,32	-1	7;48	7;41,18	7	9;55	9;54, 2	1
21	3;36	3;35,23	1	7;54	7;48, 1	6	9;56	9;55,47	0
22	3;47	3;45, 9	2	8; 0	7;54,36	5	9;56	9;57,21	1
23	3;57	3;54,51	2	8; 7	8; 1, 2	6	9;57	9;58,44	2
24	4; 8	4; 4,29	4	8;13	8; 7,19	6	9;58	9;59,56	2
25	4;19	4;14, 3	5	8;20	8;13,28	7	9;59	10; 0,57	1
26	4;29	4;23,32	5	8;25	8;19,27	6	9;59	10; 1,47	2
27	4;39	4;32,56	6	8;31	8,25,17	6	9;59	10; 2,26	3
28	4;49	4;42,16	7	8;37	8;30,59	6	9;59	10; 2,54	4
29	4;58	4;51,30	7	8;41	8;36,30	4	9;59	10; 3,11	4
30	5; 7	5; 0,39	6	8;47	8;41,53	5	9;59	10; 3,16	4

et el arredramiento de las estrellas fixas. Et es otrosy con la diuersydat del abaxamiento del moto de los signos uniuersal ct con el mudamiento de las sus eleuaçiones.

Et la causa desta es que ymaginemos una rrueda figurada en la faz de la concauydat del diámetro en los forados del exe del moto del eguador del día, que es el moto grande syn figura de ninguna estrella. Et los forados destos dos exes del moto del eguador del día son forados del moto menor et su diámetro es de 23 grados et 33 menudos. Et sobre su çircunferençia es una rrueda pequenna, el su çentro ligado a su çircunferençia. Et su diámetro es 22 menudos. Et esta rrueda faze la diuersydat general et muéuese aderredor de su çentro su mouimiento igual et lieua consigo el diámetro de la espera del eguador del día. Et este diá[219vb]metro lieua otrosy consigo el diámetro del moto de los signos et lieua consigo dos rruedas que se catan en el çentro de la espera del mundo, la una en la cabeça de Aries et la otra en la cabesça de Libra. En la parte del mouimiento son los çercos que fazen el abenimiento et el arredramiento de la cabesça de Aries et de Libra en los dos puntos de las 2 eguaçiones que en la faz de la concauydat del moto figurado

Table 6.

d	−1	0	1	2	3	4	5	6	7	8	9	10	11
R_1	0	5	9	8	2	5	7	10	5	6	13	6	4
R_2	2	6	20	10	4	10	7	11	6	5	0	1	0

en los 2 forados de los exes del moto del eguador del día. Et estos forados destos 2 exes sotienen el zodiaco. Et su diámetro es 8 grados. Et muéuense sobrestos forados destos exes mouimiento egual et traen consigo los 2 puntos de las dos eguaçiones. Et rrodean otrosy los 2 cabos del diámetro del moto del eguador del dia que son los tajamientos de los 2 puntos de las dos eguaçiones a sus tenencias de sus diámetros.

Et quando fueren los forados de los exes del moto estrellado en su arredramiento medianero en la diuersydat de la declinaçión general, fázense los dos puntos de las dos rruedas de la cabesça de Aries et del eguador del dia una differençia, la una del moto mayor syn fygura et la otra el moto estrellado de los signos. Et la rrueda que faze el abenimiento et el arredramiento non le fallan entonçe [220ra] abenimiento. Et entonçe será el moto mayor el moto estrellado un comienço. Et entonçe tírase el forado del exe escontra la parte de la longura mas çerca, creçiendo contra la longura mediana, partiéndose el punto de la cabeça de Aries del punto de la eguaçión et el punto de la eguaçión otrosy de la cabeça de Aries, la una rrueda que viene et la otra que se arriedra. Et el punto de la cabeça de Aries comiénçase de mouer desde que se parte del punto del diámetro del eguador del día medio septentrional. Et el punto del eguador del día comiença a se mouer desque se parte del punto de la rrueda del diámetro de la cabeça de Aries, arredrándose escontra la otra parte de los signos do llega el exe a la longura mediana et júntanse los dos puntos en uno, el uno la rrueda de los signos et el otro la rrueda del eguador del día. Et entonçe non ay ninguna rrueda a abenimiento nin arredramiento. Et entonçe será el comienço de la rrueda et el comienço del eguador del díia un comienço. Et entonçe se alçó el exe escontra la longura luenga yendo escontra la longura mediana onde començaron el mouymiento. Et [220rb] aquí se parten los puntos, el punto de la cabeça de Aries del punto de la eguaçión, la una rrueda que uiene et la otra que se arriedra, et el punto de la cabesça de Aries comiença de se arredrar. Et esto tal es la su declinaçión del medio oriental del punto del díametro de la eguaçión del moto del eguador del día, yendo en diuerso de lo que non los sygnos del punto de la eguaçión uiene del punto del diámetro de la rrueda oriental del punto de la cabeça de Aries commo uienen los sygnos. Et quando llegare el exe a la longura mediana onde ouo començado a mouerse ajúnta[n]se los dos puntos de la cabesça de Aries et de la eguaçión en uno onde començaron su mouimiento, et entonçe non avrá auenimiento nin arredramiento et será el comienço de los dos motos un comienço. Et quando fuere la cabesça de Aries[1] auenido será el punto de la eguaçión su opposito et será la fin del Sol con el punto de la eguaçión ante que llege al punto de la cabesça de Aries. Et quando fuere el punto de la eguaçión auenido será el punto de la cabesça de Aries su opposito et será el llegamiento del sol con el punto de la cabesça de Aries ante que llege con el [220va] punto de la eguaçión.

1. *scr. et del.* un comienço.

Table 6 (*contd.*)

12	13	14	15	16	17	18	19	20	21	22	23	24	25
1	1	0	1	0	1	0	2	0	1	0	0	1	2
1	0	1	0	2	0	0	2	0	0	2	0	0	0

References

Beaujouan, G. 1968. "Manuscrits scientifiques médiévaux de la Cathédrale de Ségovie", in *Actes du XIᵉ Congrès Internationale d'Histoire des Sciences*, Varsovie, pp. 15–18.

Chabás, J. and Goldstein, B. R. 1994. "Andalusian Astronomy: *al-zīj al-Muqtabis* of Ibn al-kammād", *Archive for History of Exact Sciences* **48**: 1–41.

Goldstein, B. R. 1964. "On the Theory of Trepidation According to Thābit b. Qurra and al-Zarqāllu and Its Implications for Homocentric Planetary Theory", *Centaurus* **10**: 232–247.

Goldstein, B. R. 1980. "Solar and Lunar Velocities in the Alphonsine Tables", *Historia Mathematica* **7**: 134–140.

Goldstein, B. R. 1992. "Lunar Velocity in the Ptolemaic Tradition", in P. M. Harman & A. E. Shapiro (eds.), *The investigation of difficult things. Essays on Newton and the history of the exact sciences*, Cambridge, pp. 3–17.

MillásVallicrosa, J. M. 1942. *Las traducciones orientales en los manuscritos de la Biblioteca Catedral de Toledo*, Madrid.

MillásVallicrosa, J. M. 1943–1950. *Estudios sobre Azarquiel*, Madrid-Granada.

MillásVallicrosa, J. M. 1962. *Las tablas astronómicas del rey Don Pedro el Ceremonioso*, Madrid-Barcelona.

Neugebauer, O. 1962a. "Thābit ben Qurra 'On the Solar Year' and 'On the motion of the Eighth Sphere' ", *Proceedings of the American Philosophical Society* **106**: 264–299.

Neugebauer, O. 1962b. *The Astronomical Tables of al-Khwārizmī*, København.

Samsó, J. 1992. *Las ciencias de los antiguos en al-Andalus*, Madrid.

Samsó, J., and Millás, E. 1994. "Ibn al-Bannā', Ibn Ishāq and Ibn al-Zarqālluh's Solar Theory", in J. Samsó, *Islamic Astronomy and Medieval Spain*, Great Yarmouth, Norfolk, X: 1–35.

Sédillot, J.-J., and Sédillot, L.-A. 1983. *Traité des instruments astronomiques des Arabes*, Paris. Repr. Frankfurt, 1984.

A note
on Copernicus' 'correction'
of Ptolemy's mean synodic month

Introduction.

In *A History of Ancient Mathematical Astronomy*[1], Neugebauer wrote that the equation

126007 (=35,0,7) days 1 hour = 4267 (=1,11,7) synodic months   (1)

was supposedly the source of Ptolemy's value (*Almagest* IV.2)[2] for the length of the mean synodic month (29;31,50,8,20^d), and he added that Copernicus[3] was the first to check Ptolemy's arithmetic to find that equation (1) in fact leads to

---

[1]   Neugebauer 1975, p. 310.

[2]   Ptolemy's text is ambiguous: "For from the observations he [= Hipparchus] set out he shows that the smallest constant interval defining an ecliptic period in which the number of months and the amount of [lunar] motion is always the same, is 126007 days plus 1 equinoctial hour. In this interval he finds comprised 4267 months, 4573 complete returns in anomaly, and 4612 revolutions on the ecliptic less about 7 ½°, which is the amount by which the sun's motion falls short of 345 revolutions (here too the revolution of sun and moon is taken with respect to the fixed stars). (Hence, dividing the above number of days by the 4267 months, he finds the mean length of the [synodic] month as approximately 29;31,50,8,20 days). He shows, then, that the corresponding interval between two lunar eclipses is always precisely the same when they are taken over the above period [126007^d 1^h]" (Toomer 1984, pp. 175-6).

[3]   *De revolutionibus* IV.4 (Nuremberg 1543, 101v): "Quapropter idem Hipparchus ulterior ista perquisiuit, nempe collatis adnotationibus, quas in eclipsibus lunaribus

1 mean synodic month = 29;31,50,8,9,20^d

and not to 29;31,50,8,20^d, a parameter that is now known to come from Babylonian system B.[4]

Later, in their *Mathematical Astronomy in Copernicus's De revolutionibus*, Swerdlow and Neugebauer pointed out that, according to Ptolemy, Hipparchus did not derive the mean motions of his lunar theory from equation (1), but merely accepted the Babylonian periods

1 synodic month = 29;31,50,8,20^d,

251 synodic months = 269 anomalistic months,

5458 synodic months = 5923 draconitic months,

adding that Copernicus, however, following Regiomontanus's implicit misunderstanding in *Epitome* IV.3[5], took the cycle as a source of Hipparchus's mean synodic month, and computed that

> diligentissime obseruauit, ad eas quas à Chaldaeis accepit: tempus in quo reuolutiones mensium et anomaliae simul reuerterentur, definiuit esse CCCLXV annos Aegyptios, LXXXII dies, & unam horam, & sub eo tempore menses IIII.CCLXVII, anomaliae uero IIII.DLXXIII circuitus compleri. Cum ergo per numerum mensium distributa fuerit proposita dierum multitudo, suntque centena vigintisex millia & vii dies atque una hora, inuenitur unus mensis aequalis dierum XXIX, scrup. primorum XXXI, secundorum L, tertiorum VIII, quartorum IX, quintorum XX. Qua ratione patuit etiam cuiuslibet temporis motus. Nam diuisis CCCLX unius menstruae reuolutionis gradibus per tempus menstruum, prodijt diarius Lunae cursus à Sole gradus [X]II, scrupula prima XI, secunda XXV, tertia XLI, quarta XX, quinta XVIII." The Basel edition of 1566 gives "vigintisex milia & xii dies"; Copernicus' autograph manuscript of *De revolutionibus* (Kraków, Biblioteka Jagiellońska, MS 10000, f. 110r) "vigintisex milia et vij dies".

[4] Cf. Kugler 1900, pp. 7, 24 and 111; Aaboe 1955; Neugebauer 1955, pp. 75-8.

[5] "Hyparchus autem quantitatem huius interualli reperit 126007 dies et horam unam et in hoc interuallo fuerunt menses lunares 4267, quod facile per numerum nouiluniorum considerare potuit. Reditiones autem in circulo diuersitatis fuerunt 4573, quod etiam per motus lune conditionatos tardum medium uelocem et medium deprehendit. Reditiones uero in orbe signorum 4612 minus septem gradibus et medietate fere. Tantum enim sol minuit in 347 reuolutionibus huius temporis, eo quod in reditionibus istis processum est in relatione ad stellas fixas. Interuallum itaque dierum diuisum per numerum mensium ostendit quantitatem unius mensis lunaris" (*Epytoma Joannis De monte regio In almagestum ptolomei* IV.3, Venice, 1496; repr. 1972, p. 116). Note however that Regiomontanus does not give in the text the result of the division 126007^d 1^h / 4267.

1 mean synodic month $= 126007^{d}\ 1^{h}\ /\ 4267 = 29;31,50,8,9,20^{d}$,   (2)

a mistake that has been repeated in modern literature, entirely independent of Copernicus. He then claims that dividing this number into 360° gives a mean daily elongation of 12;11,26,41,20,18°, although the division would correctly give 12;11,26,41,24,42°. In fact, and fortunately, Copernicus has merely rounded from the value in the Almagest, based upon the correct synodic month used by Hipparchus, that is,

$\eta^{d} = 6,0^{\circ}\ /\ 29;31,50,8,20^{d} = 12;11,26,41,20,17,59^{o/d}$.

Thus he saves himself from the error of computing what he believes to be Hipparchus's mean motions from the wrong synodic month.[6]

The object of this note is (i) to show that the parameter $29;31,50,8,9,20^{d}$ was widely considered throughout the Latin Middle Ages as the 'correct' Ptolemaic value and simply taken from Gerard of Cremona's Latin translation of the *Almagest* (depending on this point on al-Ḥajjāj ibn Maṭar's Arabic version), and (ii) to suggest that very likely Copernicus never checked the computation, but merely took these inconsistent parameters for the synodic month and the daily mean elongation from the Latin version of the *Almagest*[7].

The Arabic, Hebrew, and Latin traditions of *Almagest* IV.2.

Only two Arabic translations of the *Almagest* are extant: one dated 827/8 by al-Ḥajjāj, the other completed ca. 879-90 by Isḥāq ibn Ḥunayn, later revised by Thābit ibn Qurra (d. 901). Al-Ḥajjāj's translation has ...8,9,20^{d}

---

[6] Swerdlow and Neugebauer 1984, pp. 198-9. Pedersen (1974, pp. 162-3) also mentioned Copernicus's passage, asserting that an easy explanation for the discrepancy between Ptolemy's $29;31,50,8,20^{d}$ and the correct result in equation (2), ...8,9,20^{d}, is to assume that the parameter ...8,20^{d} was not derived from equation (1). When discussing these issues with me, B.R. Goldstein suggested a reasonable solution to the puzzle: Ptolemy had no intention of changing the Babylonian parameter, and computed $29;31,50,8,20^{d} \times 4267 = 35,0,7;2,42,38,20^{d} = 126007^{d}\ 1;5,2,41,..^{h}$, rounding this result to $126007^{d} + 1^{h}$.

[7] Copernicus used Gerard of Cremona's translation (a copy of the Venice edition, 1515, annotated by him is preserved at Uppsala) and also Trebizond's version, first published at Venice in 1528. In 1539 Rheticus brought Copernicus the 1538 edition of the Greek text with Theon's commentary, but it is assumed that the use Copernicus could make of this at so late a date was limited (Swerdlow and Neugebauer 1984, p. 92).

(MS Leiden, Or. 680, f. 50b: 6), whereas Isḥāq-Thābit's version has Ptolemy's figure ...8,20^d (MS Tunis, Bibliothèque Nationale, 07116, f. 53b:19-20; there is at this place in the manuscript a marginal note which reads "in the translation of al-Ḥajjāj nine fourths and twenty fifths", thus confirming both readings)[8]. It is likely that al-Ḥajjāj was embarrassed that equation (1) did not produce the expected result, and so he silently changed Ptolemy's text for the length of the mean synodic month, not appreciating the meaninglessness of the correction. In his *al-Qānūn al-Mas'ūdī* [9], al-Bīrūnī (973-1048) gives ...8,9,20,13^d (a value very close to the accurate result: 29;31,50,8,9,20,12,22...^{d}). Jābir ibn Aflaḥ, in his *Iṣlāḥ al-Majisṭī* (middle of the XIIth century), also translated by Gerard of Cremona and frequently cited in Western Europe, gives ...8,9,20^{d10}. Naṣīr al-Dīn al-Ṭūsī (d. 1274), in his *Taḥrīr al-majisṭī*, when discussing the mean synodic month, quotes the Babylonian parameter given in *Almagest* IV.2 and comments that instead of 29 days, 31 minutes, 50 seconds, 8 thirds, and 20 fourths, "Ḥajjāj's copy [of the *Almagest* had the value] 9 fourths, 20 fifths and 12 sixths, which was the correct [value]"[11]. A slightly different value is found in Ibn Yūnus' *al-Zīj al-Ḥākimī* (about 990)[12], where the length of Muḥarram is given as 29;31,50,8,9,24^d. The same value, ...8,9,24^d, was attributed to Ptolemy by al-Biṭrūjī (end of the XIIth c.) in his *Kitāb fī-l-hay'a*[13], a work translated into Latin by Michael Scot at Toledo in 1217 (*De motibus coelorum*) and often mentioned in thirteenth and fourteenth century Scholastic discussions of the Ptolemaic

---

[8] This information was kindly supplied by P. Kunitzsch.

[9] al-Bīrūnī 1954-6, book 7, ch. 2, vol. 2, p. 730. I wish to thank B. R. Goldstein for this information.

[10] "Cum ergo diuiserunt istos dies quos inuenerunt huic tempori reuolubili per numerum mensium qui sunt in eo, exiuit tempus mensis medij 29 dies, et 31 minutum [sic], et 50 secunda, et 8 tertia, et 9 quarta, et 20 quinta cum propinquitate..." (*Gebri Filii Affla Hispalensis Astronomi uetustissime pariter et peritissimi, libri IX de Astronomia...*, Nuremberg: Petreius, 1534, f. 49r).

[11] India Office MS No. Loth 741, f. 19v; quoted by Saliba 1987, p. 150.

[12] MS Leiden, Cod. Or. 143, p. 20 (I am grateful to B.R. Goldstein for checking this manuscript for me); see also Delambre 1819, p. 96, and Neugebauer 1979, p. 18.

[13] However, the values for the motion of the moon in longitude during a synodic month, the mean daily motion of the moon, and the daily increment in elongation given by al-Biṭrūjī agree with the corresponding parameters in the *Almagest*; cf. Goldstein 1971, 1: 145.

225   *A Note On Copernicus' 'Correction' Of Ptolemy's Mean Synodic Month*

system. It was also used by Abū Shāker in his *Ḥasāba ʿālam* (or *Chronology*; ca. 1256)[14].

Al-Khwārizmī (d. ca. 850) informs us that the Jewish calendar used a mean synodic month of 29 days and 12 793/1080 hours (each of the 1080 equal parts of an hour was called in Hebrew a *ḥeleq*, a unit ultimately of Babylonian origin)[15], that is, in sexagesimal form, $29;31,50,8,20^d$, and an identical report can be found in al-Bīrūnī[16]. Abraham Bar Ḥiyya, in his *Sefer Ḥeshbon mahlekot ha-kokabim* (*Calculation of the celestial motions*; ca. 1136), gives 12 793/1080 hours[17]; Abraham Ibn Ezra (1089-1164), in his *Liber de rationibus tabularum*[18], writes that the lunar (synodic) month is 29 days and 12;44,3,20 hours, identical to ...$8,20^d$, and, in the *Sanctification of the Moon*, Maimonides (1135-1204) also uses 29^d and 12 793/1080 hours[19]. However, in Jacob Anaṭoli's Hebrew version of the *Almagest* (ca. 1230-36) we find ...$8,9,20^d$ (MS Turin, Biblioteca Nazionale Universitaria, A.II.10, f. 40v:27), the parameter from Ḥajjāj's Arabic version[20], and this agrees with the apparent dependence of Anaṭoli's work on the Latin translation by Gerard of Cremona[21]. In subsequent astronomical literature, this 'corrected' value is always given when Ptolemy's *Almagest* is quoted. Thus, for instance, Levi ben Gerson (1288-1344) in chapter 64 of his *Astronomy* (*Milḥamot Adonai* V.1), when introducing his own value for the synodic month ($29;31,50,7,54,25,3,32^d$ or 29^d, $12;44^h$ and nearly 1/1138 of 'an hour), attributes ...$8,9,20^d$ to Ptolemy and ...12 793/1080 hours to "our ancient scholars"[22].

---

[14] Neugebauer 1987, p. 280.

[15] Kennedy 1964, p. 55. On the Babylonian origin of the division of the hour into 1080 parts, see Neugebauer 1956, p. 117.

[16] al-Bīrūnī 1879, p. 143.

[17] Millás Vallicrosa 1959, p. 55.

[18] Millás Vallicrosa 1947, p. 99: "Et hoc potest probari nam in mense lunari qui est ab adunatione solis et lune cum cursu medio suo donec iterum coniungantur sunt 29 dies et 12 hore et 44 puncta [*sic*; read: minuta] hore et medietas none minuti."

[19] Neugebauer 1949, p. 326; Neugebauer 1956, p. 114.

[20] I am grateful to M. Zonta for checking this manuscript.

[21] Zonta 1993, p. 332.

[22] "Et dico quod Ptolomeus declarauit experientijs antiquorum et suis, et Abarcas [= Hipparchus] ante eum declarauit hoc idem, scilicet, quod tempus medij mensis lunaris est $29;31,50,8,9,20^d$. Et nos inuenimus istum computum ita ueritati propinquum quod in

It is well known that Gerard of Cremona's translation of the *Almagest* was made in 1175 using the Arabic al-Ḥajjāj's version for Books I-IX, and Isḥāq-Thābit's version for Books X-XIII[23]. Consequently, Gerard of Cremona gives the 'corrected' parameter ...8,9,20^d, which we can find in the manuscript tradition as well as in the printed edition[24]. The same value is given in other widely used Latin texts, as the well known *Almagestum parvum* (or *Almagesti minoris libri VI*)[25]. George of Trebizond's

toto de cursu temporis a Ptolomeo usque ad presens non inuenitur defectus nisi 0;12°, in quibus inuenimus distantiam lune a sole in tempore nostro maiorem quam esse deberet secundum computum Ptolomei. Et iste computus quasi consentit computui cui consenserunt sapientes nostri antiqui, qui ponebant tempus mensis lunaris 29 dies, 12 horas, 793 puncta, atribuendo 1080 puncta hore cuilibet, que sunt 29;31,50,8,20^d; qui computus excedit computum Ptolomei in 0;0,0,0,10,40^d [...] Set nos in hoc considerantes subtiliter per experientias antiquorum et nostras, ut declarabitur in futuro, ista inquisitione completa inuenimus tempus medij mensis lunaris 29;31,50,7,54,25,3,32^d [...] Et secundum computum nostrum esset mensis lunaris 29 dies, 12;44^h et circa unum punctum atribuendo hore 1138 puncta" (Vat. Lat. 3098, f. 57rb). See also Mancha 1998, pp. 307-9. According to B.R. Goldstein, Levi (along with medieval Jews in general) believed that the length of the month in the Jewish calendar was already used by the rabbis of the Talmudic period, if not earlier, as attested, for example, in Judah Halevi's *Kuzari* (12th c., Spain): "The calendar, based on the rules of the revolutions of the moon, as handed down by the House of David, is truly wonderful. Though [the medieval Hebrew translation adds: "thousands and"] hundreds of years have passed, no mistake has been found in it, whilst the observations of Greek and other astronomers are not faultless. They were obliged to insert corrections and supplements every century, whilst our calendar is always free from error, as it rest on a prophetic tradition." (Hirschfeld 1905, p. 123).

[23] See, e.g., Kunitzsch 1974, pp. 99-102. Two Latin versions of the *Almagest* made from the Greek are extant (cf. Haskins 1924, pp. 103-10), which I have not checked. A fourth version, probably from the Arabic (Haskins 1924, p. 108) and of Spanish origin prior to the early thirteenth century, is preserved only in fragments.

[24] MS Memmingen F.33, f. 39v: "...per 4267 menses prouenit enim numerus dierum mensis lunaris 29 dies et 31 minuta et 50 secunda et 8 tertia et 9 quarta et 20 quinta fere..."; printed edition, Venice: Lichtenstein, 1515, f. 36r: "Et ex hoc inuenit Abrachis tempus medium mensurnum lunare, vbi diuisit numerum horum dierum per quatuor milia ducentos et sexagintaseptem menses. Prouenit enim numerus dierum mensis lunaris 29 dies et 31 minuta et 50 secunda et 8 tertia et 9 quarta et 20 quinta fere".

[25] "...et est hic numerus prefinito tempore 4573 reuersiones diuersitatis. Hijs itaque cognitis, numerus dierum et unius hore inter duas eclipses per numerum mensium diuidendus et exibit tempus equalis lunacionis, et est sicut ex premissis deprehenditur 29;31,50,8,9,20^d..." (*Almagestum parvum*, IV.3, MSS British Library, Harley 625, f. 101v, and Prague, Univ. V.A.11, f. 24r). However, the copy of this work in MS Memmingen F.33, f. 169v, has 29;31,50,8,9,25^d, very close to the value given by Ibn Yūnus, al-Biṭrūjī, and Abū Shāker (see above).

translation has also ...8,9,20^d, despite claiming to have been made directly from the Greek[26]. Once Ptolemy's Greek text was printed in 1538[27], the two values were clearly distinguished and their sources identified, as is attested in the marginal notes in some copies of printed editions of *De revolutionibus*, now attributed to Paul Wittich (ca. 1550-87) and some time ago wrongly to Tycho Brahe[28]. So, in the copy of *De revolutionibus* preserved at the University Library at Prague[29], in the margin of f. 101v, next to the passage where Copernicus asserts that Ptolemy's mean synodic month is 29;31,50,8,9,20^d, there is the annotation: *Sic habet translatio Arabica, sed graeca sic: 29;31,50,8,20^d, qua et usus Ptolemaeus hinc colligit diurnum motum distantiae lunae a solis 12;11,26,41,20,18°* [30], and in the text, the value for the elongation of the Moon, 12;11, 26,41,20,18$^{o/d}$, rounded from Ptolemy's result of the subtraction of the mean daily motion of the Sun from the mean daily motion of the moon in longitude, namely

$$13;10,34,58,33,30,30^{o/d} - 0;59,8,17,13,12,31^{o/d} = 12;11,26,41,20,17,59^{o/d}$$

is corrected to 12;11,26,41,24,42$^{o/d}$, which results from the consistent calculation

$$\frac{360° + 0;59,8,17,13,12,31° \times 29;31,50,8,9,20^d}{29;31,50,8,9,20^d} = 13;10,34,58,37,54,41^{o/d}$$

and, therefore,

$$13;10,34,58,37,54,41^{o/d} - 0;59,8,17,13,12,31^{o/d} = 12;11,26,41,24,42,10^{o/d}.$$

[26] Venice: Junta, 1528, f. 33r.

[27] *Claudii Ptolemaei Magnae Constructionis, id est Perfectae coelestium motuum pertractationis, Libri XIII*, Basel: Hervagius, 1538.

[28] Gingerich and Westman 1981.

[29] Basel: Petrina, 1566, shelf-mark No. 14 B 16.Tres M 11; a facsimile of it was published in Horský 1971.

[30] At the bottom of the folio Wittich adds: *Mensis Synodicus Iuxta Hipparchum 29;31,50,8,9,20,12^d.*

Regiomontanus, who owned a Greek copy of the *Almagest* and intended to publish a Latin translation, was surely aware of the discrepancy between the original text and Gerard of Cremona's version from the Arabic, and perhaps his perplexity in confronting the dilemma ('wrong' Ptolemy *versus* 'correct' translation) may explain his refusal to give us the exact result deriving from equation (1).

Bibliography

Aaboe, A. 1955. "On the Babylonian origin of some Hipparchian parameters", *Centaurus*, 4:122-5.

al-Bīrūnī, 1954-56. *al-Qānūn al-Mas'ūdī*. Hyderabad. 3 vols.

al-Bīrūnī, 1879. *The Chronology of Ancient Nations. An English Version of the Arabic Text of the Athâr ul-Bâkiya of Albīrūnī or the "Vestiges of the Past"*. Transl. C. E. Sachau, London.

Delambre, J.B.J. 1819. *Histoire de l'Astronomie du Moyen Age*. Paris.

Gingerich, O. and Westman, R.S. 1981. "A Reattribution of the Tychonic Annotations in Copies of Copernicus's *De Revolutionibus*", *Journal for the History of Astronomy*, 12:53-54.

Goldstein, B.R. 1971. *Al-Biṭrūjī: On the Principles of Astronomy*. New Haven-London. 2 vols.

Haskins, C.H. 1924. *Studies in the History of Mediaeval Science*. Cambridge, Mass.

Hirschfeld, H. 1905. *Judah Hallevi's 'Kitab al Khazari'*, London.

Horský, Z. 1971. *Nicolai Copernici De revolutionibus orbium coelestium libri sex (editio Basileensis) cum commentariis manu scriptis Tychonis Brahe*. Praga.

Kennedy, E.S. 1964. "Al-Khwārizmī on the Jewish Calendar", *Scripta Mathematica*, 27:55-59.

Kugler, F.X. 1900. *Babylonischen Mondrechnung*. Freiburg.

Kunitzsch, P. 1974. *Der Almagest. Die Syntaxis Mathematica des Claudius Ptolemäus in arabisch-lateinischer Überlieferung*. Wiesbaden.

Mancha, J.L. 1998. "The Provençal Version of Levi ben Gerson's Tables for Eclipses", *Archives Internationales d'Histoire des Sciences*, 48:269-352.

Millás Vallicrosa, J.M. 1947. *Abraham Ibn Ezra. El libro de los fundamentos de las tablas astronómicas*. Madrid.

Millás Vallicrosa, J.M. 1959. *La obra Séfer Ḥešbón mahlekot ha-kokabim (Libro del cálculo de los movimientos de los astros) de R. Abraham Bar Ḥiyya ha-Bargeloni*. Madrid.

Neugebauer, O. 1949. "The Astronomy of Maimonides and Its Sources", *Hebrew Union College Annual*, 22:321-363.

Neugebauer, O. 1955. *Astronomical Cuneiform Texts*. 3 vols. London.

Neugebauer, O. 1956. *The Code of Maimonides. Book Three, Treatise Eight. Sanctification of the New Moon*. Trans. by S. Gandz, with supp. and intr. by J. Obermann, and astronomical commentary by O. Neugebauer. New Haven.

Neugebauer, O. 1975. *A History of Ancient Mathematical Astronomy*. Berlin-New York.

Neugebauer, O. 1979. *Ethiopic Astronomy and Computus*. Österreische Akademie der Wissenschaften, Philosophisch-Historische Klasse, Sitzungberichte, 347. Wien.

Neugebauer, O. 1987. "The Chronological System of Abū Shāker (A.H. 654)", in King, D.A. and Saliba, G. (eds.), *From Deferent to Equant: A Volume of Studies in the History of Science in the Ancient and Medieval Near East in Honor of E.S. Kennedy*. Annals of the New York Academy of Sciences, 500. New York.

Pedersen, O. 1974. *A Survey of the Almagest*. Odense.

Saliba, G. 1987. "The Role of the *Almagest* Commentaries in Medieval Arabic Astronomy: A Preliminary Survey of Ṭūsī's Redaction of Ptolemy's *Almagest*", *Archives Internationales d'Histoire des Sciences*, 37:3-20 (reprinted in *A History of Arabic Astronomy. Planetary Theories during the Golden Age of Islam*, New York-London, 1994, pp. 143-60).

Swerdlow, N.M. and Neugebauer, O. 1984. *Mathematical Astronomy in Copernicus's De Revolutionibus*. Berlin-New York.

Toomer, G.J. 1984. *Ptolemy's Almagest*. London.

Zonta, M. 1993. "La tradizione ebraica dell'*Almagesto* di Tolomeo", *Henoch*, 15:325-350.

Al-Biṭrūjī's Theory of the Motions of the Fixed Stars

Communicated by J. D. NORTH

Quarum causas in orbibus sub suppremo collocatis indagantes, in suppremo enim uniformitas semper cernitur, exploratum habuerunt id prouenire ex motibus giratiuis lulabinis appellatis, factis quidem a permistione motus orbis super suis polis cum motu eiusdem super polis alterius, itaque ex multis motibus simul collectis unus fit motus. Quae quidem theorica phisicis conformis rationibus cunctis ueteribus ad Aristotelem philosophorum principem usque uigebat, quin immo sui summi acie ingenii eam 2 de coelo textu commentario 35 teste Auerroe innuere non desinit.

Qalo Qalonymos

1. Introduction

The problem which this paper deals with is the following: were the homocentric models of Eudoxus alluded to by Aristotle (*Metaphysics*, xii.8) preserved in al-Biṭrūjī's *Kitāb fi'l-hay'a*, as argued by Kennedy in his reviews of Carmody's edition of Scot's Latin version and Goldstein's English translation based on the Arabic and Hebrew versions of the work[1] – or, as asserted by Goldstein, was the essence of al-Biṭrūjī's reform to place Ptolemaic models on the surface of a sphere, influenced by the tradition of the trepidation theory and especially by Ibn al-Zarqāllu's contributions, without knowledge whatever of Eudoxus?[2] To answer this question, this paper presents an analysis and new translation of the second chapter of al-Biṭrūjī's work, devoted to the motion of the

---

[1] F. J. Carmody, *al-Biṭrūjī's De motibus celorum*, Berkeley and Los Angeles, 1952; E. S. Kennedy, Review of Carmody, *op. cit.*, *Speculum* 29 (1954), pp. 246–251; E. S. Kennedy, "Alpetragius's Astronomy", *Journal for the History of Astronomy* 4 (1973), pp. 134–136.

[2] B. R. Goldstein, *Al-Biṭrūjī's: On the Principles of Astronomy*, New Haven and London, 1971; see also *id.*, "On the Theory of Trepidation according to Thābit b. Qurra and al-Zarqāllu and its implications for Homocentric Planetary Theory", *Centaurus* 10 (1964), pp. 232–247. Carmody's view was also that al-Biṭrūjī's models were only "Ptolemy in a spherical projection" (*op. cit.*, pp. 12–13); "this system is no more than a projection of Ptolemy's eccentrics and epicycles into the inner surface of the sky" (*id.*, "The planetary theory of Ibn Rushd", *Osiris* 10 (1952), pp. 556–586, on p. 557).

stars,[3] based on the two extant Latin versions.[4] From them it follows that the model described in this chapter is indeed a Eudoxan couple,[5] whose hippopede accounts for the motions of the stars: in al-Biṭrūjī's words, a true motion in declination and another, merely apparent, in accession and recession.

2. The phenomena

Tunc istud est quod accidit celo stellarum fixarum ex accidentibus facientibus nobis apparere diuersitatem in motu longitudinis. Et diuersitas in latitudine eius est salua et apparens et uisa.

al-Biṭrūjī

---

[3] The clue for this new interpretation of al-Biṭrūjī's text was provided by the discovery that Ptolemy's method to compute right ascensions was used in Levi ben Gerson's *Astronomy* as a procedure to deal with the geometry of the curve which we call 'hippopede' (see J. L. Mancha, "Right Ascensions and Hippopedes. Homocentric Models in Levi ben Gerson's Astronomy. I. First Anomaly", to appear in *Centaurus*). It is worth mentioning that the theory of trepidation is by no means mentioned in Levi's *Astronomy*, which suggests that it was unknown to him, and from which we can at least infer that in the Hebrew version of al-Biṭrūjī's work available to Levi the second chapter was lacking or abridged, and consequently that al-Biṭrūjī was not Levi's source concerning the use of the right ascensions method in homocentric theory. Levi does not claim originality for himself, and so I guess that the method was, so to say, a standard and anonymous procedure. It will be obvious from this paper that this use of Ptolemy's procedure was not original either in al-Biṭrūjī, who described it but did not understand very well the whole subject. As far as I know, the only modern account of Eudoxus' spheres close to this right-ascensions approach is found in A. Pannekoek, *A History of Astronomy*, London 1961 (repr. 1989), pp. 109–110. The underlying idea (the relations between equal arcs of the ecliptic and their projections on the equator) can be already found in Theodosius' *Sphaerica*, iii.6 (see H. Mendell, "The trouble with Eudoxus", in P. Suppes, J. M. Moravcsik, and H. Mendell, *Ancient and Medieval Traditions in the Exact Sciences*, Stanford, 2000, pp. 59–138).

[4] The first resulted from the collaboration of the Jew Abū Dawd and Michael Scot at Toledo in 1217. The second was due to Qalo Qalonymos, who translated Moshe ibn Tibbon's Hebrew rendering from the Arabic original made in 1259. This version was printed in 1531 with the title *Alpetragii Arabi planetarum theorica physicis rationibus probata, nuperrime latinis litteris mandata a Calo Calonymos Hebreo Neapolitano* (Venetiis in aedibus Luceantonii Iunte Florentini anno Domini MDXXXI mense Ianuario). According to the colophon, Qalonymos translated the text in 1528. On Qalonymos, also known as Qalonymos b. David, see M. Steinschneider, *Die Hebraeischen Übersetzungen des Mittelalters und die Juden als Dolmetscher*, Berlin, 1893 (repr. Graz, 1956), p. 984.

[5] Despite the plausibility of some current sceptical criticisms on Simplicius' report on Eudoxus and the subsequent historiographical troubles about Schiaparelli's classical interpretation of Aristotle's and Simplicius' passages, I shall use for convenience the expression 'Eudoxan couple' to denote 'a pair of concentric spheres moving with equal velocity in opposite directions about different axes'. We know that some medieval scholars (e. g., Ibn al-Haytham, Ibrāhīm b. Sinān, Levi ben Gerson) understood the working of Eudoxan couples; however, how they conceived Eudoxus' and Aristotle's astronomical theories is another matter, for Simplicius' commentary on *De caelo* was apparently never translated into Arabic.

What, according to al-Biṭrūjī's, were the phenomena to be accounted for?[6] Besides the daily motion, observation reveals changes in stellar declinations and longitudes; nevertheless, astronomers do not agree in the description of this apparent motion in longitude. A summary of the history of their discrepancies is given in sentences 65–80, although al-Biṭrūjī does not mention his source(s). The oldest astronomers – as Hermes and the Masters of Talismans[7] – believed that the stars have a forward and backward motion, but others (for instance, the Babylonians) rejected both trepidation and precession (sentences 66–67). Ptolemy, about 265 years after Hipparchus and agreeing with him, asserted that the stars move eastward about the poles of the ecliptic at an uniform rate of 1° each 100 years (sentences 3, and 70–71). Timocharis' and Aristyllus' observations, reported by Ptolemy (*Almagest*, vii.2-3)[8], are mentioned in sentences 69–71, but al-Biṭrūjī's source was probably corrupt for the dates are sometimes wrong.[9] A combination of trepidation (amplitude of 8°) and precession (1° /100 years) is attributed to Theon of Alexandria in sentence 73,[10] and so al-Biṭrūjī agrees with other Islamic writers,[11] who understood Theon's account of trepidation in this way. Al-Battānī's suggestion of a variable precession[12] is explicitly mentioned in sentence 75, and probably sentence 4 also refers to him, but his rate for uniform precession of 1° /66 years is never mentioned. Ibn al-Zarqālluh is mentioned twice: in sentence 7 it is said that he wrote a treatise entitled *On the motion of accession and recession*, and al-Biṭrūjī informs us in sentence 78 that the motion attributed by Ibn al-Zarqālluh to the pole of the sphere of the fixed stars around the pole of the equator[13] was a source of inspiration for his own model.

---

[6] See sentences 2–8, 55–59, 65–80, 99, and 140–141 of the translation in section 4 of this paper.

[7] οἱ παλαιοὶ τῶν ἀποτελεσματικῶν in Theon's Greek text (cf. A. Thion, *Le "Petit Commentaire" de Théon d'Alexandrie aux tables faciles de Ptolémée*, Roma, 1978, pp. 236–237 (French translation, p. 319); an English translation in O. Neugebauer, *A History of Ancient Mathematical Astronomy*, Berlin-New York, 1975, p. 632.

[8] G. J. Toomer, *Ptolemy's Almagest*, London, 1984, pp. 327–338.

[9] The date of Menelaus' observations given in the text (the 845th year from Nabonassar) is correct, but Timocharis' observations reported by Ptolemy are from years 466th, 465th, and 454th of Nabonassar era. The date attributed to Hipparchus (400 years after Alexander's death) is clearly wrong.

[10] On Theon's theory, see F. Jamil Ragep, "Al-Battānī, Cosmology, and the Early History of Trepidation in Islam", in *From Bagdad to Barcelona. Estudios sobre Historia de las Ciencias Exactas en el Mundo Islámico en honor del Prof. Juan Vernet*, Barcelona, 1996, pp. 268–298).

[11] For instance, al-Battānī (cf. Ragep, *op. cit.*), and in Muslim Spain, Ṣāʿid al-Andalusī and Ibn al-Zarqālluh (J. Samsó, "On the Solar Model and the Precession of the Equinoxes in the Alphonsine Zīj and its Arabic Sources", in G. Swarup, A. K. Bag, and K. S. Shukla (eds.), *History of Oriental Astronomy*, Cambridge, 1987, pp. 175–183, on p. 178 (reprinted in *id.*, *Islamic Astronomy and Medieval Spain*, Aldershot, Hampshire, Variorum, 1994).

[12] For chapter 52 of al-Battānī's *Zīj*, see C. A. Nallino, *al-Battānī sive Albatenii Opus astronomicum*, Publicazioni del reale Osservatorio di Brera in Milano, 3 vols. 1899–1907, vol. 1 (1903): pp. 126–128 (Latin), and vol. 3 (1899): pp. 190–192 (Arabic), and especially Ragep, *op. cit.*, which includes an English translation of chapter 52.

[13] *K* 12r28–29: quod poli huius orbis moueantur super duobus circulis aequidistantibus aequinoctiali: itaque motus earum subsequatur ad motum horum duorum polorum, *C* X18: quod duo

Echoing *Almagest*, vii.3, al-Biṭrūjī considers the changes in declination of the stars to be a well established fact[14], from which astronomers inferred their motion in longitude, whose rate and direction remain at his time still uncertain (sentences 55–58). Without discussing in detail models, parameters, or observations, he repeats at several places (sentences 8 and 79, for instance) that the appearances agree with the motion of accession and recession assumed by Ibn al-Zarqālluh,[15] adding that sometimes the stars appear stationary (a phenomenon compatible with trepidation, but not with uniform or variable precession):

> ... the distance covered by the motion of the fixed stars has been found by observation to vary at different times – for sometimes the retardation of this slow motion is found to be less and at other times greater, and sometimes following the order of the signs and at other times in the opposite direction (namely that of the motion of the universe), and sometimes, during a long period of time, no motion at all could be determined from observations by people living then... [16]

The observed changes in the declination can result from a *true* northward and southward motion of the stars with respect to the celestial equator; however, the eastward component of trepidation must be an *apparent* motion, since all the spheres in the

---

poli huius celi mouentur super duos circulos, et erit motus stellarum sequens motum istorum polorum [Hereafter, references to the Latin versions will be abbreviated as follows: "*K* **9v**22" denotes folio 9 verso, line 22, in Qalonymos' printed edition, and "*C* **IX**5" refers to chapter IX, sentence number 5, in Carmody's edition of Scot's text]. The only known copy of Ibn al-Zarqālluh's work on the fixed stars (preserved in Hebrew in Paris, BnF, Heb. 1036), was reproduced with a Spanish translation in J. M. Millás Vallicrosa, *Estudios sobre Azarquiel*, Madrid-Granada, 1943–1950, pp. 250–342. On Ibn al-Zarqālluh's model for the variation of the obliquity of the ecliptic, see Goldstein, "On the theory of trepidation". Ibn al-Zarqālluh's *Treatise on the fixed stars* (Millás, *op. cit.*, esp. pp. 274–278) could be the source for at least a part of the information given by al-Biṭrūjī in sentences 65–75.

[14] See sentence 56: plurimum enim ex quo cognouerunt motum stellarum fixarum fuit cum uiderint declinationem earum in latitudine *K* **11v**13–14, quia plus quod induxit nos ad motus istarum stellarum fixarum est earum extractio in latitudine *C* **X**5.

[15] See sentence 8: et hic quidem motus ut apparet est sicut posuit Auoashac *K* **9v**29–30, et iste motus prout posuit Azarkel apparet in re *C* **IX**4 [79]: et uerificatus est nunc ille motus sicut posuit Alzarkala predictus ut scilicet quod uidetur de diuersitate motus stellarum fixarum sit accessus et recessus *K* **12r**33–34, quiescat [*ithbāt/thabat*] ergo motus secundum quod dixit Abu Isac Azarkel, quod illud scilicet quod apparet de diuersitate motus stellarum est motus accessus et recessus *C* **X**19.

[16] Sentence 55: et quoniam spacium motus stellarum fixarum reperitur in obseruatione diuersum secundum diuersitatem temporum, nam aliquando reperitur hic motus tardus parua retardatione et aliquando maiori retardatione et aliquando secundum successionem signorum et aliquando contra ordinem signorum et ad partem motus uniuersi, et quandoque magno temporis spacio non apprehenditur in eis motus ex obseruatione hominum illius aetatis *K* **11v**8–13, et quia illud quod inuenitur de motu stellarum fixarum per considerationes diuersatur in quantitate secundum diuersitatem temporum, quia inueniunt istum motum cum sua tarditate in uno tempore maiori, in alio tempore minori, et in uno tempore secundum signa, in alio contra signa ad partem motus generalis, et remanent tempore longo sine motu sensibili quoad considerationes hominum in illo tempore *C* **X**4–5 (I have slightly modified Carmody's punctuation).

universe move from east to west, and an actual motion in the direction opposite to that of the daily motion is impossible. Moreover, al-Biṭrūjī does not exclude that trepidation can be combined with precession:

> This agrees with what was accepted by recent astronomers concerning both motions of accession and recession; although they also admitted a motion in the direction of the signs, which would result from some lagging still remaining in the orb [of the fixed stars after these two motions], noticeable only after a long time – this is possible, although its amount would not be as Ptolemy asserted.[17]

Thus, the model for the fixed stars must account for the observed changes in the declination of the stars – whose maximum reaches the maximum solar declination[18] – and also for an apparent forward and backward longitudinal motion, using only concentric spheres which move from east to west around different axes.

3. The model

> Sed non fuit intentio nostra nisi expergefacere in qualitatibus motus uerificati qui facit mutationes diuersas multis modis, et posuimus astrologiam celi possibilem, et radices probabiles loco earum Tholomei que sunt difficilis imaginationis, que sustinet omnem partem celi similem in suo motu cum toto et totum motum unum sine motibus diuersis.
>
> al-Biṭrūjī

---

[17] Sentence 99: et hoc profecto est iuxta id quod constituerunt recentiores de ambobus motibus accessus et recessus, licet cum hoc habuerunt etiam motum secundum ordinem signorum et est quod quidam defectus remanserit orbi et apparebit longinquitate temporis et potest esse quidem licet non fuerit spacium eius id quod dixerit Ptolemaeus *K* **13r**7–10, sed si cum hoc habuit motum secundum signa, erit ita quod remaneat semper post istos duos motus incurtatio modica remanens in celo, et apparebit istud in longitudine temporis; et est istud res que potest esse, sed erit minus quam quod nominauit Tholomeus *C* **XI**13 (I have modified Carmody's punctuation). See also sentence 8: et hic quidem motus ut apparet est sicut posuit Auoashac, nisi quod motus in longitudine cum hoc motu accessus et recessus possibile est ut sit secundum ordinem signorum, licet adhuc non peruenerint ad ueritatem eius ex obscruatione *K* **9v**29–32, et iste motus prout posuit Azarkel apparet in re, sed suus motus tantum in longitudine cum motu accessus et recessus potest esse secundum signa, licet non possimus figi super eius rectificationem *C* **IX**4, and sentence 141: Habet autem alium defectum preter hunc quem diximus itaque erit stellis fixis motus in longitudine secundum ordinem signorum ut dixit Ptolemaeus et alii pristini adeo quod existimauerunt ex hoc hunc orbem moueri contra motum uniuersi esse id quod magis uidetur et quod propinquius sit possibilitati *K* **14v**2–5, sed si sit incurtatio preter hoc quod nos diximus, ita quod per eam habeant stelle fixe mutationem in longitudine secundum signa, sicut nominauit Tholomeus et alii antiqui, in tantum quod opinati sunt quod istud celum mouetur motu diuerso motui generali, hoc quidem dubium est; sed tunc magis apparens in potentia *C* **XI**45.

[18] See sentence 18: et erit declinatio stellarum quae incedunt per medium orbis ut declinatio solis ab aequinoctiali et hoc quidem ut pristini dixerint absque praecisione illius spacii *K* **10r**15–17, tunc erit declinatio istius stelle similis declinationi solis; et istud est quod dixerunt antiqui uerbo grosso non rectificante *C* **IX**9.

What is the model to account for these phenomena? The irregular apparent motions of the stars result from the compounding of circular and uniform motions of the ninth and the eighth spheres.[19] The ninth or highest sphere (*al-falak al-a^c lā, orbis suppremus*), the first mover, is the fixed system of reference, with the vernal equinox as zero point, against which the motion of the stars, situated on the eighth, takes place. This supreme orb rotates with the daily westward motion, and transmits it to all the spheres placed below it, although the power (*quwwa, uirtus*) received by these spheres will diminish as their distance from it one increases. This diminution produces a loss of motion of each sphere, and in consequence a reduction in the number of degrees traversed by it, as compared with the number of degrees traversed by the highest one; so, the ninth sphere moves 360° in twenty-four hours, and the eighth moves only $(360 - \alpha)°$ in this time. These α degrees that it lags behind the ninth one are called by al-Biṭrūjī the *taqṣīr* of the sphere (a word rendered into Latin by Scot as *incurtatio*, and as *defectus* by Qalonymos),[20] and it increases as one proceeds from the outermost sphere toward the center of the world. However, the *taqṣīr* is compensated by a second westward motion of each sphere about its own poles, due to its desire (*shawq, desiderium*) of perfection (*kamāl, complementum, perfectio*) and assimilation with the prime mover; so, all the spheres below the highest one move by their own motion and seek their aim either by nature (*bi-l-ṭab^c*) or by desire to imitate (*tashabbuh*) it.[21]

In Fig. 1, let O and S be the poles of the ninth sphere, about which it performs a revolution in twenty-four hours, circle $ACKGM$ its equator, and T the earth. Let circle $ABGD$ be the ecliptic on the ninth sphere; thus, point A is the autumnal equinox and the distance $AZ = 90°$. Let us also assume that at the initial position, at time t_0, the pole of the sphere of the fixed stars (the eighth) coincides with point Z. The angular distance between the poles of the ninth and the eighth spheres, $OZ = i$, is equal to the obliquity of the ecliptic. The daily motion of the ninth sphere is transmitted to the eighth, so that its poles rotate around the poles of the ninth one, tracing around O and S, respectively, circles ZPF and ELN – called by al-Biṭrūjī 'circles of the path of the poles' (*mamarr*

---

[19] See sentences 9–54 and 81–139. On the philosophical and cultural background of al-Biṭrūjī see A. I. Sabra, "The Andalusian revolt against Ptolemaic astronomy", in E. Mendelsohn (ed.), *Transformation and tradition in the sciences. Essays in honor of I. Bernard Cohen*, Cambridge, 1984, pp. 133–153; on the influence of Neoplatonic dynamics on al-Biṭrūjī, see J. Samsó, *Las ciencias de los antiguos en al-Andalus*, Madrid, 1992, on pp. 344–348, and *id.*, "On al-Biṭrūjī and the *hay'a* tradition in al-Andalus", in *Islamic Astronomy and Medieval Spain*, XII, pp. 1–13.

[20] With verbal forms *incurto* < *curto* and *deficio*, respectively; besides *lagging* or *defect*, acceptable translations for *incurtatio* would also be *shortening* or *curtailement* (etymologically related to *curto*). Although the denotation of *incurtatio/defectus* is clearly spatial, there is a passage in Scot (**XI**12) where time is connoted, and *festinum* (faster) is given as antonym of *incurtatum* (*maior* and *minor*, respectively, in Qalonymos, **13r**5). There are, however, at least two passages where Scot translated *taqṣīr* as *defectus* (**VIII**13: ad complendum defectum primi motus, and **XIV**23: ibi recedens a modo mediocritatis ad finem tarditatis et ad defectum).

[21] omnes orbes qui sunt sub eo mouentur ad motum eius et quilibet perficit intentionem suam siue natura siue desiderio ad se assimilandum ei K **8r**44–**8v**1, omnes celi inferiores eo mouentur per motum suum et intendunt ire uia sua, aut per naturam aut per desiderium imitandi ipsum C **VII**18; cf. Goldstein, 1971, vol. 1: p. 75, and Samsó, "On al-Biṭrūjī . . .", p. 10.

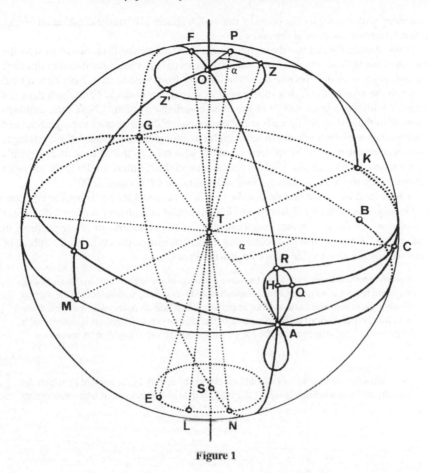

Figure 1

al-quṭbayn; *circuli transitus polorum* and *circuli cursus polorum* in Scot's and Qalony-mos' versions, respectively).[22]

However, when at time t_n the ninth sphere completes a number k of revolutions (from east to west), the pole of the eighth sphere does not return to point Z, for due to the *taqṣīr* the sphere lags behind the ninth and the pole only reaches point P (so that the point of the eighth sphere that at t_o did coincide with A, will coincide at t_n with point C);

---

[22] Sentence 16: hii duo poli huius orbis facient in hoc suo defectu duos circulos super quos prolixitate temporis mouentur, quos appellabimus circulos cursus duorum polorum *K* **10r**7–8, cum ista incurtatione paulatim paulatim in fine longissimi temporis complebuntur illi duo parui circuli et mutabuntur loca polorum super ipsos; et tunc dicemus illos duos circulos transitus polorum *C* **IX**8. 'Rotation of poles (around poles)', an expression often used to describe al-Biṭrūjī's system, is obviously equivalent to 'motion of concentric spheres around different poles (or different axes)'.

of course, with the *taqsīr* the poles of the eighth sphere also traverse the small circles called 'circles of the path of the poles'.[23]

Thus, despite the fact that the eighth sphere is indeed moved from east to west by the motion of the highest one, due to the *taqsīr* it seems to have been moved during this time through angle α in the opposite direction, namely from west to east. Note that at time t_n, when the pole of the eighth sphere is at P, its equator is circle CH, which does not coincide with the ecliptic $ABGD$ of the fixed coordinate system.[24] Note that, although throughout the text al-Biṭrūjī calls the intersections of the equator of the eighth sphere and the equator of the ninth one "points similar to the equinoctial points", these passages have been misinterpreted by modern scholars, who refer to the equator of al-Biṭrūjī's eighth sphere as "the ecliptic"; indeed, the equator of the eighth is similar to the movable ecliptic of the trepidation models. Obviously, at t_n arc CP is equal to $90°$.

The second westward motion that the sphere of the fixed stars performs to compensate its *taqsīr* is called by al-Biṭrūjī *istīfā'* ('motion of completion' = *completio, motus completionis* in Scot's version; *complementum, motus complementi* or *supplementi* in Qalonymos). Probably due to its closeness to the first mover, the *istīfā'* of the sphere of the fixed stars is equal to the *taqsīr*. Al-Biṭrūjī writes:

> this orb is moved about its poles in the same direction as the highest one, in order to bring to completion what could not be accomplished and was lacking with respect to the motion of the highest orb, and [this motion] is performed around its poles (which do not move with this proper motion) until the motion of the highest orb is caught up, namely until it [= the eighth orb] arrives at the place at which the highest orb completed its revolution, [25]

and also

> the pole [of the eighth orb] lags behind [the ninth orb] exactly by the amount by which the [eighth] orb lagged behind; however, the stars on it do not lag behind, as later astronomers

---

[23] See sentence 19: Et ambo isti circuli profecto cursus amborum polorum sunt idem duo circuli quos faciunt ambo poli cum perfectione reuolutionis orbis suppremi *K* **10r**17–18; this sentence is absent from *C*.

[24] See sentence 85: et erunt intersectiones huius circuli medii in zona huius orbis cum circulo aequinoctiali super duobus punctis similibus duobus punctis amborum aequinoctiorum et due maxime distantie inter ea sunt similes duobus punctis amborum solsticiorum *K* **12v**14–17, erit stella que erit in medio huius spere describens circulum inclinatum super equatorem diei secantem ipsum per duo media, sicut est circulus signorum ... et erit sectio huius circuli super duo puncta que sunt similia duobus punctis equalitatis, et finis longitudinis est similis finibus duarum mutationum *C* **XI**4–5.

[25] Sentence 13: mouetur hic orbis super polos suos, ut perficiat id quod non potuit et defecit a motu suppremi, ad partem quidem motus suppremi et perfecit sese super polos suos, qui sunt fixi huic motui sibi proprio, quoad perueniat ad motum suppremi, scilicet quousque applicet ad locum postquam perfecerit suppremus reuolutionem suam *K* **9v**41–**10r**1, mouetur istud celum super proprios polos ad complendum illam incurtationem a motu superiori, et tunc mouetur super proprios polos iam fixos ut possit acquirere complementum simile superiori *C* **IX**6.

believed, for they accomplish with the motion of completion of their orb what the orb lagged behind [the highest one];[26]

that is, in order to imitate the ninth sphere, the eighth is also moved around pole P (Fig. 1), so that the point of the eighth sphere placed by the *taqsīr* at point C will move with the *istīfā'* on circle CH through angle α. Therefore, at time t_n, the eighth sphere has completed two westward motions: with the power of the ninth sphere it moves with a lagging motion around pole O through $(k \cdot 360 - \alpha)°$, and with its own motion of completion it moves around pole P through $\alpha°$; so, at time t_n, it has been moved by an amount equal to

$$(k \cdot 360 - \alpha)° + \alpha° = k \cdot 360°,$$

that is, equal to that of the ninth sphere – a concept expressed in a rather clumsy way by al-Biṭrūjī, who writes that the eighth orb catches up with the ninth one and "arrives at the place at which the highest orb completed its revolution".

It is a trivial matter to transform al-Biṭrūjī's model in a Eudoxan couple: the outer sphere (the ninth one) carries first the inner one (the eighth) from west to east about pole O through angle α (so that Z moves to P, and A to C) and then the eighth one moves also through α in the opposite direction about pole P (so that C moves to Q). Consequently, a point at A, on the equator of the eighth sphere, will generate with these combined motions Simplicius' hippopede; its length can therefore account for changes in declination of a star placed at A at t_o, whereas its width can account for eastward and westward oscillations in longitude. So, al-Biṭrūjī's ingenuity merely consisted in replacing the physically not allowed eastward motion of the eighth sphere by a westward motion lagging behind the ninth one, in conformity with the accepted principles.

Since the word "hippopede" was in all probability not available to al-Biṭrūjī, we can now ask how exactly he formulated the geometrical consequences of his model, i.e., how he described the path of a point placed on the equator of the eighth sphere at A, which results from these two motions.

Al-Biṭrūjī asserts that the lagging motion of the orb (namely, the different positions of the poles of the orb on the circles of their paths) accounts for the changes in stellar declination:

Since the distance of the stars from both poles [of their orb] is always the same, the stars seem to incline in the direction towards which the poles move when they revolve with their lagging motion. It is called motion in declination; for the poles, due to their lagging, move backwards in the circles of their paths and lie at different times in different directions

---

[26] Sentence 21: et deficit polus tantum quantum defecit orbis: stellae tamen quae sunt in eo non deficiunt, et hoc quidem sicut fecerunt posteriores huius scientiae, sed perficit sibi ipsi motu sui orbis motum complementi id quod ab eo prius defecerat orbis ipse *K* **10r**21–24; the sentence is drastically abbreviated in *C* **IX**10: et incurtat solus polus istud quod incurtabat celum (reading *solus* instead of Carmody's *Solis*).

with respect to the poles of the highest orb, and for this reason the declination of the stars change... ;[27]

and later:

When the highest orb begins to move from point A, the pole of the orb of the stars will follow it from point Z; but when point A returns to its place in the figure, point Z will not have completed the circle of its path, for it lags, and the lagging will be equal to a part of arc ZF, as if point Z moved toward F by the amount of this lagging. When point Z reaches point F with the lagging motion, the maximum declination of the star (which was on point A) to the north of the celestial equator is reached, and [once it has been moved with its motion of completion] the star (which preserves its position in longitude) will be placed at the same distance [from the equator] as point D on the oblique circle; thus, the star will rise from horizon OAS at a distance from A equal to the length of arc OF and equal to the distance of point M from point D (which is similar to the solstitial point), for its distance from pole Z is always the fourth part of the circle. In this way the motion of the pole from point F to the point opposite to point Z on the circular path will take place, and the star will return from its maximum northerly declination to the point similar to the other equinoctial point, namely point G.[28]

In other words: at t_0, the pole of the eighth sphere is at Z and the star at A, and its declination is zero (Fig. 1); but when, due to the *taqṣīr*, the pole is moved through 90° to F, the star will be at K, and when the sphere moves again through 90° about F due to the *istūfā'* the star will be at R, and its declination equal to arc AR = arc OF = arc MD. When the lagging leads the pole to Z', so that arc $ZZ' = 180°$, the star will be at G, and after moving again through 180° about Z' due to the *istūfā'*, the star will return to A, with zero declination.

As for the way by which the combined motions of the two spheres (or, properly, the combined motions of the eighth around different axes) produce apparent oscillations in

---

[27] Sentences 23–24: Et quoniam diatantia stellarum ab uno quoque amborum polorum <non> est eadem semper, ideo apparet stellis declinatio loci ad quem mouentur poli cum reuoluuntur defectiue, et appellatur motus in latitudine nam poli propter defectum suum postponuntur in circulis cursus suis et erunt diuersis temporibus in partibus diuersis polorum supremi, unde stellae declinantur... *K* **10r**25–29; Et quia longitudo cuiuslibet stelle est semper salua ad polos, uidentur stelle quasi impelli ad partem motus polorum quando rotantur poli cum sua incurtatione; et illud est quod nominatur motus latitudinis, quia poli cum sua incurtatione posteriorant super duos circulos et distabunt a polis superioris in temporibus diuersis in partibus diuersis. Et ideo uidentur stelle iste quasi errare... *C* **IX**11.

[28] Sentences 111–114: nam cum inceperit orbis supremus moueri ab A et subsecutus fuerit eum polus orbis stellarum fixarum a Z et reuersus fuerit punctus A ad suum locum figurae, non compleuit quidem polus circulum cursus sui ipsius Z, deficit enim ab eo et defectus erit parte arcus ZP, quasi ipse punctus Z motus fuerit ad partem P, ut est illa pars quam defecit. Et cum peruenerit punctus Z defectiue ad P, tunc perfecta est declinatio stellae ad septentrionem a circulo aequinoctiali, scilicet qui fuerat super puncto A, et peruenerit stella ad distantiam puncti D in circulo obliquo cum continuitate stellae suo loco in longitudine; et erit A stellae ascendenti in orizonte UAS [= OAS] in tali distantia ab A ut est spacium arcus UP [= OF] et est aequalis distantie M a puncto D, qui est similis puncto solsticii, nam distantia eius a polo Z semper quarta circuli. Et sic etiam erit motus poli a puncto P [= F] ad punctum similem punctum Z circuli cursus et stella reuertitur ab

longitude, al-Biṭrūjī's description is somewhat awkward and, as we will see, not without errors.[29]

Al-Biṭrūjī asserts that point A generates a circle when the sphere has a single motion around axis OS, and also when the sphere rotates uniquely around axis ZE.[30] However, the motion of the point is more complicated when the sphere revolves around its two axes (namely, with its lagging or shortened motion, and with its proper motion or *istīfā'*), and the characteristics of the resulting figure are described twice in the text: in sentences 35–54 (where it is called *leuleb*), and then in sentences 81–139 (where its adequacy to represent a motion in accession and recession as well as the method to calculate it are explained in detail).

Is there some single procedure to deal with the geometry of Simplicius' hippopede without reconstructing it as the intersection of the sphere and a cylinder? An elementary problem of spherical astronomy, the computation of right ascensions, provides a fair approximation. In Fig. 2, the observer is at T, the celestial equator AB goes through the zenith Z, and the circle of declination SAC coincides with the horizon; ε is the obliquity of the ecliptic and N the north pole. In the situation represented in the figure, the vernal point B has risen by an arc AB at the moment when the sun is in the horizon rising at C (the diurnal rotation takes place from A to B). Since time is measured by the uniform rotation of the equator, the arc $\alpha = AB$ represents the time it took the ecliptic arc $\lambda = BC$ to rise. Ptolemy's procedure to compute α as function of λ required the use of a Menelaus theorem twice,[31] and it is equivalent to the modern formula

$$\tan \alpha = \cos \varepsilon \tan \lambda. \tag{1}$$

Ptolemy's procedure can be now applied to al-Biṭrūjī's model. In Fig. 3, T and N are again the observer and the north pole of the celestial equator, respectively, and P will represent different positions of the pole of the eighth sphere (the diurnal rotation taking now place from B to A). At time t_0, a given star is at A, and the pole of the eighth sphere

---

ultima declinatione sua in septentrionem ad punctum similem puncto aequinoctii alterius, scilicet puncto C [= G] K **13v**10–21; quia quando incipiet celum superius cum motu puncti A, et sequitur eam [?] polus stellarum fixarum a puncto Z, et reuertetur punctum A ad suum proprium locum forme, polus Z non complebit circulum qui transit per Z, quia Z incurtat ipsum; tunc incurtat partem arcus ZF, et apparebit quod punctum Z mutabatur ad partem F secundum quantitatem illius partis quam incurtabat. Tunc quando applicabitur punctum Z cum incurtatione ad F, finiebatur inclinatio stelle septentrionalis ab equatore diei, uolo dicere quod stella que apparuit super punctum A reuertebatur super longitudinem puncti D in circulo inclinato; tunc stella ascendit in orizonte OAC [= OAS] super longitudinem puncti A equalem arcui OF et equalem arcui MD, quod D est [simile] loco puncti mutationis diei. Et sic erit mutatio poli a puncto F ad punctum quod est in oppositione puncti Z in circulo transitus, et stella reuertetur a fine inclinationis sue septentrionalis ad punctum quod est [simile puncto] equalitatis secunde, uolo dicere punctum G C **XI**25–27.

[29] This strongly suggests that, as Kennedy (1973, p. 136) stated, "he obtained the theory at second or third hand and put it down as best he could, garbling it in so doing".

[30] See sentences 28–34.

[31] The results are tabulated twice in the *Almagest* (i.16 and ii.8; see G. J. Toomer, *Ptolemy's Almagest*, London, 1984, pp. 74 and 100; see also O. Neugebauer, *A History of Ancient Mathematical Astronomy*, Berlin-New York, 1975, pp. 31–32).

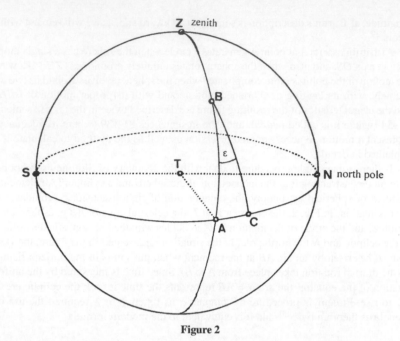

Figure 2

at P_o; let circle NAC be the horizon. Let us assume that the *taqṣīr* amounts to 45° ; it leads the pole to P_1 and the star to B_1, so that $AB_1 = \alpha = 45°$. The *istūfā'* moves then the star from B_1 along circle B_1C around pole P_1 through 45° ; once finished this motion of completion, where is the star placed on circle B_1C? From Ptolemy's procedure it follows that, when $\alpha < 90°$, $\alpha < \lambda$; consequently, we know that with the *istūfā'* the star will be moved on circle B_1C through a distance equal to α, namely from B to D, so that arc $B_1D =$ arc AB_1, and arc $DC = \lambda - \alpha$. So, as seen from T, the star seems to move in the opposite direction, as if it receded with respect to the motion of the universe or it was moved opposite to that motion:

> and this lagging is called [by astronomers] accession because it takes place in the direc-
> tion of the signs... although it actually occurs unlike [what they asserted], for accession
> according to them is a motion opposite to the motion of the universe, and recession accord-
> ing to them is a motion in the direction of the motion of the universe; however, according
> to the truth, it occurs the reverse.[32]

It follows from (1) that the distance DC increases for values of α between 0° and about 45°, when it reaches its maximum, and then decreases until $\alpha = 90°$, for then the *taqṣīr* leads the pole to P_2 and the star to B_2, and once it is moved with the *istūfā'* the star will be again on circle NAS, at E, for now arc $AB_2 =$ arc $B_2E = 90°$, and so $\lambda = \alpha$. Therefore, when $0° < \alpha < 90°$, the star will seem to be placed to the east of point A, as if it

---

[32] See sentences 94 and 79.

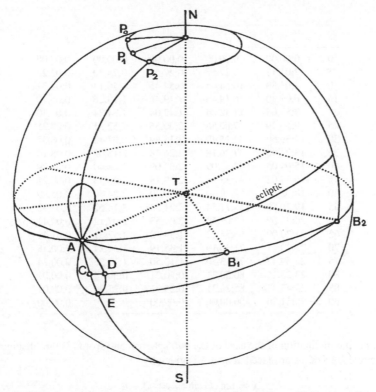

Fig. 3. Arc $AB_1 = 45°$, arc $B_1 C >$ arc AB_1; arc $AB_2 = 90°$, arc $B_2 E =$ arc AB_2

were moved in the direction of the signs (despite its sphere has been moved with its two motions the same number of degrees as the ninth orb), and to the west of point A when $90° < \alpha < 180°$. It is also clear that the star will exhibit a southerly declination for values of α between $0°$ and $180°$, which reaches its maximum when $\alpha = 90°$. The pattern will be repeated for values of α between $180°$ and $360°$, with northern declinations in this second half, and "the circle will be so completed and two accessions and two recessions will occur along it" (sentences 98 and 139).

Thus, let A be the intersection on the ninth sphere of the celestial equator and the ecliptic, and let x be the distance from the celestial equator of a star on the equator of the eighth sphere placed at t_0 at A. This distance can be computed from

$$\sin x = \sin i \sin \alpha. \tag{2}$$

The arc of the equator of the eighth sphere rising with the equatorial arc α follows from

$$\tan \lambda = \tan \alpha / \cos i, \tag{3}$$

Table 1

α	x	λ	y_b	y_a	$y_b - y_a$
0	0;00,00	0;00,00	0;00,00	0;00,00	0;00,00
5	2;01,12	5;27,52	0;27,52	0;25,30	0;02,22
10	4;01,38	10;54,45	0;54,45	0;50,14	0;04,31
15	6;00,30	16;19,46	1;19,46	1;13,26	0;06,20
20	7;57,03	21;42,04	1;42,04	1;34,24	0;07,40
25	9;50,29	27;00,56	2;00,56	1;52,31	0;08,25
30	11;39,59	32;15,49	2;15,49	2;07,12	0;08,37
35	13;24,48	37;26,18	2;26,18	2;18,02	0;08,16
40	15;04,05	42;32,09	2;32,09	2;24,40	0;07,29
45	16;37,02	47;33,18	2;33,18	2;26,53	0;06,25
50	18;02,53	52;29,49	2;29,49	2;24,40	0;05,09
55	19;20,50	57;21,54	2;21,54	2;18,02	0;03,52
60	20;30,09	62;09,53	2;09,53	2;07,12	0;02,41
65	21;30,09	66;54,12	1;54,12	1;52,31	0;01,41
70	22;20,11	71;35,19	1;35,19	1;34,24	0;00,55
75	22;59,42	76;13,50	1;13,50	1;13,26	0;00,24
80	23;28,16	80;50,21	0;50,21	0;50,14	0;00,07
85	23;45,33	85;25,31	0;25,31	0;25,30	0;00,01
90	23;51,20	90;00,00	0;00,00	–	0;00,00

where i is the inclination of the axes of the two spheres, and from (3) the 'distance'[33] y from circle NAS of a star placed at t_0 at A is given by

$$y = \tan^{-1}(\tan \alpha / \cos i) - \alpha \qquad (4)$$

or

$$y = \lambda - \alpha. \qquad (5)$$

Thus, a table giving λ as a function of α allows easily to compute the difference $\lambda - \alpha$ (as in Table 1).[34]

The occurrences of the expressions *figura leuleb / figura laulabina* in the text deserves special attention. The interpretation usually adopted is that *leuleb* means "spiral" and refers to the curve drawn by a heavenly body (star or planet) when its slow eastward and its fast diurnal rotations are compounded.[35] Nevertheless, there is a passage in the text (sentences 46–48) which suggests that al-Biṭrūjī uses it to denote *the four parts into*

---

[33] Obviously, arc DC is not perpendicular to circle NAS.

[34] In Table 1 column I (α) displays the values for the argument at 5° -intervals; column II (x) the length of the hippopede computed from (2); column III (λ) is computed from (3); column IV (y_b) displays the difference between columns III and I, i.e., the width of the hippopede computed with the right ascensions method; column V (y_a), the width of the hippopede computed from the correct formula $\sin y = \frac{1}{2}(1 - \cos i) \sin 2\alpha$, and column VI the differences between the corresponding entries in columns IV and V. Note that y_b max does not occur when $\alpha = 45°$.

[35] Carmody (1952, pp. 52–54) and Goldstein (1971, vol. 1, pp. 23 and 26) agree on this point.

which the hippopede is divided by its two axes of symmetry, namely circles *OAS* and *AKGM* in Fig. 1, or circles *NAS* and AB_1B_2 in Fig. 2. The *leuleb* figure that a star on the equator of the eighth sphere at *A* at t_0 will generate when moved with both motions (the lagging one around the axis of the ninth sphere, and the *istīfā'* around the axis of the eighth) is alluded to in sentences 35, 36, and 39, and described with some detail in the "proof" given in sentences 40–48.[36] Al-Biṭrūjī repeatedly claims that the rising of point *A* – namely, the rising of a star on the equator of the eighth sphere at t_0 at *A*, which apparently slides eastward along circle *AKGM* with the *taqṣīr* and moves westward on inclined circles with the *istīfā'* – occurs each daily revolution at a different place (with respect to the fixed reference system), and that a *leuleb* figure results from all the circles generated by the rising positions of the star (due to the *istīfā'*) when it is moved by the *taqṣīr* between A and K in Fig. 1, or between A and B₂ in Fig. 3. (In this interpretation leuleb figures would be arcs AQR and ADE in Figs. 1 and 3, respectively.) The same will occur when the pole of the eighth sphere is moved due to the *taqṣīr* between *F* and *Z'* (and the star from K to G)

> for it will likewise generate a figure similar to the first one, and in the two remaining quadrants two figures similar to the first two ones will result, and point A will return to its initial place. Thus, the combination of all these two motions [namely, all the motions that take place about these two different axes] will produce four figures...[37]

I will now consider some passages which demonstrate that al-Biṭrujī's had no clear understanding of the model he was describing. We can distinguish in it (1) the arc α of the celestial equator rising at time t at horizon h, which measures the *taqṣīr* of the sphere, (2) the arc λ of the equator of the eighth sphere which rises simultaneously with α, and (3) the arc γ of the equator of the eighth sphere traversed by the star with its motion of completion, which is always equal to α. These variables are related, and we know that (i) λ and γ are functions of α, and (ii) when $\alpha < \lambda$, $\lambda > \gamma$, and when $\alpha > \lambda$, $\lambda < \gamma$. Let us consider the following sentences:

non erit illud quod ascendit de equatore diei cum eo quod de suo celo inclinato equale semper ei quod secuit stella de celo inclinato (sentence 50: *C* **X**2)	id quod ascendit ex gradibus illius spacii quos pertransiit stella in circulo suo obliquo non erit aequale semper ei quod ascendet cum eis ex gradibus circuli aequinoctialis (*K* **11r** 40–42)
diuersantur quantitates quas secant stelle in suis circulis inclinatis a quantitatibus equatoris diei que ascendunt cum ipsis (sentence 89: *C* **XI**7)	differunt propter hoc spacia que transeunt stellae sui circuli obliqui a spaciis que fuerint in frontespicio et eo quod ascendit cum eis circuli aequinoctialis (*K* **12v**25–26)

---

[36] See also Fig. 6.
[37] Sentences 47–48; see also Fig. 7.

quia gradus quos secat stella cum suo motu proprio **diminuuntur** a gradibus qui ascendunt cum eo de equatore diei, tunc ascendunt cum ea de equatore diei qui sunt plures gradus, tunc apparet quod stella multiplicat motum suum...et istud es quod numerauerunt recessum (sentence 95: *C* **XI**10)

tunc gradus quos pertranseunt stellae motu suo proprio **sunt pauciores** [plures] illis qui sunt in frontespicio circuli aequinoctialis qui sunt plures [pauciores] et apparet ex hoc additamentum citra motum uniuersi... et hoc est quod appellant recessum (*K* **12v**43–13r2)

et istud quia arcus quem secat stella cum motu sui celi qui est proprius ei et latet nos ascendit cum arcu equatoris diei **maiore** ipso. Tunc apparet stelle additio et precessio propter gradus equatoris ascendentes cum eo, et uidebitur precedere motum superioris (sentences 137-138: *C* **XI**41)

nam arcus quem pertransit stella ad motum sui orbis sibi proprium nobis occultum ascendit cum **maiori** arcu aequinoctialis et uidebitur stellae excessus gradibus aequinoctialis ascendentibus cum ea et apparet precedens motum suppremi (*K* **14r**38–40)

As we can see, the expression "arc [of the equator of the eighth sphere] traversed by the star" (literally γ, which is always equal to α, for arcs due to the the *istīfā'* are always equal to arcs due to the *taqsīr*) is used to refer to arc λ (which is sometimes greater and sometimes smaller than α), and so the sentences are literally false. Consequently, we should expect for sentences 50 and 89 something as: *gradus que transit stella in circulo inclinato cum motu complementi sunt equales gradibus equatoris que incurtat stella cum incurtatione polorum, sed differunt a gradibus circuli inclinati que ascendunt cum illis gradibus equatoris*; however, since both Latin versions agree, the error was very likely found in al-Biṭrūjī's original.[38]

A second problematic passage concerns al-Biṭrūjī's determination of the stationary points of the star on the hippopede. Sentences 93–94 assert that, for values of α between $315°$ and $45°$ (i.e., when the star traverses with the *taqsīr* the quadrant of the celestial equator whose middle point is the equinoctial point) arcs λ are greater than arcs α rising with them, and that, in this interval, the star seems therefore to move in the direction of the signs. It is also said (sentences 95–96), that for values of α between $45°$ and $135°$ (i.e., when the star traverses the next quadrant of the equator), arcs λ are smaller than arcs α rising with them; so, as the star moves on the equator of the eighth sphere through an arc equal to α, it seems then to move in the opposite direction, ahead the motion of the universe. Al-Biṭrūjī here misunderstands his sources, for (see Fig. 4a) when α ranges between $315°$ and $45°$, the star moves from west to east, though in this interval $\alpha > \lambda$ between $315°$ and $360°$, and $\lambda > \alpha$ between $0°$ and $45°$; similarly, it is true the star moves from east to west for values of α between $45°$ and $135°$, though $\lambda > \alpha$ between $45°$ and $90°$, whereas $\alpha > \lambda$ between $90°$ and $135°$. In sentence 98 al-Biṭrūjī correctly asserts that in a complete revolution the star accomplishes two accessions and two recessions (it moves in accession, using the traditional terminology, between $315°$ and $45°$, and between $135°$ and $225°$, whereas it moves in recession between $45°$ and $135°$, and between $225°$ and $315°$).

---

[38] Latin sentences 50 and 89 indeed agree with the Arabic text reproduced in Goldstein (1971, vol. 2). I am grateful to J. Samsó for this information.

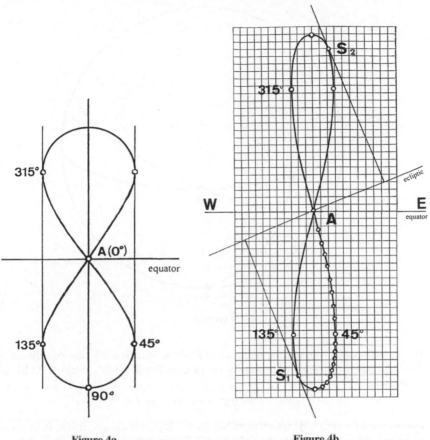

Figure 4a   **Figure 4b**

However, it is evident that Fig. 4a cannot represent the model, for stationary points on the hippopede are determined by tangent arcs perpendicular to the ecliptic, and taking into account its length and width. We can see in Fig. 4b (which preserves the ratio length/width for the hippopede generated when $i = 23;51,20°$, as in al-Biṭrūjī's text) that the star indeed moves in accession from S_1 to S_2 (namely between 110° and 250°, approximately), and in recession from to S_2 to S_1 (namely between 250° and 110°, approximately). It is also clear that great changes in declination and latitude (23;51,20° and about 22°, respectively) are required to obtain relatively small changes in longitude (the maximum amplitude is about 9;39°).

Sentences 128–133 contain the only section in the chapter where Scot's and Qalonymos' versions differ, and they provide further evidence strongly suggesting that al-Biṭrūjī did not invent the model. In Scot's version, we read (Fig. 5) that, since arcs BK (the maximum declination or the obliquity of the ecliptic) and AL (= 45°) are known, arcs LT and AT will be known, according to the procedure explained in the *Almagest* (i.14 and

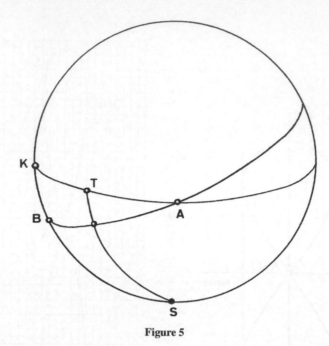

Figure 5

16). In Qalonymos' version, the reference to Ptolemy is replaced by the description of a method based on Jābir's trigonometry (in fact, on Ibn Muʿādh's rules)[39]. Al-Biṭrūjī asserts that

$$\sin AB : \sin BK = \sin AL : \sin LT,$$

whence arc $LT = 16;37,21°$ (taken from *Almagest*, i.15, confirming so that al-Biṭrūjī uses Ptolemy's value for the obliquity of the ecliptic;[40] accurately computed: $16;37,2°$), and that

$$\sin LB : \sin LS = \sin TK : \sin TS,$$

whence $LS = 73,22,39°$ (acc. $73;22,58°$), $TK = 47;31°$ (acc. $47;33,18°$), and $AT = 42;29°$ (acc. $42;26,42°$).[41] Although in his model arcs λ are a function of arcs α, al-Biṭrūjī seems to be unaware of it and he follows slavishly Ptolemy and Jābir, who explained how right ascensions α can be computed as a function of ecliptic arcs λ.

Finally, it is worth mentioning that, as suggested by sentences 99, 102, and 141, al-Biṭrūjī considered it was sufficient, for accounting for a combination of precession

---

[39] See *Gebrii filii Affla Hispalensis de astronomia libri IX*, Nuremberg, 1534, ii, De scientiis particularibus; on Ibn Muʿādh, see J. Samsó, *Las ciencias de los antiguos...*, pp. 139–144.

[40] Toomer, *op. cit.*, pp. 72 and 71 (critical apparatus).

[41] As indicated by Goldstein (1971, vol. 1, p. 25), $42;29°$ also results from linear interpolation in the table in the *Almagest* (ii.8) for the right ascension of $45°$.

and trepidation, to make the *istīfā'* of the sphere of the stars (arc γ) slightly smaller than its *taqṣīr* (arc α).[42]

4. The text

There is an essential agreement between Scot's and Qalonymos' versions with respect to the content, although style and vocabulary of the translators are quite different. Scot's language is more concise and clear, and Qalonymos' seems to be a slightly enlarged version.[43] Accepting Carmody's and Goldstein's claims on the relationship between the Arabic, Tibbon's and Qalonymos' versions,[44] the text which follows translates Qalonymos' text, although I have systematically collated it with Carmody's edition of Scot's version.[45]

My purpose has been to provide a translation as close as possible to the Latin original, avoiding simplifications (clarifying, for instance, al-Biṭrūjī's obscure expression of some concepts) and anachronisms (modifying his terminology). Editorial insertions and sentence numbers are given between square brackets. In order to facilitate comparative analysis, correspondences (folio and line numbers) with Qalonymos' printed version as well as Carmody's edition of Scot's version and Goldstein's English trans-

---

[42] There is some incertainty on this matter in the text for, as we have above seen, sentence 55 (which explicitly claims that during some periods no motion in longitude at all can be observed) excludes a variable precession; however, sentences 102 and 99 (where it is also said that, if precession really exists, its rate is different from Ptolemy's), are ambiguous. Indeed, the text does not allow us to think that al-Biṭrūjī distinguished variable precession from variable trepidation. We could get a variable precession from al-Biṭrūjī's model if the *taqṣīr* of the eighth sphere is made equal to $p + \alpha$, and the *istīfā'* equal to α (where p and α are the precessional and trepidational rates, respectively), with $p > \alpha$. I will deal with this model in a forthcoming paper on al-Biṭrūjī's planetary theory.

[43] For the divergences, see Carmody, *op. cit.*, pp. 16–17. Carmody suggested three possible explanations: "either all the Latin manuscripts belong to a corrupt sub-family, or they represent a shortened version of Scot's making, or they translate a second Arabic version". In my opinion, it is likely that the common ancestor of the preserved Arabic, Tibbon's and Qalonymos' versions resulted from a revision by al-Biṭrūjī himself of a first redaction of the work (preserved only in Scot's version); be that as it may, in what concerns al-Biṭrūjī's astronomy these divergences are not significant at all, and geometrical errors, for instance, remain unchanged in both versions.

[44] Carmody, *op. cit.*, p. 16: "The Latin version of Calonymos is much closer to the Arabic than is that of Michael Scot; in general, in the variants of the present edition, a quotation from Calonymos automatically serves to represent the original of al-Biṭrūjī"; Goldstein 1971, vol. 1: p. 49: "It became apparent that the Hebrew version is a word for word translation which, in general, agrees very well with the Arabic text". Qalonymos' version was not used by Goldstein for his translation except for the figures.

[45] Unfortunately, Carmody's critical apparatus contains more information on the differences between Qalonymos and Scot's versions than on the variant readings from the eleven Latin manuscripts used to establish the edited text.

lation are indicated at the beginning of each paragraph.[46] In the few places where I have preferred to translate Scot's readings (as, for instance, in sentences 93, 127, and 135), both Latin texts are given in footnotes. The figures for the text given below (Figs. 6, 7, and 8) are tridimensional redrawings of those that we find in the manuscripts and the printed editions.[47]

Second Chapter

9v14,**IX**1, §63 [1] We must deal next with the orb that succeeds it, which is the orb of the fixed stars: according to the order [of matters] we have proposed we must now begin the chapter on the apparent motion of the stars existing in this orb (that which follows the highest one, and is nearest and contiguous to it) and about the arrangement [producing it].

[2] We say that the stars of this orb seem to have two different motions besides the diurnal one, as it appears when we observe them; the one in longitude and the other in declination, in longitude with respect to the signs, in declination with respect to the north and the south.[48] [3] According to Ptolemy, the one which takes place in longitude goes from west to east and on the oblique orb. [4] However, astronomers closer to us said that they had found that this motion was not uniform throughout, but irregular, for it sometimes increases and sometimes decreases.

9v22, **IX**2*, §64 [5] Nevertheless, others before them said that the motion of the fixed stars goes sometimes forward, in the direction of the order of the signs, and sometimes backward, in the opposite direction, and that the stars with this motion do not complete

---

[46] For references to Qalonymos' and Carmody's texts I will use the procedure explained in note 13. "**IX**23*" means that the beginning of the paragraph does not coincide with the beginning of sentence 23 in Carmody's chapter IX. "§66" denotes the paragraph of the same number in Goldstein's translation. When my subdivisions into paragraphs differs from Goldstein's, his paragraph number is given in the text between square brackets.

[47] These plane figures are meaningless without an appropriate reconstruction, and once they have been properly reconstructed, they are superfluous. In the Latin text the position of points F, Q (Fig. 6), and H (Fig. 8) – the clue for a correct understanding of the text – "is in doubt in the manuscripts" (Carmody, 1952, p. 162), and the same occurs in the Arabic text reproduced by Goldstein (1971, vol. 2, pp. 3–431) and in the printed editions. Figure 6 corresponds to Figs. 1–2 in Carmody (1952, p. 96), and it combines two figures of Qalonymos' version (ff. 10v and 11r), corresponding to Figs. 2.1 and 2.2 in Goldstein (1971, vol. 1, pp. 82 and 84; see also vol. 2, pp. 155 and 161). Figures 7 and 8 correspond to the figure in f. 13r of Qalonymos' version, Figs. 3–4 in Carmody (1952, p. 104), and Fig. 2.3 in Goldstein (1971, vol. 1, p. 95; see also vol. 2, p. 195).

[48] *with C*: et unus modus est secundum longitudinem et alius secundum latitudinem: secundum longitudinem quoad signa, secundum latitudinem quoad meridiem et septentrionem; *K reads*: quorum motuum unus est in longitudine et est contra motum uniuersi quem appellant secundum ordinem signorum, alius uero in latitudine et est qui uidetur procedere ad septentrionem et meridiem [= *of these motions, the one takes place in longitude and in the direction opposite to that of the motion of the universe (that they call 'following the order of the signs'), the other one in declination, which seems to take place to the north and the south*].

the circle of the signs;[49] [6] this motion was called by them motion of accession and recession, and it was accepted by subsequent astronomers, although it is still uncertain. [7] The master Abū Isḥāq al-Zarqāla composed on it his tract *On the motion of accession and recession,* for which, as well as for the variation in declination of the solar circle due to it, more recent astronomers composed tables. [8] This motion, as it appears to us, is as Abū Isḥāq posited, [§65] except that it is possible that with this motion of accession and recession a motion in longitude occurs following the direction of the signs, although they have not still reached the truth of this matter from observations; knowledge on it cannot be reached except after a very long period of time with continual observations through it – and it is so extremely close to the truth.

9v34, **IX5** [9] We will now describe how this motion occurs and which the arrangement [of orbs producing it] is. [10] When the highest orb is moved with the diurnal motion about its two poles (which never move with this motion), the succeeding orb is also moved with this motion. [11] Now, as it is carried by the highest one and its poles are different from the poles of the highest one, for they are inclined with respect to them, it is necessary that these two poles are moved and they trace with their motion two circles whose poles are the poles of the universe.

9v39, **IX6** [12] Although this orb is moved with the motion of the highest one, it lags behind this motion, and therefore its two poles also lag behind so that they do not complete these two circles in the time in which the highest one accomplishes its revolution. [13] For this reason, this orb is moved about its poles in the same direction as the highest one, in order to bring to completion what could not be accomplished and was lacking with respect to the motion of the highest orb, and [this motion] is performed around its poles (which do not move with this proper motion) until the motion of the highest orb is caught up, namely until it arrives at the place at which the highest orb completed **[10r]** its revolution.[50] [14] The motion of this orb about its poles is named 'motion of completion'. [15] When this orb brings to completion the revolution and the highest orb is caught up,[51] as it was above said, then the fixed stars bring also to completion their motion in longitude, due to what they add with the motion of their orb,[52] and so of the

---

[49] et quod ipse stellae non perficient circulum signorum motu suo ad finem eius *K; in C (IX.2) the text corresponding to sentences [4]–[5] is*: et postquam inuenerunt alii posteriores quod iste motus diuersatur in uelocitate et tarditate secundum modos diuersos; uocauerunt ipsum motum antecessionis et recessionis eo quod anterioratur et posterioratur secundum sucessum signorum [= *later, other subsequent astronomers found that this motion varies in swiftness and slowness in different ways; they called it motion of accession and recession because it goes forward and backward in the direction of the signs*].

[50] *and* [this motion] ... *completed its revolution*: et perficit sese super polos suos qui sunt fixi huic motui sibi proprio quoad perueniat ad motum suppremi, scilicet quousque applicet ad locum postquam perfecerit suppremus reuolutionem suam *K; C IX.6*: et tunc mouetur super proprios polos iam fixos ut possit acquirere complementum simile superiori [= *it is then moved about its own poles, which are at rest with respect to this motion, so that it can obtain a complement similar to the highest one*].

[51] *C*: tunc dico quod in isto motu incurtatio erit polis rotantibus, complementum autem [*instead of Carmody's* aut] erit ex celo.

[52] *C*: et stelle in longitudine complebunt motum suum qui est motus completus.

lagging of the orb only remains the lagging of the poles (for the motion of completion takes place around them and they are at rest with respect to it), and their lagging takes necessarily place in the opposite direction, that is, opposite to the motion of the universe.

10r6, IX8, §66 [16] As the distance of these poles from the poles of the highest orb is the same and immutable, these two poles of this orb trace with their defect two circles on which they are moved with the course of time; we name these circles 'circles of the path of the two poles', and they are on this orb, as we said, parallel to the equatorial circle, and their poles are the poles of the equator. [17] The amount[53] [of the radius] of this circle is equal to the amount of the [maximum] declination of the stars placed on the middle of this orb to the north or the south of the equator, and the distance between their maximum declinations to the north and to the south is twice the radius of this circle – that is, the arc of a great circle traced to the poles of the highest orb from the circle of the path of the pole of this orb is twice the inclination of the solar circle to the equator.[54] [18] The [maximum] declination of the stars on the middle of this orb will be equal to the [maximum] declination of the sun to the equator, as the most ancients asserted, although they did not precise that amount. [19] The two circles of the path of the two poles are identical to the two circles traced by both poles when a revolution of the highest orb is accomplished.[55]

10r18, **IX**10, §67 [20] As this orb lags behind the motion of the universe and it brings to completion this lagging with its own motion around its poles (and the pole does not move with this additional motion), the stars existing in this orb maintain all together their positions in longitude, but not in declination. [21] The pole [of this orb] lags behind exactly by the amount by which the orb lagged behind; however, the stars on it do not lag behind, as later astronomers believed, for they accomplish with the motion of completion of their orb what the orb lagged behind [the highest one]. [22] Thus, both poles revolve on these two circles of their path, although they do not move with the motion by which the orb is moved around them.[56]

10r25, **IX**11, §68 [23] Since the distance of the stars from both poles [of their orb] is[57] always the same, the stars seem to incline in the direction towards which the poles move when they revolve with their lagging motion. [24] They call it motion in declination; for the poles, due to their lagging, move backwards in the circles of their paths and lie at different times in different directions with respect to the poles of the highest orb, and for this reason the declination of the stars changes when they follow their poles, as their distance from them is necessarily the same. [25] Thus, a star that [at a certain time]

---

[53]   *C* quantitas, *K* spacium.

[54]   *with C; in K*: est quasi duplum declinationis <et est fere duplum declinationis> circuli solaris ab aequinoctiali (*the part of the sentence between angle brackets is, in my opinion, superfluous and misleading, but it translates the preserved Arabic text; cf. Goldstein, 1971, vol. 2, p. 149. I am grateful to J. Samsó for this information.*

[55]   *sentence [19] is absent from C.*

[56]   however... [22]... around them: *absent from C.*

[57]   *with C; K*: quoniam distantia stellarum ab unoquoque amborum polorum non est eadem [= *since the distance of the stars from each one of both poles is not the same*]; *this reading is also possible, but in this case the poles mentioned in the sentence are those of the ninth orb. According to J. Samsó the Arabic sentence (Goldstein, 1971, vol. 2, p. 151) is affirmative.*

coincides with the celestial equator does not remain stationary on it; it inclines from it depending on the positions of the poles on the circles of their paths and on the directions in which they lie with respect to the poles of the highest one. [26] Therefore, the star will be sometimes north of the equator and sometimes south of it,[58] according to the inclination of both rotating poles with respect to the poles of the highest orb; the other stars [which at that time did not coincide with the equator][59] will be likewise moved from their positions, and they will sometimes approach the equator and sometimes recede from it.

10r35, IX13, §69 [27] Thus, how can the stars move with these two motions of accession and recession that astronomers closer to us described and confirmed by observation? We must say that this motion necessarily appears in the stars due to the revolution of both poles on both circles of their paths and from the combination of the lagging of the two poles (namely, their lagging motion due, as above said, to the fact that they do not accomplish a revolution in the time in which the highest orb does it) with the motion of the orb of the stars around its own poles – for this motion is not visible to us, and we can only perceive its accidental [and apparent] consequences, as we will explain later.

10r41, IX14 [28] Let us now make some preliminary remarks that are necessary to help us to conceive this motion. We assert that, if there is a sphere whose two poles are moved on two small circles (a motion with which the sphere itself is moved around the pole of another sphere placed above it), and a point **[10v]** is marked on its surface (namely on the surface of the sphere whose poles are moved), then that point (due to the motion by which it is moved with the motion of its poles revolving on both small circles) will trace a true circle.[60]

10v3, IX15, §70 [29] Let, for instance, AB be the sphere, whose poles (which are G and E) revolve on two small circles GD and EZ around the two poles of another sphere. [Fig. 6] Let A be a point marked on the surface of the sphere AB and let the sphere AB be moved with the motion with which its poles revolve on both circles GD and EZ, and let it carry out its revolution until it returns to its [starting] place.[61] [30] It must be said that with the motion of the sphere point A will trace a true circle.[62]

10v12, IX16 [31] The proof of this is as follows: let us imagine that points G and E of the two circles are the two revolving poles of this sphere, and let us draw through them and point A an arc of a great circle; thus, this arc will necessarily be a semicircle. Point A is marked in the middle of it,[63] so that its distances from both poles, G and E, will be known. [32/§71] When the sphere rotates with the revolution of both poles on the two circles GD and EZ, arc GAE will move with them, and its motion will generate an imaginary sphere when the poles complete a revolution on the two circles GD and EZ.

---

[58] on the directions... [26] ... south of it: et partes secundum quas respicient polum superioris ad dextram scilicet et ad sinistram; tunc erit stella aliquando a dextro equatoris et aliquando a sinistro *C*.

[59] *the sentence between square brackets*, que non sunt super equatorem, *is absent from K.*

[60] circulum perfecte circulationis *K*, circumferentia non spiralis *C*.

[61] ad locum suum *K*, ad locum a quo inceperunt *C*.

[62] circulum perfecte circulationis *K*, circulum completum *C*.

[63] in medietate eius *K*; *the text in C* (et punctus A in hoc dimidio est notum) *does not require to place point A in the middle of this semicircle.*

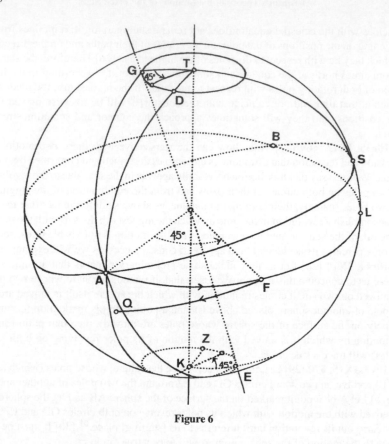

Figure 6

Let us imagine that this sphere is at rest, and let G and E be its poles. [33] Thus, since the distance of A from each of the poles G and E is the same in both directions[64] throughout every revolution, and the distances AG and AE are equal throughout the motion of point A in every rotation,[65] and the distance AG, as well as the distance AE, remain without change throughout the motion of point A, points G and E are therefore the two poles of the circle traced by point A with the motion of the sphere on which it is marked[66] – sphere which is [also] moved with the motion of its revolving poles on both circles GD and EZ); [34] and it is certainly a true circle,[67] for the arcs traced in both directions from

---

[64] in ambabus partibus *K*, in utraque parte *C*.

[65] in toto circulo *K*, in tota rotatione *C*.

[66] *reading in K* ad motum orbis super quo designatum est cum mouetur *instead of the textual* ad motum orbis super quo designatum est moueri.

[67] et est quidem [circulum] perfecte circulationis *K*, et saluatur circumferentia *C*.

the poles of the sphere to the circumference[68] are equal. This is what was intended to explain.

10v27, IX20, §72 [35] Moreover we assert: if we consider a sphere whose poles revolve as we said, and in addition to this motion the sphere is moved with another motion around its own poles (which do not move with this additional motion),[69] and these motions are combined, and a point is marked on the surface of the sphere, and the sphere is simultaneously moved with these two motions, then the marked point will not trace with these two motions a true circle,[70] [all the points of which lie] on the same plane, and it will not return after a complete revolution of the sphere at the place of the circle from which it started, but it will incline to it, and the revolved point will generate something like a gyrating figure, called *leuleb*;[71] [36] that is, the circle will begin at a given point and once completed a revolution it will end at another point in another plane, and when the sphere has been revolved a number of revolutions, the circles generated by that point will be like a gyrating revolution, called *leuleb*.[72]

10v37, IX22, §73 [37] Let, for instance, be AB the sphere whose poles G and E revolve on the two circles GD and EZ, and let the sphere be moved with their motion, namely with the motion of the poles around the poles of both circles, which are [points] T and K. [38] Let this sphere be also circularly moved at the same time with another motion different from the previous one about poles G and E (which do not move [with this motion]), and let point A be marked on the surface of the sphere AB. [39] We must say that point A, with the motion of the sphere about poles G and E combined with its other motion with which it moves with the motion of its revolving poles, will generate a gyrating circle and a *leuleb* figure.[73]

10v44, IX23* [40] The proof of it is as follows: let us draw a great circle through the poles of the two **[11r]** small circles on which the poles of the sphere revolve; this is circle TSLK, intersecting circles GD and EZ at points E and G.[74] It is evident that point E is opposite to point G so that the line drawn between them is a diameter of the sphere. [41/§74] Let us imagine the circle generated by point A when the sphere is moved about this diameter: if the sphere did not have another motion (with which the diameter is moved on the two circles[75] GD and EZ), it would be circle ASB, and the

---

[68] nam arcus protracti a polis sphaerae ad circumferentiam undique ad quamlibet duarum partium *K*, quia arcus qui extrahuntur a polis spere ad illam circumferentiam *C*.

[69] *with C*: et habebit spera cum isto motu alium motum super polos fixos isti motui; *K reads*: et fuerit huic sphaerae cum illo motu alius motus super polis <duorum circulorum> quiescentibus <in motu amborum polorum> (*between angle brackets, the words which, in my opinion, are meaningless, although they literally translate the Arabic text; I am grateful to J. Samsó for this information*).

[70] circulum perfecte circulationis *K*, circulum uerum *C*.

[71] per modum figurae giratiue dicte laulabine *K*, sicut figura leuleb, id est, spiralis *C*.

[72] ut reuolutio giratiua dicta laulabina *K*, punctum giratum, et hoc est leuleb *C*.

[73] faciet circulum giratiuum et figuram laulabinam *K*, facit leuleb *C*.

[74] intersecting ... and G: *absent from C*.

[75] *reading in K* ad motum [respectu] huius diametri [qui mouetur] super duobus circulis *and in C* propter motum [respectu] axis sui [qui mouetur].

revolving motion [of point A] perfectly circular,[76] as we previously explained. [42] Let us now imagine the circle generated by point A when the sphere AB is moved about poles K and T: if the sphere had not another motion, it would be circle ALB (and this circle, namely ALB, is necessarily parallel to circles EZ and GD).[77] [43] Let us draw horizon TAK through the points of intersection of these two great circles and through poles T and K. Since the rising of point A is placed on the horizon at the intersection of both circles, when pole E is constrained by its motion on the circle [EZ] toward Z, and the sphere inclines due to its constraint, point A will therefore move[78] in the direction of the inclination of the pole. [44] When the sphere is moved so with the other motion about poles T and K, point A will be moved from its place inclining from circle ASB (for it retains the same distance from pole E which is moved toward Z), and it arrives so, for instance, to point F.[79] [45/§75] When the sphere is rotated about poles G and E,[80] and point F[81] rises [later] from horizon TAK, it will not rise at its original place, but at point Q in this case;[82] similarly as long as pole E is moved toward Z, and when [pole E] reaches point Z, point A will reach [point] L. [46] Thus, after each revolution it will rise from a point different from its rising point in the succeeding revolution, and it will rise from that point and will end with its revolution at another point, and non true circles[83] will be always generated, for every one of these circles, as we asserted, is placed on a different plane, and a gyrating figure, called *leuleb*,[84] will result from all of them. [47] This will also be the behaviour of point A when the pole is moved from point Z to the point opposite to point E on the circle [of the path of the poles] and with this motion point A is moved from L to B, for it will likewise generate a figure similar to the first one, and in the two remaining quadrants two figures similar to the first two ones will result, and point A will return to its initial place. [48] Thus, the combination of all these two motions [namely, all the motions that take place about these two different axes] will produce four figures, as we asserted, and this is what we intended to represent.

11r32, X1, §76 [49] We also assert that the circle traced by the heavenly bodies – the fixed stars as well as the planets – with the motion of their orbs about their poles is inclined to the celestial equator according to the distance of the pole of the orb of that planet or the fixed stars from the pole of the highest one, that is, from the pole of the celestial equator, and the more the poles of the equator are far from the poles of the orb of the heavenly bodies, the more the circle on which is [placed] the body inclines to the equator, and this inclination is as great as the distance of the pole [of the orb] to the pole of the highest one – this is evident.

11r37, X2 [50] We also assert that when whichever of these orbs is moved by itself about its own poles in the direction of the motion of the universe, namely of the diurnal

---

[76] circulationem perfectam *K*, rotatio circularis sana *C*.

[77] and this circle ... and GD: *absent from C.*

[78] prosequitur *K*, trahetur *C*.

[79] ad puntum B *KC*.

[80] *K*: super polis T K; C: super duos polos T K.

[81] punctum B *K,* punctum F *C*.

[82] eius exempli gratia *K*, per similitudinem *C*.

[83] circuli non perfecte circulationis *K*, circulos incompletos *C*.

[84] proueniet figura omnium giratiua dicta laulabina *K*, et sit figura eorum figura leuleb *C*.

motion, and the heavenly body fixed on that orb is moved with that motion along some distance on its oblique circle (although this motion is not perceived by us), the rising degrees of the distance traversed by the star on its oblique circle will not be always equal to the number of degrees of the equatorial circle which rises with them,[85] [51] for sometimes it will differ according to the inclination of the part of the oblique circle[86] traversed [by the body] to the north or to the south of the celestial equator or [when it is] at the intersections; this has been already explained in the *Almagest*.

11r44, X2*, §77 [52] It has been also shown that **[11v]** when the maximum declination is known, the declination of any degree marked on the oblique circle will be known, and the number of degrees of the equatorial circle rising with this degree in the assumed horizon will be also known. [53] It has been also explained there that (concerning the number of ascending degrees) the rising times of a complete quadrant of the oblique circle (of those included between the two equinoctial points and the two solstitial points) are equal to the rising times of the complete quadrant of the equatorial circle facing it; [54] however, they are unequal when the degrees [of these arcs] are less than a quadrant of circle, and so the rising times [of arcs of the oblique circle] are sometimes greater than the rising times [of arcs of the equator facing them] and sometimes they are lesser; we will deal with this matter in its place, God willing.

11v8, X4, §78 [55] As the distance covered by the motion of the fixed stars has been found by observation to vary at different times – for sometimes the retardation of this slow motion is found to be lesser and at other times greater, and sometimes following the order of the signs and at other times in the opposite direction, namely that of the motion of the universe, and sometimes, during a long period of time, no motion at all could be determined from observations by people living then[87] –, [56] what mainly led them to know the motion of the fixed stars was the variation in their declination[88] that they noticed when they observed the stars situated on the equator; [57] namely they observed that some star[s] situated on the equator were removed from it, and that those which were in the past to the north of the equator were [later] to the south, and those which were to the south, to the north of their places; so, they concluded that this motion takes place on the inclined circle. [58] Moreover, as this motion with its [changes in] declination recedes, they inferred that this motion, which moves the stars on the oblique circle, takes place in the direction opposite to that of the motion of the highest orb. [59/§79] Thus, since the seven planets vary greatly in their motion in longitude, but their variation in

---

[85] id quod ascendit ex gradibus illius spacii quos pertransiit stella in circulo suo obliquo non erit aequale semper ei quod ascendet cum eis ex gradibus circuli aequinoctialis K, non erit illud quod ascendit de equatore diei cum eo quod de suo celo inclinato equale semper ei quod secuit stella de celo inclinato C.

[86] circuli declinationis K, de circulo inclinato C.

[87] quandoque magno temporis spacio non apprehenditur in eis motus ex obseruatione hominum illius aetatis K, et remanent tempore longo sine motu sensibili quoad considerationes hominum in illo tempore C.

[88] declinationem in latitudine K, extractio in latitudine C.

latitude is small, they posited (with respect to the different motion) that the motion of all of them follows this oblique orb[89] and [takes place] about its poles.

11v22, X7 [60] Once they assumed that the orb of the fixed stars is oblique, I do not know why they supposed that it is placed immediately below another higher orb without stars – for it would have been sufficient for them to posit the oblique orb, and there is no need to work on the hypothesis of a higher orb.

11v28, X7 [61] Moreover, why did they not posit this oblique orb (which moves these orbs about their poles with a motion different from the motion of the highest one, namely the motion of the universe) to be placed in a position opposite to that of the highest one – for instance, below the orb of the moon? [62] For they have no proof from astronomy that this orb is [really] placed below the highest one and above all the others, and it would have been more reasonable to assume that it is placed below the orb of the moon, moving first the lunar orb, explaining so that the orb of the moon is the swiftest one among all the orbs moved with this motion because it is the closest one to its mover – therefore, the closer an orb would be to the moon, the swifter it would be, and this would explain why the orb of the fixed stars is the slowest one, and so all the orbs would be properly ordered with respect to this motion.

11v32, X8*, §80 [63] Moreover, why did they not attribute the eccentricity of all the orbs exclusively to the oblique orb? If so, the diversity of these orbs in their swift and slow motions would result from the eccentricity of the oblique circle and from the fact that the motion of all the remaining orbs would take place around it.[90] [64] Instead they concede and admit some separation between these orbs and some division of them into parts that differ each other with respect to the motion, and at the same time some reciprocal participation and combination with respect to motion and substance.

11v37, X10, §81 [65] As the ancients did not attain a certainty on this motion [of the fixed stars] admitted by them, they had many doubts about it and many troubles concerning the truth of their motion. [66] For the oldest ones – such as Hermes and the Masters of Talismans after him – assert that the stars move sometimes following the order of the signs and sometimes in the opposite direction, as if it were evident for them or conceded by them.[91] [67] Those who came after them – the Chaldeans and those who observed the stars[92] before the time of Nabonassar in order to establish the truth of what the oldest ones discovered – did not find a motion in the stars and rejected this motion stated by their ancestors, for they did not compose tables for them **[12r]** or a theory[93] that could explain how this motion is possible, [68] and their view was that the orb of the

---

[89] motum omnium simul sequi hunc orbem obliquum quoad motum diuersum *K*, motus omnium est sequens celum inclinatum in motu diuerso *C*. *The expression* motus diuersus *seems to allude here to the motion of the Different* (*as opposed to the motion of the Same*) *in the two-sphere model of Plato* (Republic *X 617A, and* Timaeus *35–40*) *and Aristotle* (Metaphysics Λ 8).

[90] from the fact ... around it *is a tentative translation; in both versions this sentence is obscure*: et eius moti omni orbium super se *K*, et suus motus celorum istorum super illud *C*.

[91] quasi hoc esset eis quid per se notum seu traditum a maioribus *K*, et uidetur quod ista res fuit scita siue concessa eis *C*.

[92] ut caldei et qui inuenerunt motum harum stellarum *K*, ut alkademein de illis qui inspexerunt has stellas *C*.

[93] computum nec theoricam *K*, tabulas neque astrologiam *C*.

fixed stars is the orb which moves [the remaining ones] with the diurnal motion and that the orb of the signs (which is the inclined circle of the sun) intersects the equatorial circle at two points – one of them is named the vernal equinox, the other the autumnal equinox (which are the beginnings of Aries and Libra)– and [they believed] these intersections are immutable.

12r5, X13, §82 [69] Afterwards, those who came after them not long[94] before the time of Alexander – as Hipparchus posited from observations made by Timocharis and Aristyllus in the 450th year of the era of Nabonassar, and also from observations made by the geometer Menelaus in the 845th year of the era of Nabonassar, and by himself about 400[95] years after the death of Alexander – asserted to have found from observations made by the people living at these times that the stars are moved following the order of the signs, and they carefully considered their motions and concluded that the motion of this orb takes place only in the direction of the signs.

12r25, X15 [70] Later, Ptolemy made observations almost 266[96] years after Hipparchus and found that the motion of the fixed stars takes always place in the direction of the signs. Hipparchus computed this motion and asserted that it takes place about the poles of the orb of the signs and in their direction one degree each 100 years. [71] Later, Ptolemy found the motion of the stars in agreement with Hipparchus' reckoning and confirmed this motion.[97] [72/§83] However, astronomers who came after Ptolemy were truly amazed by these earlier observations and – since they observed these stars, compared their positions determined by observation with the positions which resulted from that rate, and found them in disagreement – did not accept this motion. [73] One of these who came after Ptolemy, Theon of Alexandria, believed that the stars have a motion of accession and a motion of recession, each one of 8 degrees, and that they also have a motion in the direction of the signs of one degree each 100 years. [74] Others who came after him rejected this motion when they did not find that the observed positions of the stars differed (sometimes by some additional amount and sometimes by some negative amount) from the positions at which they ought to be placed according to the computation made by the ancients.[98] [75/§84] Later, al-Battani asserted that the fixed

---

[94] et postea succedentes eis neque multo tempore ante alexandrum *K*, et illi qui uenerunt post istos per magnum tempus ante tempus alexandri *C*.

[95] *sic in both versions*: atque demum ex obseruatione ipsius yparchi post obitum alexandri fere 400 annis *K*, et postea considerauit Abrachis per se post mortem Alexandri fere anno 400 *C*.

[96] ferme 266 annis *K*, in anno 265° *C*.

[97] *with C*: et postquam inuenit Tholomeus progressum stellarum in uniformitate numerationi Abrachis, firmauit et sententiauit istum motum secundum illud; *K reads*: et cum inuenerit Ptolomeus id quod perscrutatus fuit ille ex locis stellarum iuxta quidem eandem uiam constituit computum illius motus et subtilius quidem.

[98] *with C*: et illi qui uenerunt post expulerunt istud, qui non inuenerunt sus loca per uisum alia a suis locis per equationem computationis quam fecerunt antiqui aliquando diuersa per additionem aliquando diuersa per diminutionem; *K reads*: quem quidem motum posteriores reiecerunt inuenientes loca earum secundum obseruationem in locis praeter loca in quibus erant situate in locatione sua priori nam aliquando addunt et aliquando diminuunt iuxta tempora determinata eis.

stars move away from the point of the vernal equinox with different speeds in equal times, and thus he left aside this motion.[99]

12r25, X18 [76] When[100] Abū Isḥāq al-Zarqāla considered these different motions, he found a way of combining them according to his views, although he did not determine perfectly the truth of the motions of the stars. [77] He constructed for them tables and a theory[101] according to which the poles of this orb are moved on two circles parallel to the celestial equator, so that the motion of the stars follows the motion of these two poles. [78] What he said induced us to discover something never considered by others,[102] namely a motion doing really so – this is indeed the motion of this orb (which is placed immediately below the highest one and follows its motion) about its poles in order to complete the lagging by which it remains behind the motion of the highest one, so that [due to this motion] it can be known[103] and distinguished from it. [79/§85] Thus, it is true that the mentioned motion is as al-Zarqāla posited, that is, what appears to us as irregular in the motion of the fixed stars is a motion of accession and recession,[104] although it takes actually place unlike [what they asserted], for accession according to them is a motion opposite to the motion of the universe, and recession according to them is a motion in the direction of the motion of the universe; however, according to the truth, it occurs the reverse, as it will be evident. [80] So, [the appearance of] the motion opposite to the motion of the universe that Ptolemy posited is saved with[105] accession and recession (as in the motions of the planets: the appearance of retrogradations and stations is saved with their motion opposite to the motion of the universe); however, its amount is still not known.

12r40, XI1 [81] In order to verify what has been said, let us return to the description of the motion of this orb, that is, the orb of the fixed stars; we will present later a figure so that our representation of this motion could be more adequate and its truth more accessible. [82/§86] We assert that when this orb is moved about its poles with its proper motion[106] following the motion of the highest one (motion that we call 'motion of completion'[107]), and its poles revolve opposite to the motion of the universe on the two circles of their paths (due to their lagging with respect to the motion [12v] of the highest orb, for both poles lag behind it, although the stars do not lag behind, for their motion of completion takes place about these poles, which do not move with it), the

---

[99] praetermisit hunc motum K, dimisit rem suam in hoc C (*lit.*: he left the matter here; *reading* dimisit *from mss DIKP instead of* diuisit).

[100] *reading* quando (*as in* C) *instead of* quoniam.

[101] astrologia C, theorica K.

[102] quod latuit ipsum C.

[103] eo cognoscatur *om.* C.

[104] quiescat ergo motus secundum quod dixit Abu Isac Azarkel, quod illus scilicet quod apparet de diuersitate motus stellarum est motus accessus et recessus C, et uerificatus est nunc ille motus sicut posuit alzarcala predictus ut scilicet quod uidetur de diuersitate motus stellarum fixarum sit accessus et recessus K.

[105] stat cum K, est saluus cum C.

[106] C *add.* motum qui est per se.

[107] motus complementi C, motus supplementi K.

motion[108] of the stars which takes place in longitude about the poles of this orb in the direction of the motion of the universe is combined with the motion of both poles[109] on the two circles of their paths (this [second cause] acts by accident,[110] for the distance of the stars to the two poles is immutable). [83] Thus, when the two poles are moved in some direction due to the change of their position, the stars existing in this orb are moved [with respect to the equator] in the direction towards which [the poles] remove them, according to the distance of the two poles to the poles of the equator; [84] therefore, the stars of this orb incline (although they brought to completion the motion in longitude of the highest orb and, due to the motion by which this orb is moved about its poles, they caught up with it – except for a little amount [if a motion of the stars in the direction of the signs really exists], something still not determined by observation[111]), and in this way their declinations change with the motion of the two poles.

12v9, XI4, §87 [85] Since this motion of the stars – by which the motion of the universe is followed to bring to completion [their lagging] – does not take place on circles parallel to the celestial equator, but on circles inclined to it, as we pointed out in our introduction, the stars moving on the girth of this sphere will generate a circle inclined to the celestial equator, which cuts it into two halves, like the circle of the signs, and its inclination will be equal to the distance of the poles of this orb from the poles of the highest one. [86] The mean circle on the girth of this orb[112] will intersect the celestial equator at two points similar to the two equinoctial points, and the two points placed [on the circle] at the maximum distance between them are similar to the two solstitial points. [87] The motion of the fixed stars in the middle of this orb[113] takes place on this oblique circle; the remaining stars on the girth of this orb revolve on circles parallel to this oblique [circle] – except that the diurnal motion of all of them takes place on circles parallel to the celestial equator.

12v19, XI6*, §88 [88] This motion on this oblique circle in the direction of the motion of the universe, named by us 'motion of completion', was discovered by none of my predecessors; since it was unknown to them, they fell into error, for they believed that the orbs below the highest one were moved opposite to its motion and resisting it[114] with their natural motion; this led them to complications and far from the truth on these motions and the arrangement of orbs producing them.[115]

---

[108] motus stellarum *K*, mutatio motus stelle *C*.

[109] cum eo quo mouentur ambo poli *K*, cum hoc quod mouet ipsam mutatio duorum polorum *C*.

[110] quod est agens per accidens: *om. C*.

[111] et peruenerint per motum quo mouetur orbis super polis suis praeter paululum quod adhuc non est re uera perceptum *K*, et alkaneauit [= *pertingit*; from Ar. *laḥaq*] illud cum motu sui celi super duos polos preter parum uel multum, et est res que usque adhuc non est uerificata *C*.

[112] in zona huius orbis *om. C*.

[113] *with C*: motus stellarum fixarum que sunt medie in isto celo; *K*: et super hoc circulo obliquo erit motus stellarum fixarum medius et zona huius orbis.

[114] moueri contra motum eius et oppugnare eum motu suo naturali *K*, mouentur motu contrario eius motui *C*.

[115] in errorem extractionis ueritatis in suis rebus et sua astrologia *C*, exitu a ueritate rei et theoricae earum *K*.

12v23, **XI**7*, §89 [89] As the proper motion of the orb of these stars is uniform and similar to the motion of the highest one and in the same direction – except that it is inclined to it, namely to the celestial equator –, the distances traversed by the stars on its oblique circle differ from the distances facing them on the circle of the celestial equator and rising with them.[116] [90] This was explained in the *Almagest*, and we will explain it as an example, for the ascending degrees of this oblique circle are not always equal to the degrees of the circle of the celestial equator which rise with them, for they are greater or less or equal.[117] [91] Nevertheless, this motion (namely the motion of completion) is not perceived by our senses, but we observe some traces [of it] sometimes in the increase in longitude of their positions, sometimes in their decrease, and also in their changes in declination. [92/§90] These accidents revealed to us this motion, for if this motion were not, these accidents could not be [perceived] in the stars, since they are affixed to their orb [and remain] continuously at their places. [93] Consequently, [for] the stars placed near the points of intersection of the two circles (the oblique one and the celestial equator), when they are about 45 degrees in either direction – namely, within the two quadrants bisected by the two points of intersection [of the equator and the ecliptic] –, the number of degrees of its inclined heaven traversed by the star is smaller than the [rising] arc of its heaven because of the degrees of the equator equal to them, since the degrees [of the inclined circle] rise with a smaller number of degrees of the equator.[118] [94] For this reason, all [the stars] seem to move in the opposite direction, as if they receded with respect to the motion of the universe or they were moved opposite to that motion, and this lagging is called accession because it takes place in the direction of the signs.[119]

12v40, **XI**10 [95] When the stars placed on the middle of their orb[120] are moved on the degrees that follow the aforementioned degrees (namely from 45 degrees [counted]* from the point of intersection to the point where 135 degrees are completed, or approximately – that is, the next quadrant, bisected by the point which is similar to the solstitial point), the degrees traversed by the stars with their own motion are less[121] than the degrees facing them in **[13r]** the celestial equator (which are more[122]). [96] For this

---

[116] ideo differunt propter hoc spacia que transeunt stellae sui circuli obliqui a spaciis que fuerint in frontespicio et eo quod ascendit cum eis circuli aequinoctialis *K*, diuersantur quantitates quas secant stelle in suis circulis inclinatis a quantitatibus equatoris diei que ascendunt cum ipsis *C*.

[117] and we will explain . . . or equal: *om. C.*

[118] *with C:* gradus quos secat stella de suo celo inclinato incurtantur a complemento sui celi propter gradus equales illis de equatore, quia illi ascendunt cum paucioribus se de equatore; *K*: gradus quidem quos transeunt circuli obliqui sunt minus gradibus qui sunt in frontespicio, scilicet gradibus aequalibus eis circuli aequinoctialis et ascendunt propter hoc cum eis minus ex illis.

[119] est enim ad partem secundum ordinem signorum *om. C.*

[120] with *C:* et quando incipient stelle moueri (que sunt in medio celi); *K*: et cum inceperint hae stellae existentes in zona orbis signorum designati in orbe suo moueri.

[121] diminuuntur *C*, sunt pauciores *K* (*corrected to* plures *in the list of errata*).

[122] qui sunt plures *CK* (*corrected to* pauciores *in the list of errata of K*; these corrections are required by Qalonymos' wrong translation of sentence 93 – see note 70).

reason, it appears in the stars, with their motion of accession, an increase over[123] the motion of the universe, and this is what they called recession. [97/§91] Therefore, the fixed stars seem to be moved in these two quadrants with two different motions, although they are indeed moved only with an uniform motion – that is, when they are on the quadrant bisected[124] by the point of intersection, their motion seems to be lesser, but in the next quadrant their motion seems to be greater;[125] [98] the circle will be so completed and two accessions and two recessions will occur along it, despite both motions in longitude (i.e., the motion of the highest orb and the motion of the orb of the fixed stars) being equal. [99] This agrees with what was accepted by recent astronomers concerning both motions of accession and recession; although they also admitted a motion in the direction of the signs, which would result from some lagging still remaining in the orb [after these two motions],[126] noticeable only after a long time – this is possible, although its amount would not be as Ptolemy asserted.[127]

   13r10, XI14, §92 [100] Since this motion (namely the proper motion of the fixed stars) takes place in the direction of the motion of the universe and follows it, when it begins at the point similar to the summer solstice it goes toward the point similar to the vernal equinox, and from it toward the point similar to the winter solstice and from it toward the point similar to the autumnal equinox, opposite to the motion in the direction of the signs they assumed; this motion is opposite to that they attributed to the circle of the signs (along which the motion of the sun takes place) and opposite to the motion of the sun.[128] [101] The difference between the degrees of the oblique circle of this orb and the degrees of the celestial equator facing them and rising with them is known by the method established by Ptolemy; it was explained there that once the maximum declination is known, the lengths of whatever arcs of great circle drawn between the oblique circle and the celestial equator [and perpendicular to it] are also necessarily known.

   13r19, XI17, §93 [102] Let us use a figure with circles and letters as an example, to make the explanation clearer. We will leave aside the lagging (of this orb with respect to the highest one) still remaining once this orb has been moved with its own motion with

---

[123] *reading* citra motum uniuersi *instead of* contra motum uniuersi *in K; C*: additionem motui generali.

[124] *reading* in cuius medio est *instead of* in cuius medio extra.

[125] erit illud quod apparet ex motu eius incurtatum ; et in quarta que sequitur in cuius medio est punctum mutationis erit illud quod apparet ex motu eius multiplex festinus *C*.

[126] *C*: sed si cum [*instead of Carmody's* sicut] hoc habuit motum secundum signa, erit ita quod remaneat semper post istos duos motus incurtatio modica remanens in celo; post istos duos motus *is omitted in K*.

[127] licet non fuerit spacium eius id quod dixerit Ptolemaeus *K*, sed erit minus quam quod nominauit Tholomeus *C*.

[128] *with C*: econtrario illi motui quem ipsi posuerunt scilicet secundum signa. Et iste motus est contrarius ei quem posuerunt circulo signorum (super quem habet motum Sol) et contrarius motui Solis; *K*: contra id quod posuerint in circulo signorum quoad motum quem putauerunt esse soli super eo.

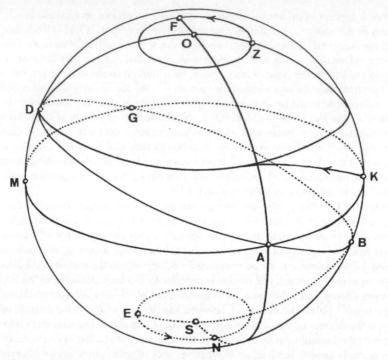

Fig. 7. Arc OAS: horizon; arc ESBKZODM: meridian; arc AMG: visible part of the equator

which follows the highest one,[129] and we will consider that it completes its own motion and catches up with it, as more recent astronomers did[130]. [103] Let circle AKGM be the celestial equator, and circle ABGD the mid-circle of this orb (which is the circle described by any of the stars placed on the middle of it with the motion of the orb about its poles). Let these two circles intersect at points A and G, and let points S and O be the poles of the universe. [Fig. 7] [104] Let also points E and Z be the poles of the orb of the fixed stars which revolve around the two poles that do not move with their motion, namely around S and O, and let the two circles EN and ZF be the two circles of their path (on which they are moved), as in this figure. [105] We draw circle ESBKZODM through the poles of the celestial equator (and the poles of this orb[131]). Since the two circles ABGD and AKGM intersect at the two points A and G, which are similar to the

<hr />

[129] *with C*: et dimittamus incurtationem quam facit hoc celum a superiori post motum suum, qui est per ipsum solum, quo (*instead of Carmody's* quod) sequitur superius; *K*: et incipiemus recitare defectum quo deficit iste orbis in se totum a suppremo post motum suum sibi proprium quo subsequitur eum.

[130] ut fecerunt recentiores: *absent from C*.

[131] et super duos polos huius celi: *absent from K*.

two equinoctial points of the orb of the signs, the two points B and D are thus similar to the two solstices. [§94] Let semicircle OAS be the horizon at *sphaera recta*.

13r37, XI21 [106] As the orb of the fixed stars lags behind the motion of the highest one, as we said, its poles necessarily lag behind on the two circles of their paths, receding from their [initial] positions with respect to the highest one[132] (although their distance to the poles of the highest one, i.e., the distance from points E and Z to points S and O, [respectively], is always the same); [107] thus, their lagging motion opposite to the motion of the highest orb and about its poles will necessarily take place on two circles opposite and parallel to the celestial equator,[133] namely on circles EN and ZF. Poles E and Z are moved on these two circles in the direction opposite to the motion of the universe. [108] When one of the stars existing in this orb is placed at one of the points of intersection (i.e., the two points A and G), **[13v]** and both poles recede on their circles from the intersections of the two circles EN and ZF with circle ESB (i.e., the meridian circle[134]), pole E will be moved toward the part [of the orb] hidden for the people [living] at the assumed horizon, and pole Z toward the part visible to them. [109] Both points E and Z will be moved toward N and F, respectively, i.e., in the direction opposite to that of the motion of the universe; consequently, a star on point A necessarily inclines from circle AKGM (i.e., the celestial equator) and digresses from it on its oblique circle in the direction toward which both poles incline, for the distance [of point A] from them always remains the same.

13v7, XI25, §95 [110] Let point O, for instance, be the north pole, and let the motion of the highest orb take place from point A toward point M on the celestial equator, and the motion of pole Z [when it is carried by the highest orb take place] in the same direction, but its lagging toward point F. [111] When the highest orb begins to move from point A, the pole of the orb of the stars will follow it from point Z; but when point A returns to its place in the figure, point Z will not have completed the circle of its path, for it lags, and the lagging will be equal to a part of arc ZF, as if point Z moved toward F by the amount of this lagging. [112] When point Z reaches point F with the lagging motion, the maximum declination of the star (which was on point A) to the north of the celestial equator is reached, and [once it has been moved with its motion of completion] the star (which preserves its position in longitude) will be placed at the same distance [from the equator] as point D on the oblique circle; [113] thus, the star will rise[135] from horizon OAS at a distance from A equal to the length of arc OF and equal to the distance of point M from point D (which is similar to the solstitial point), for its distance from pole Z is always the fourth part of the circle. [114] In this way the motion of the pole from point F to the point opposite to point Z[136] on the circular path will take place, and the star will return from its maximum northerly declination to the point similar to

---

[132] tunc mutauerunt se a suo directo quod prius habebant in superiori ad aliud directum in eodem C.

[133] circulis oppositis et aequidistantibus ab aequinoctiali *K*, circulos equidistantes et equidistantes equinoctiali C.

[134] *with C:* uolo dicere circulum medie diei; *K*: scilicet circulo aequinoctiali.

[135] *with C:* tunc stella ascendit; *K*: erit A stelle ascendenti.

[136] *with C:* ad punctum quod est in oppositione puncti Z; *K*: ad punctum similem puncto Z.

the other equinoctial point, namely point G. [115] In the two remaining quarters, the relative positions[137] of the pole and the star will be as in these two quadrants, each quadrant with its corresponding quadrant (namely, the quadrant of the circle of the path on which the pole is moved and the quadrant of the oblique circle on which the star revolves.[138]

13v23, XI28, §96 [116] We have discussed the arrangement[139] producing this motion, to which we were induced with the help of God; it is the motion that the orb of the stars has by itself, by means of which it follows the motion of the highest one and brings to completion what was lacking with respect to the universal motion – [117] a motion natural to it, by which it achieves its form, and by which it differs and is distinguished from the motion of the highest one, although it is not different from it nor contrary to it – [118] [namely] the motion of this orb around its own poles (which takes place following the motion of the highest one and combining with it,[140] but not around the poles of the highest one), which do not move with this motion. [119] For this reason the poles of this orb lag, although the whole orb and the stars on it do not lag, for they catch up with the highest orb by the motion of completion.[141] [120] Nevertheless, this motion is not perceived by our senses; our reason necessarily concludes it[s existence] from the motion in declination of the stars and from their apparent motion in accession and recession which result from the completion of the motion[142] with respect to [the motion of] the highest orb.

13v32, XI30, §97 [121] Thus, any star of those affixed to this orb on circle ABG is moved with the motion of its orb about poles Z and E through the distance by which it completes the lagging of its orb with respect to the highest one. [122] When it achieves this motion of completion, there remains only a change in declination, and this because [the star] inclines in the direction toward which the pole (that does not complete this lagging, but lags behind) inclines – [123] for the stars are moved with the motion of the orb about their poles, but the poles [themselves] do not move with this motion, for they are at rest [with respect to it], thence the lagging of the pole will produce only a motion in declination in the stars themselves and [with it] they incline sometimes to the north and sometimes to the south.[143]

13v39, XI32, §98 [124] Let us place one of these stars at some point on circle ABG, for instance, at point L, and let the distance of this point from A (which is the point similar to the equinoctial point) be 45 degrees. [Fig. 8] [125] We assert that when the star on point L traverses these degrees (i.e., from point L toward point A) with the motion of completion (which is its proper motion by which it is moved with the motion of its orb about its poles), the arc of the celestial equator, namely circle AKGM, facing them and rising with them is smaller, **[14r]** and we form our judgement on the longitude [of

---

137 dispositio *K*, modus *C*.
138 *with C*: rotatur stella; *K*: reuoluitur planeta.
139 'amr *Arabic*, res *C*, dispositio *K*.
140 copulans ei *K*, associat motum superiorem *C*.
141 cum motu complementi *C*, per motum supplementi *K*.
142 cum complemento motus *CK*.
143 aliquando ad dextrum aliquando ad sinistrum *C*.

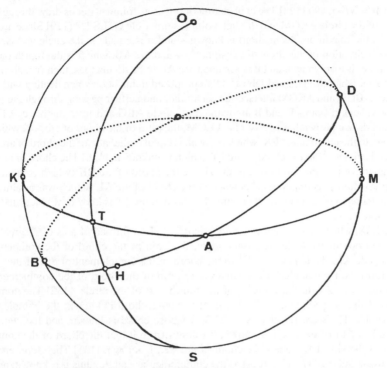

Fig. 8. Arc AT = arc AH. According to the text (sentences 124–135), A is the point "similar to the equinoctial point"; therefore, the true equinoctial point is T

the star] with respect to this [circle], not with respect to the oblique circle,[144] for the motion of the star on the oblique circle cannot be perceived. [126] When it occurs so, the star seems to be retarded and receding in longitude with respect to the motion of the highest orb by an amount equal to the difference between the degrees of the oblique circle and the degrees of the celestial equator [ascending with them], although its motion is always uniform; [127] but due to the inclination [of the pole], arc AL is reduced to an arc equal to the arc of the equator rising with it, for it rises with a smaller arc; the star seems then to recede from its [initial] position, although it has completed its proper motion.[145]

---

[144] *with C*: et est respectu cuius iudicatur [*instead of Carmody's* indicatur] longitudo non secundum circulum inclinatum; *K*: est id cuius distantia debemus adherere non autem circulo obliquo.

[145] *with C*: arcus AL incurtatur suum equale in ascensu cum eo de equatore, et ascendit cum minus ipso; tunc uidebitur stella retrocedere a suo loco licet compleuit locum suum; *K*: arcus AL deficit in ascensionibus ab eo qui est ei aequalis et non obliquus et apparet stellae tarditas sui loci licet perfecerit motum suum.

14r8, **XI**35, §99 [128] The explanation of this is as follows. Let us draw through the poles of the circle AKGM and through point L semicircle OTLS.[146] [129] Since arc BK (that is the maximum declination) is known – and it is equal, as the ancients asserted, to the inclination of the circle of the signs[147] – and arc AB, which is the fourth part of the circle, is also known, and it is assumed arc AL to be 45 degrees, then (according to what Abū Isḥāq Muḥammed Jābir b. Aflaḥ proposed in his book when dealing with two circles ABGD and AKGM intersecting each other and neither passing through the poles of the other), if points L and B are marked on circle ABG and from them arcs LT and BK are drawn perpendicular to circle AKGM, the ratio of the chord of arc AB, which is known, to the chord of arc BK, which is given, is equal to the ratio of the chord of arc AL, which is given,[148] to the chord of arc LT, which is unknown. [130] The chord of arc AB is known for it is the fourth part of the circle, the chord of arc BK (which is the arc of the [maximum] declination) is known, and the chord of arc AL (which was assumed to be 45 degrees) is also known; therefore, the chord of arc LT is known; whence arc LT is known to be 16;37,21°.[149]

14r18, §100 [131] Moreover, since triangle ATL is formed of arcs of[150] great circles and one of its angles is a right angle, the ratio of the chord of the complement of the side AL, the hypotenuse,[151] to the chord of the complement of one of the sides that form the right angle, TL, is equal to the ratio of the chord of the complement of the other side, AT, to the chord[152] of the fourth part of the circle. [132] The complement of the side AL is arc LB, that is 45°, whose chord is known; the complement of the side TL is arc LS, that is 73;22,39°, whose chord is known, and the chord of the fourth part of the circle is known; it follows from it that the chord of the complement of the side TA, namely the chord of arc TK, is known. [133] Therefore, arc TK is known, and it is 47;31°. Arc AT is the complementary angle; thus it is known, and is 42;29°. Consequently, it is smaller than arc AL, marked on the oblique circle, which rises with it.

14r27, **XI**39*, §101 [134] When the star traverses with its proper motion an arc equal to arc AT (let AH be this arc), it appears to us as receding from its [former] position

---

[146] *instead of sentences [129]–[133] C gives the following text:* et quia arcus BK, qui est maior inclinatio, est scitus, et arcus AL est scitus, erit arcus LT notus, et arcus AT notus, secundum quod explanatum est in Almagesti. Et erit scitum quod arcus AT erit minor arcui AL [= *since arc BK (that is the maximum declination) is known, and arc AL is known, arc LT and arc AT will be also known, according to what was explained in the* Almagest. *It will be also known that arc AT will be smaller than arc AL*]. *I have slightly modified Carmody's wording and punctuation.*

[147] *K add.* et est circulus BC et AB.

[148] *reading* positi *instead of* posite.

[149] *K:* 16;37,11°.

[150] *reading* ex arcubus *instead of* et arcubus.

[151] *reading* oppositae recto *instead of* remanentis recto.

[152] *reading* ad cordam *instead of* corde.

in longitude by the amount of arc LH,[153] since[154] there is no difference between the horizon and semicircle OTLS, for if semicircle OTLS is moved, there is no change in the rising times of any of their parts.[155] [135] For this reason, when the star traverses on arc AL the number of degrees of the equator [rising with that arc], it seems to lag behind, and from it they believe that the star goes back and moves following the order of the signs.[156] [136] This appearance of retardation or postposition from its [former] location in its orb occurs [because its motion is perceived] with respect to the celestial equator[157] – that is, because its proper motion in longitude is hidden from our senses, although [its motion] in declination is evident.

14r35, XI41 [137] The same occurs [for the star] in the succeeding degrees of this quadrant (i.e., 45° to either side of point A); but in the next quadrant (in the middle of which is the solstitial point)[158] it occurs the opposite of what we said; [138] for the arc that the star traverses by the proper motion of its orb, which is hidden from us, rises with an arc of the equator greater than it, and the star seems to have an excess with respect to the degrees of the celestial equator rising with it; thus, it appears to us as preceding the motion of the highest orb, and from it we believe it moves ahead the motion of the universe. [139] In the remaining two quadrants, it will occur as in the two preceding ones. The first lagging is what they call accession or excess, but they call the second increment over the motion of the universe recession, although neither accession nor recession [really] occur[159].

14r43, XI44 [140] This is what happens in the orb of the fixed stars concerning the accidents from which results the apparent variation in the motion in longitude. **[14v]** However, the variation in declination is true and apparent to our senses.[160] This is what we intended to explain.

14v2, XI45, §102 [141] Whether [this orb] has another lagging besides the afore-mentioned one, and thus the fixed stars have a motion in longitude in the direction of the

---

[153] tunc quando secabit stella (cum motu qui est per eam) arcum equalem arcui AT (et sit arcus LE), uidebitur posteriorari in loco suo in longitudine cum quantitate arcus AE *C*, et cum pertransierit stella motu suo proprio arcum aequalem arcui AT ac si esset arcus AH [apparet stella] recedens a loco suo in longitudine spacio arcus AH *K*. AE *and* LE *must be interchanged in C; in* K *the second* AH *must be replaced by* LH.

[154] quia *C*, tunc *K*.

[155] *C add.* sed simul ascederet totum.

[156] *with C*: et propter hoc uidetur stella quando secabit arcum AL quod incurtat similitudinem graduum equatoris diei, tunc credunt propter hoc quod stella reuertitur et mouetur secundum signa; *K*: quare apparet stella cum pertransiret arcum AL deficiens a totidem partibus sequinoctialis et existimatur ex hoc stella regredi et moueri contra [*sic*] ordinem signorum.

[157] quia aspicitur secundum equatorem diei *C*, est respectu equinoctialis *K*.

[158] *sic*: punctus solsticii *K*, punctus mutationis *C*.

[159] et primus quidem defectus est ille quem appellant accessum seu excessum et additamentum ante motum uniuersi secundum uero ipsi appellant recessum et tamen re uera non est accessus nec recessus *K*, et incurtatio prima est que nominatur elewel et est accessus secundum Tholomeum, et precessio ad motum generalem nominatur ab eis alidber, et est secundum Tholomeum recessus, et secundum ueritatem neque est accessus neque recessus *C*.

[160] diuersitas uero in latitudine est re uera et apparet sensu *K*, et diuersitas in latitudine res eius est salua et apparens et uisa *C*.

signs – as Ptolemy asserted, and other ancient [astronomers], for they thought that this orb was moved opposite to the motion of the universe –, this seems to be in accordance with the appearances and most likely, because from this motion could follow the great changes that occur in this generable and corruptible lower world, and the transformation of inhabited areas into uninhabited areas and vice versa. This is what we think on the motion of this orb.

Acknowledgments. I am grateful to Professors John D. North and Julio Samsó for their comments on a draft of this article, and to C. Mancha for his help with the figures.

INDEX OF NAMES

Abraham: V 17
Abraham Bar Ḥiyya: VI 307, 317, 351;
VII 281, 283; X 225, 229
Abraham Ibn ʿEzra: V 50; VI 307–8,
316–17, 351; VII 281, 283; X 225,
228
Abū Dawd (Abraham ibn Daud): XI 144
Abū al-Qāsim ʿAbdallāh ibn Amājūr: II 284
Abū Shāker: VI 309, 352; X 225–6, 229
Abūl-Ḥasan ʿAlī b. ʿUmar al-Marrākushī,:
IX 1
Adelard of Bath: I 33
Albert of Brudzewo: VIII 80–81, 87
Alchabitius: VI 316; see also al-Qabīṣī
Alexander: XI 145, 171
Alfonso X, Alfonsine: II 283; IX 2, 6
Alhazen, see Ibn al-Haytham
ʿAlī ibn Khalaf, Abū-l-Ḥasan: IX 7
Alpetragius: XI 144; see also al-Biṭrūjī
Aly Aben Ragel, ʿAlī ibn Abī-l-Rijāl: IX 6
Amico, G.B.: VII 265, 280–81
Anaritius: I 34; see also al-Narīzī
Antiphon: I 23
Apianus, P.: II 284, 297
Archimedes: I 33, 34
Aristarchus of Samos: VII 281
Aristotle: I 1, 3, 23, 33, 34; II 284; IV 2;
V 17, 23, 35; VI 305; VII 264, 274,
280–82; VIII 78, 82; XI 143, 170
Aristyllus: XI 145, 171
Augustine: III 2, 16, 23; IV 3
Avendauth: III 14, 23; see also Abū Dawd
Averroes, Ibn Rushd: IV 2; V 20, 49;
VI 306; VII 280, 282–3; XI 143
Avicenna, Ibn Sīnā: III 14; VI 352;
VIII 82, 88
Azarquiel, see Ibn al-Zarqāllu

Bacon, F.: I 33
Bacon, R.: I 1, 20, 33, 34; VI 304, 349
Bartholomeus Anglicus: VI 278
al-Battānī: III 9, 18; IV 5–6, 8, 13, 19;
V 28–31, 41, 44, 49; VI 276, 278,
301, 304, 313, 317, 331, 352;

XI 145, 171
Benedict XII: III 16; IV 3; VI 271, 278,
303
al-Bīrūnī: II 284, 297; VI 309, 350;
X 224–5, 228
al-Biṭrūjī: IV 1, 22; V 36; VI 309, 351;
VII 265, 282–3; X 224, 226;
XI passim
Bianchini, G.: VIII 82, 88
Blasius of Parma: I 1; II 289
Bonjorn: VI 312, 313, 317
Bryson: I 23

Callippus: VII 274, 281–2
Capuanus, J.B.: VIII 86–8
Cardan, G.: V 18, 48
Clement VI: III 1–2, 6, 14–15; IV 3, 16, 17;
VI 269–70, 272, 278, 303, 349
Copernicus: V 19; VI 309, 352; VIII 81;
X 221–3, 228–9

Dionysius Exiguus: VI 310
Dominicus Gundissalinus: III 14, 23
Duns Scotus: VI 269

Egidius of Baisiu: I passim; II 298; III 25;
VI 351
Esculeus, see Hypsicles
Euclides: I 1, 3, 4, 7–8, 12, 15–16, 33–4;
II 285; III 21; V 14, 18–19, 20
Eudoxus: VII 264–5, 274, 276, 279, 281–2;
VIII 70; XI 143, 144
Eutocius: I 33

al-Fārābī: V 13, 49
al-Farghānī: VIII 79, 84
Fibonacci, Leonardo: V 14, 17, 48
Firmin of Belavall: IV 3; VI 269
Fracastoro, G.: VII 265

Gaston II, count of Foix: VI 278
Gemma Frisius: II 275, 276, 297
George of Trebizond: VI 309; X 223, 226
Gerard of Cremona: I 28, 34; VI 309;

X 223–6, 228
al-Ḥajjāj ibn Yūsuf ibn Maṭar: VI 309;
    X 223–6
Hasdai Crescas: VI 352
Henry of Hesse, Henry of Langenstein: I 1,
    2; II 286, 287, 288, 289, 297;
    VIII 73–5, 77–8, 81–3, 87
Hermes: VII 283; XI 145, 171
Hervagius, Iohannes: X 227
Hipparchus: V 19; XI 145, 171; VI 307,
    308, 349; X 221, 225–8
Hippocrates: I 23
Hyppolitus: VI 310
Hypsicles: I 6, 12, 28, 33

Ibn al-Bannā al-Marrākushī: IX 2, 5–6, 8
Ibn al-Hā'im, Abū Muhammad ʿAbd al-
    Ḥaqq: IX 6
Ibn al-Haytham: I 1, 2, 3, 16–17, 21–3, 28,
    33, 34, 35; II 289, 297; V 22, 49;
    VI 306, 352; VII 264; XI 144;
    VIII 70–73, 80–2, 88
Ibn al-Kammād, Abū Yaʿfar Ahmad b.
    Yūsuf: IX passim
Ibn al-Muthannā: VI 317, 350
Ibn al-Raqqām, Abū ʿAbd Allāh
    Muḥammad: IX 2, 5–8
Ibn al-Shāṭir: VI 305
Ibn al-Zarqāllu, Abū Isḥāq: III 14; IX 1–2,
    5, 7–8; XI 143, 145–6, 163, 172
Ibn Gabirol, Solomon ibn Yehuda: III 14
Ibn Isḥāq al-Tūnisī: IX 5–6, 8
Ibn Muʿādh al-Jayyānī: XI 160
Ibn Rushd, see Averroes
Ibn Sīnā, see Avicenna
Ibn Yūnus: II 284; VI 309, 349; X 224, 226
Ibrāhīm ibn Sinān: VII 264; XI 144
Isaac Qimḥi: III 6; IV 3
Isaiah: VI 280, 290, 299
Isḥāq ibn Ḥunayn: VI 309; X 223–4, 226
Iunta, L.A.: XI 144; v. Junta

Jābir ibn Aflaḥ: V 20, 36, 50; X 224;
    XI 160, 180
Jacob Anatoli: VI 309; X 225
Jacob ben David, see Bonjorn
Jacob ben Makhir ibn Tibbon: III 14, 23
Jafar Indus: VI 273
Jean de Venette: VI 350
Jeanne of Navarre: II 277
Job: VI 281, 291
Johannes Brixiensis: III 14, 23
John of Croyba: VIII 87
John of Dumpno: IX 1
John of Genoa: IX 2–3

John of Montfort: IX 2–3
John of Murs: II 283, 298; IV 473; V 17,
    49; VI 269, 278, 349
John of Palermo: V 14
John of Sicily: II 278
Joseph ibn Kaspi: VII 283
Juan Gil: IX 1, 3, 8
Judah Halevi: X 226, 228
Julmann: VIII 78, 81, 84
Junta, L.A.: VI 309; X 227; see also Iunta
al-Jūzjanī, Abū ʿUbayd: VI 306, 352;
    VIII 82, 88

Kepler, J.: I 1, 2, 3, 19, 23–4, 32, 34; II 277,
    295, 297; V 20,49; VII 280
al-Khwārizmī, Muḥammad ibn Mūsa: V 17;
    VI 317, 352; VIII 85; X 225, 228
al-Kindi: I 33

Levi ben Gerson: I 29, 30–32, 34; II 277–8,
    285–6, 289–97; III, IV, V, VI passim;
    VII 265, 274, 276–82; X 225;
    XI 144
Lupis, de: VI 273

Maestlin, M.: II 276
Maimonides: V 35, 49; VI 303, 306, 308,
    351–2; X 225, 229
Marie of Brabant: II 277
Masters of Talisman: XI 145, 171
Matthew of Miechów (Maciej z Mie-
    chowa): I 34
Maurolico, F.: I 1, 2, 19, 34
Menelaus of Alexandria: XI 145, 153, 171
Messahala (Māshāllāh): I 3
Michael, bishop: VI 273
Michael Scot: X 224; XI 144, 146, 148,
    149, 150, 159, 161, 162
Moses: VI 281, 291
Moshe ibn Tibbon: VII 286; XI 144, 161
Munster, S.: VII 281

Nabonassar: XI 145, 170, 171
al-Narīzī, al-Faḍl ibn Hātim al-Nayrīzī: I 34

Oresme, Nicole: III 1; 25; IV 3; VI 270

Pecham, John: I 1, 3, 14–18, 20–28, 31,
    33–4; II 286, 289
Peter of Saint-Omer: III 16, 25
Petit de Nyons, see Isaac Qimḥi
Petrarch, F.: VI 270
Petrina, H.: X 227
Petrus of Alexandria: I 30; II 290;
    III passim; IV 3, 15; V 14, 17;

VI 270–71, 278
Petrus Philomena de Dacia: III 25
Peurbach, Georg: II 298; VIII 77, 80–81,
    86–8
Philip III: II 277
Philip IV: II 277
Philippe de Vitry: III 1; IV 3, 14; VI 270
Plato: XI 170
Pliny: V 19
Profatius (Judaeus): III 23; see also Jacob
    ben Makhir
Ptolemy: I 6, 12, 28, 34; II 296–7; III
    4, 7–8, 17; IV 1, 7–8, 12–14, 17,
    19, 22; V 18–24, 26–9, 31, 35–6,
    40–41, 44, 49, 50; VI 272, 281,
    291, 299–309, 317, 352; VII 264–5,
    276–7, 280–81; VIII 70, 72–4, 78,
    81, 82, 85–8; X 221, 224–9; XI 144,
    147, 153–4, 160–62, 171, 175

al-Qabīsī, Abū al-Saqr Abd al-Aziz ibn
    Uthman: IX 6
Qalo Qalonymos (Calonymos, Qalonymos
    b. David): XI 143, 144, 146, 148,
    149, 150, 159, 160, 161, 162, 174
Qalonymos ben Qalonymos: III 6; VII 283

Regiomontanus: II 276, 298; V 22, 49;
    VIII 82, 88; X 222
Reinhold, E.: II 275, 278, 298; VII 265,
    281; VIII 86, 88
Rheticus, J.: X 223
Risner, F.: I 23, 33; II 297
Robert d'Anjou: III 6; IV 3
Roder, Christian: VIII 88
Roger, Pierre: IV 17; VI 272, 278
Roger of Hereford: II 275, 278

Ṣāʿid al-Andalusī, Abū-l-Qasim: XI 145
al-Samawʾal, Ibn Yahyā al-Maghribī: V 14
Santbech, Daniel: II 276, 298
Scaliger, J.: VI 273
Simplicius: VII 264–5, 274, 282; XI 144,
    151, 153
Solomon ben Gerson: III 2, 14; VI 278
Speier, Jacob von: VIII 88

Thābit b. Qurra: II 277; VI 309; VIII 77, 84;
    IX 5–6; X 223–4, 226; XI 143
Thales of Mileto: I 33; VII 280
Theodosius of Tripoli: IV 16; VII 268, 272,
    276; XI 144
Theon of Alexandria: VIII 86; XI 145, 171;
    X 223
Thomas Aquinas: VI 269, 352
Tideus (Diocles): I 33
Timocharis: III 9, 18; IV 7, 12; V 44;
    VI 311–12; XI 145, 171
al-Ṭūsī, Naṣīr al-Dīn: V 49; VII 281;
    VIII 73, 82, 88–9; X 224, 229
Tycho Brahe: I 34; II 275, 276, 297;
    X 227–8

Viète, F.: V 20

William of Moerbeke: VII 282
William of Saint-Cloud: II 275, 277, 278,
    283–6, 289, 298; IV 19; VI 317,
    351
Witelo: I 1, 33, 34, 35; II 277, 297
Wittich, Paul: X 227

Yahyā ibn Abī Manṣūr: IX 4
Yehudah b. Moshe ha-Kohen: IX 6
Yomtob Poel: v. Bonjorn

INDEX OF MANUSCRIPTS

Brescia, Biblioteca Civica Queriniana
A.IV.11: VI 303
Bruxelles, Bibliothèque Royale 2962–2978:
III 22, 35–6

Cues, Stiftsbibliothek 215: II 278–81

Firenze, Conv. soppr. J.X.19: II 288

India Office, MS n. Loth 741: X 224

Klagenfurt, Bischöfliche Bibliothek
XXX.b.7: III 2, 15–17, 23; VI 303
Kraków, Biblioteca Jagiellońska
569: I 2, 3–8
10000: X 222

Leiden, Universiteitsbibliotheek
Add. 26921: IV 10; VI 271, 274–5,
310–11, 333
Cod. Or. 143: X 224
Cod. Or. 680: X 224
Harley 625: X 226
London, British Library
Or. 10725: VI 271, 275, 331
Scaligerani 46: IV 4; VI 273, 279,
280–90, 320–33
Scaligerani 46*: IV 4; VI 273, 318–20
Jews Collage, Heb. 135: IX 1, 3
Lyon, Bibliothèque Municipale 326: II 290;
III 2; V 47; VI 333–49; VII 265–70,
282

Madrid, Biblioteca Nacional 10023: IX 1,
3, 4, 7
Melk, Stiftsbibliothek 51: VIII 83–4
Memmingen, F.33: X 226
Milano, Ambrosiana D 327: II 290; III 2;
V 47; VI 333, 335–49; VII 265–70,
282
Munich, Bayerische Staatsbibliothek
CLM 8089: III 2, 5, 15–17, 23
CLM 26667: VIII 84–6
Heb. 314: VI 271, 275, 316, 318, 320,

322, 331, 332–3

Napoli, Biblioteca Nazionale, III F.9: IV 9,
13; VI 271

Oxford, Bodleian Library
Ashmol. 192: III 1
Ashmol. 393: III 1
Bodley 300: VIII 83
Digby 168: II 1
Digby 176: III 1

Paris, Bibliothèque nationale
Hebr. 724: II 295; III 23; IV 2, 9, 13;
VI 271, 275, 277, 318, 331, 333
Hebr. 725: II 295; III 23; IV 9, 13;
VI 271, 275, 318, 331, 333
Hebr. 1036: XI 146
Lat. 7281: II 278–81
Lat. 7293: III 2–5, 15; IV 3, 16; VI 270,
333
Lat. 7378A: III 1
Lat. 15171: II 1
Lat. 16401: I 2; VIII 83–4
n. a. l. 1242: II 278–81; VI 317
Praga, Knihovna Metropolitní Kapitoly
1272: VIII 82–4
Princeton, Garret 95: VIII 83

Ripoll, Biblioteca Lambert Mata 21: IX 1, 3
Roma, Biblioteca Apostolica Vaticana
Barberini 350: VIII 83
Hebr. 391,3: VI 271
Lat. 3098: I 29–30; II 290–92, 294–5;
III 2–5, 8–13, 16–24; IV passim;
V 15, 21–4, 28, 31–6, 40–41,
44, 47–8; VI 277, 299, 300–308,
311–12, 317, 325, 327–8, 330–31,
333–49; VII 265–70, 282; X 225–6
Lat. 3380: II 290–92; III 2; V 47;
VI 333–4; VII 265–70, 282
Lat. 4082: VIII 83
Ottobon. 1906: III 1, 17
Segovia, Catedral 115: IX 1, 6–8, 10

Torino, Bibl. Nazionale Universitaria
  A.II.10: X 225
Tunis, Bibliothèque Nationale 07116:
  X 224

Utrecht, Universiteitsbibliotheek 725:

II 278–81; VIII 83
Wien, Nationalbibliothek
  5203: VIII 83
  5277: III 2, 5, 15–17, 23; VI 302
  5292: VIII 84–6
Wrocław, Ossol. 759: VIII 87